Analytical Chemistry

The INSTANT NOTES series

Series editor
B.D. Hames
School of Biochemistry and Molecular Biology, University of Leeds, Leeds, UK

Animal Biology
Biochemistry 2nd edition
Chemistry for Biologists
Developmental Biology
Ecology 2nd edition
Genetics
Immunology
Microbiology
Molecular Biology 2nd edition
Neuroscience
Plant Biology
Psychology

Forthcoming titles
Bioinformatics

The INSTANT NOTES Chemistry series
Consulting editor: Howard Stanbury

Analytical Chemistry
Inorganic Chemistry
Medicinal Chemistry
Organic Chemistry
Physical Chemistry

Analytical Chemistry

D. Kealey

School of Biological and Chemical Sciences
Birkbeck College, University of London, UK
and
Department of Chemistry
University of Surrey, Guildford, UK

and

P. J. Haines

Oakland Analytical Services,
Farnham, UK

First published 2002 (ISBN 1 85996 189 4)

A CIP catalogue record for this book is available from the British Library.

ISBN 1 85996 189 4

BIOS Scientific Publishers Ltd
9 Newtec Place, Magdalen Road, Oxford OX4 1RE, UK
Tel. +44 (0)1865 726286. Fax +44 (0)1865 246823
World Wide Web home page: http://www.bios.co.uk/

Distributed exclusively in the United States, its dependent territories, Canada, Mexico, Central and South America, and the Caribbean by Springer-Verlag New York Inc, 175 Fifth Avenue, New York, USA, by arrangement with BIOS Scientific Publishers, Ltd, 9 Newtec Place, Magdalen Road, Oxford OX4 1RE, UK

Production Editor: Nadine Séveno
Typeset by Phoenix Photosetting, Chatham, Kent, UK
Printed by Biddles Ltd, Guildford, UK, www.biddles.co.uk

CONTENTS

ABBREVIATIONS

AAS atomic absorption spectrometry
ADC analog-to-digital converter
AFS atomic fluorescence spectrometry
ANOVA analysis of variance
ATR attenuated total reflectance
BPC bonded-phase chromatography
CC chiral chromatography
CGE capillary gel electrophoresis
CI confidence interval
CIEF capillary isoelectric focusing
CL confidence limits
CPU central processing unit
CRM certified reference material
CZE capillary zone electrophoresis
DAC digital-to-analog converter
DAD diode array detector
DMA dynamic mechanical analysis
DME dropping mercury electrode
DSC differential scanning calorimetry
DTA differential thermal analysis
DTG derivative thermogravimetry
DVM digital voltmeter
ECD electron-capture detector
EDAX energy dispersive analysis of X-rays
EDTA ethylenediaminetetraacetic acid
EGA evolved gas analysis
FA factor analysis
FAES flame atomic emission spectometry
FFT fast Fourier transform
FID flame ionization detector or free induction decay
GC gas chromatography
GLC gas liquid chromatography
GSC gas solid chromatography
HATR horizontal attenuated total reflectance
HPLC high-performance liquid chromatography
IC ion chromatography
ICP inductively coupled plasma
ICP-AES ICP-atomic emission spectrometry
ICP-OES ICP-optical emission spectrometry
ICP-MS ICP-mass spectrometry
IEC ion-exchange chromatography
ISE ion-selective electrode
LVDT linear variable differential transformer
MEKC micellar electrokinetic chromatography
MIR multiple internal reflectance
MS mass spectrometry
NIR near infrared
NMR nuclear-magnetic resonance
NPD nitrogen-phosphorus detector
PAH polycyclic aromatic hydrocarbons
PC paper chromatography
PCA principal component analysis
PCR principal component regression
PDMS polydimethylsiloxane
PLS partial least squares
QA quality assurance
QC quality control
RAM random access memory
RF radiofrequency
RI refractive index
ROM read only memory
RMM relative molecular mass
SCE saturated calomel electrode
SDS sodium dodecyl sulfate
SDS-PAGE SDS-polyacrylamide gel electrophoresis
SE solvent extraction
SEC size-exclusion chromatography
SHE standard hydrogen electrode
SIM selected ion monitoring
SPE solid phase extraction
SPME solid phase microextraction
SRM standard reference material
TCD thermal conductivity detector
TG thermogravimetry
TIC total ion current
TISAB total ionic strength adjustment buffer
TLC thin-layer chromatography
TMA thermomechanical analysis

PREFACE

Analytical chemists and others in many disciplines frequently ask questions such as: What is this substance?; How concentrated is this solution?; What is the structure of this molecule? The answers to these and many other similar questions are provided by the techniques and methods of analytical chemistry. They are common to a wide range of activities, and the demand for analytical data of a chemical nature is steadily growing. Geologists, biologists, environmental and materials scientists, physicists, pharmacists, clinicians and engineers may all find it necessary to use or rely on some of the techniques of analysis described in this book.

If we look back some forty or fifty years, chemical analysis concentrated on perhaps three main areas: qualitative testing, quantitative determinations, particularly by 'classical' techniques such as titrimetry and gravimetry, and structural analysis by procedures requiring laborious and time-consuming calculations. The analytical chemist of today has an armoury of instrumental techniques, automated systems and computers which enable analytical measurements to be made more easily, more quickly and more accurately.

However, pitfalls still exist for the unwary! Unless the analytical chemist has a thorough understanding of the principles, practice and limitations of each technique he/she employs, results may be inaccurate, ambiguous, misleading or invalid. From many years of stressing the importance of following appropriate analytical procedures to a large number of students of widely differing abilities, backgrounds and degrees of enthusiasm, the authors have compiled an up-to-date, unified approach to the study of analytical chemistry and its applications. Surveys of the day-to-day operations of many industrial and other analytical laboratories in the UK, Europe and the USA have shown which techniques are the most widely used, and which are of such limited application that extensive coverage at this level would be inappropriate. The text therefore includes analytical techniques commonly used by most analytical laboratories at this time. It is intended both to complement those on inorganic, organic and physical chemistry in the *Instant Notes* series, and to offer to students in chemistry and other disciplines some guidance on the use of analytical techniques where they are relevant to their work. We have not given extended accounts of complex or more specialized analytical techniques, which might be studied beyond first- and second-year courses. Nevertheless, the material should be useful as an overview of the subject for those studying at a more advanced level or working in analytical laboratories, and for revision purposes.

The layout of the book has been determined by the series format and by the requirements of the overall analytical process. Regardless of the discipline from which the need for chemical analysis arises, common questions must be asked:

- How should a representative sample be obtained?
- What is to be determined and with what quantitative precision?
- What other components are present and will they interfere with the analytical measurements?
- How much material is available for analysis, and how many samples are to be analyzed?
- What instrumentation is to be used?
- How reliable is the data generated?

These and related questions are considered in Sections A and B.

Most of the subsequent sections provide notes on the principles, instrumentation and applications of both individual and groups of techniques. Where suitable supplementary texts exist, reference is made to them, and some suggestions on consulting the primary literature are made.

We have assumed a background roughly equivalent to UK A-level chemistry or a US general chemistry course. Some simplification of mathematical treatments has been made; for example, in the sections on statistics, and on the theoretical basis of the various techniques. However, the texts listed under Further Reading give more comprehensive accounts and further examples of applications.

We should like to thank all who have contributed to the development of this text, especially the many instrument manufacturers who generously provided examples and illustrations, and in particular Perkin Elmer Ltd. (UK) and Sherwood Scientific Ltd. (UK). We would like also to thank our colleagues who allowed us to consult them freely and, not least, the many generations of our students who found questions and problems where we had thought there were none!

DK
PJH

A1 ANALYTICAL CHEMISTRY, ITS FUNCTIONS AND APPLICATIONS

Key Notes

Definition Analytical chemistry is a scientific discipline used to study the chemical composition, structure and behavior of matter.

Purpose The purpose of chemical analysis is to gather and interpret chemical information that will be of value to society in a wide range of contexts.

Scope and applications Quality control in manufacturing industries, the monitoring of clinical and environmental samples, the assaying of geological specimens, and the support of fundamental and applied research are the principal applications.

Related topics

Analytical problems and procedures (A2)
Chemical sensors and biosensors (H1)
Automated procedures (H2)
Computer control and data collection (H3)
Data enhancement and databases (H4)

Definition

Analytical chemistry involves the application of a range of techniques and methodologies to obtain and assess qualitative, quantitative and structural information on the nature of matter.

- **Qualitative analysis** is the identification of elements, species and/or compounds present in a sample.
- **Quantitative analysis** is the determination of the absolute or relative amounts of elements, species or compounds present in a sample.
- **Structural analysis** is the determination of the spatial arrangement of atoms in an element or molecule or the identification of characteristic groups of atoms (functional groups).
- An element, species or compound that is the subject of analysis is known as an **analyte**.
- The remainder of the material or sample of which the analyte(s) form(s) a part is known as the **matrix**.

Purpose

The gathering and interpretation of qualitative, quantitative and structural information is essential to many aspects of human endeavor, both terrestrial and extra-terrestrial. The maintenance of, and improvement in, the quality of life throughout the world, and the management of resources rely heavily on the information provided by chemical analysis. Manufacturing industries use analytical data to monitor the quality of raw materials, intermediates and finished products. Progress and research in many areas is dependent on establishing the chemical composition of man-made or natural materials, and the monitoring of toxic substances in the environment is of ever increasing importance. Studies of biological and other complex systems are supported by the collection of large amounts of analytical data.

Scope and applications

Analytical data are required in a wide range of disciplines and situations that include not just chemistry and most other sciences, from biology to zoology, but the arts, such as painting and sculpture, and archaeology. Space exploration and clinical diagnosis are two quite disparate areas in which analytical data is vital. Important areas of application include the following.

- **Quality control (QC)**. In many manufacturing industries, the chemical composition of raw materials, intermediates and finished products needs to be monitored to ensure satisfactory quality and consistency. Virtually all consumer products from automobiles to clothing, pharmaceuticals and foodstuffs, electrical goods, sports equipment and horticultural products rely, in part, on chemical analysis. The food, pharmaceutical and water industries in particular have stringent requirements backed by legislation for major components and permitted levels of impurities or contaminants. The electronics industry needs analyses at ultra-trace levels (parts per billion) in relation to the manufacture of semi-conductor materials. Automated, computer-controlled procedures for process-stream analysis are employed in some industries.
- **Monitoring and control of pollutants**. The presence of toxic heavy metals (e.g., lead, cadmium and mercury), organic chemicals (e.g., polychlorinated biphenyls and detergents) and vehicle exhaust gases (oxides of carbon, nitrogen and sulfur, and hydrocarbons) in the environment are health hazards that need to be monitored by sensitive and accurate methods of analysis, and remedial action taken. Major sources of pollution are gaseous, solid and liquid wastes that are discharged or dumped from industrial sites, and vehicle exhaust gases.
- **Clinical and biological studies**. The levels of important nutrients, including trace metals (e.g., sodium, potassium, calcium and zinc), naturally produced chemicals, such as cholesterol, sugars and urea, and administered drugs in the body fluids of patients undergoing hospital treatment require monitoring. Speed of analysis is often a crucial factor and automated procedures have been designed for such analyses.
- **Geological assays**. The commercial value of ores and minerals is determined by the levels of particular metals, which must be accurately established. Highly accurate and reliable analytical procedures must be used for this purpose, and referee laboratories are sometimes employed where disputes arise.
- **Fundamental and applied research**. The chemical composition and structure of materials used in or developed during research programs in numerous disciplines can be of significance. Where new drugs or materials with potential commercial value are synthesized, a complete chemical characterization may be required involving considerable analytical work. Combinatorial chemistry is an approach used in pharmaceutical research that generates very large numbers of new compounds requiring confirmation of identity and structure.

A2 ANALYTICAL PROBLEMS AND PROCEDURES

Key Notes

Analytical problems — Selecting or developing and validating appropriate methods of analysis to provide reliable data in a variety of contexts are the principal problems faced by analytical chemists.

Analytical procedures — Any chemical analysis can be broken down into a number of stages that include a consideration of the purpose of the analysis, the quality of the results required and the individual steps in the overall analytical procedure.

Related topics

Analytical chemistry, its functions and applications (A1)
Sampling and sample handling (A4)
Chemical sensors and biosensors (H1)
Automated procedures (H2)
Computer control and data collection (H3)
Data enhancement and databases (H4)

Analytical problems

The most important aspect of an analysis is to ensure that it will provide useful and reliable data on the qualitative and/or quantitative composition of a material or structural information about the individual compounds present. The analytical chemist must often communicate with other scientists and nonscientists to establish the amount and quality of the information required, the time-scale for the work to be completed and any budgetary constraints. The most appropriate analytical technique and method can then be selected from those available or new ones devised and validated by the analysis of substances of known composition and/or structure. It is essential for the analytical chemist to have an appreciation of the objectives of the analysis and an understanding of the capabilities of the various analytical techniques at his/her disposal without which the most appropriate and cost-effective method cannot be selected or developed.

Analytical procedures

The stages or steps in an overall analytical procedure can be summarized as follows.

- **Definition of the problem**. Analytical information and level of accuracy required. Costs, timing, availability of laboratory instruments and facilities.
- **Choice of technique and method**. Selection of the best technique for the required analysis, such as chromatography, infrared spectrometry, titrimetry, thermogravimetry. Selection of the method (i.e. the detailed stepwise instructions using the selected technique).
- **Sampling**. Selection of a small sample of the material to be analyzed. Where this is heterogeneous, special procedures need to be used to ensure that a genuinely representative sample is obtained (Topic A4).

- **Sample pre-treatment or conditioning**. Conversion of the sample into a form suitable for detecting or measuring the level of the analyte(s) by the selected technique and method. This may involve dissolving it, converting the analyte(s) into a specific chemical form or separating the analyte(s) from other components of the sample (the **sample matrix**) that could interfere with detection or quantitative measurements.
- **Qualitative analysis**. Tests on the sample under specified and controlled conditions. Tests on reference materials for comparison. Interpretation of the tests.
- **Quantitative analysis**. Preparation of standards containing known amounts of the analyte(s) or of pure reagents to be reacted with the analyte(s). Calibration of instruments to determine the responses to the standards under controlled conditions. Measurement of the instrumental response for each sample under the same conditions as for the standards. All measurements may be replicated to improve the reliability of the data, but this has cost and time implications. Calculation of results and statistical evaluation.
- **Preparation of report or certificate of analysis**. This should include a summary of the analytical procedure, the results and their statistical assessment, and details of any problems encountered at any stage during the analysis.
- **Review of the original problem**. The results need to be discussed with regard to their significance and their relevance in solving the original problem. Sometimes repeat analyses or new analyses may be undertaken.

A3 ANALYTICAL TECHNIQUES AND METHODS

Key Notes

Analytical techniques	Chemical or physico-chemical processes that provide the basis for analytical measurements are described as techniques.
Analytical methods	A method is a detailed set of instructions for a particular analysis using a specified technique.
Method validation	A process whereby an analytical method is checked for reliability in terms of accuracy, reproducibility and robustness in relation to its intended applications.
Related topic	Quality in analytical laboratories (A6)

Analytical techniques

There are numerous chemical or physico-chemical processes that can be used to provide analytical information. The processes are related to a wide range of atomic and molecular properties and phenomena that enable elements and compounds to be detected and/or quantitatively measured under controlled conditions. The underlying processes define the various **analytical techniques**. The more important of these are listed in *Table 1*, together with their suitability for qualitative, quantitative or structural analysis and the levels of analyte(s) in a sample that can be measured.

Atomic and **molecular spectrometry** and **chromatography**, which together comprise the largest and most widely used groups of techniques, can be further subdivided according to their physico-chemical basis. **Spectrometric techniques** may involve either the **emission** or **absorption** of **electromagnetic radiation** over a very wide range of energies, and can provide qualitative, quantitative and structural information for analytes from major components of a sample down to ultra-trace levels. The most important atomic and molecular spectrometric techniques and their principal applications are listed in *Table 2*.

Chromatographic techniques provide the means of separating the components of mixtures and simultaneous qualitative and quantitative analysis, as required. The linking of chromatographic and spectrometric techniques, called **hyphenation**, provides a powerful means of separating and identifying unknown compounds (Section F). **Electrophoresis** is another separation technique with similarities to chromatography that is particularly useful for the separation of charged species. The principal separation techniques and their applications are listed in *Table 3*.

Analytical methods

An analytical method consists of a detailed, stepwise list of instructions to be followed in the qualitative, quantitative or structural analysis of a sample for one or more analytes and using a specified technique. It will include a summary and

Table 1. Analytical techniques and principal applications

Technique	Property measured	Principal areas of application
Gravimetry	Weight of pure analyte or compound of known stoichiometry	Quantitative for major or minor components
Titrimetry	Volume of standard reagent solution reacting with the analyte	Quantitative for major or minor components
Atomic and molecular spectrometry	Wavelength and intensity of electromagnetic radiation emitted or absorbed by the analyte	Qualitative, quantitative or structural for major down to trace level components
Mass spectrometry	Mass of analyte or fragments of it	Qualitative or structural for major down to trace level components isotope ratios
Chromatography and electrophoresis	Various physico-chemical properties of separated analytes	Qualitative and quantitative separations of mixtures at major to trace levels
Thermal analysis	Chemical/physical changes in the analyte when heated or cooled	Characterization of single or mixed major/minor components
Electrochemical analysis	Electrical properties of the analyte in solution	Qualitative and quantitative for major to trace level components
Radiochemical analysis	Characteristic ionizing nuclear radiation emitted by the analyte	Qualitative and quantitative at major to trace levels

Table 2. Spectrometric techniques and principal applications

Technique	Basis	Principal applications
Plasma emission spectrometry	Atomic emission after excitation in high temperature gas plasma	Determination of metals and some non-metals mainly at trace levels
Flame emission spectrometry	Atomic emission after flame excitation	Determination of alkali and alkaline earth metals
Atomic absorption spectrometry	Atomic absorption after atomization by flame or electrothermal means	Determination of trace metals and some non-metals
Atomic fluorescence spectrometry	Atomic fluorescence emission after flame excitation	Determination of mercury and hydrides of non-metals at trace levels
X-ray emission spectrometry	Atomic or atomic fluorescence emission after excitation by electrons or radiation	Determination of major and minor elemental components of metallurgical and geological samples
γ-spectrometry	γ-ray emission after nuclear excitation	Monitoring of radioactive elements in environmental samples
Ultraviolet/visible spectrometry	Electronic molecular absorption in solution	Quantitative determination of unsaturated organic compounds
Infrared spectrometry	Vibrational molecular absorption	Identification of organic compounds
Nuclear magnetic resonance spectrometry	Nuclear absorption (change of spin states)	Identification and structural analysis of organic compounds
Mass spectrometry	Ionization and fragmentation of molecules	Identification and structural analysis of organic compounds

Table 3. Separation techniques and principal applications

Technique	Basis	Principal applications
Thin-layer chromatography	Differential rates of migration of analytes through a stationary phase by movement of a liquid or gaseous mobile phase	Qualitative analysis of mixtures
Gas chromatography		Quantitative and qualitative determination of volatile compounds
High-performance liquid chromatography		Quantitative and qualitative determination of nonvolatile compounds
Electrophoresis	Differential rates of migration of analytes through a buffered medium	Quantitative and qualitative determination of ionic compounds

lists of chemicals and reagents to be used, laboratory apparatus and glassware, and appropriate instrumentation. The quality and sources of chemicals, including solvents, and the required performance characteristics of instruments will also be specified as will the procedure for obtaining a representative sample of the material to be analyzed. This is of crucial importance in obtaining meaningful results (Topic A4). The preparation or pre-treatment of the sample will be followed by any necessary standardization of reagents and/or calibration of instruments under specified conditions (Topic A5). Qualitative tests for the analyte(s) or quantitative measurements under the same conditions as those used for standards complete the practical part of the method. The remaining steps will be concerned with data processing, computational methods for quantitative analysis and the formatting of the analytical report. The statistical assessment of quantitative data is vital in establishing the reliability and value of the data, and the use of various statistical parameters and tests is widespread (Section B).

Many **standard analytical methods** have been published as papers in analytical journals and other scientific literature, and in textbook form. Collections by trades associations representing, for example, the cosmetics, food, iron and steel, pharmaceutical, polymer plastics and paint, and water industries are available. Standards organizations and statutory authorities, instrument manufacturers' applications notes, the Royal Society of Chemistry and the US Environmental Protection Agency are also valuable sources of standard methods. Often, laboratories will develop their own **in-house methods** or adapt existing ones for specific purposes. **Method development** forms a significant part of the work of most analytical laboratories, and **method validation** and periodic revalidation is a necessity.

Selection of the most appropriate analytical method should take into account the following factors:

- the purpose of the analysis, the required time scale and any cost constraints;
- the level of analyte(s) expected and the detection limit required;
- the nature of the sample, the amount available and the necessary sample preparation procedure;
- the accuracy required for a quantitative analysis;
- the availability of reference materials, standards, chemicals and solvents, instrumentation and any special facilities;
- possible interference with the detection or quantitative measurement of the analyte(s) and the possible need for sample **clean-up** to avoid matrix interference;

- the degree of selectivity available – methods may be **selective** for a small number of analytes or **specific** for only one;
- quality control and safety factors.

Method validation

Analytical methods must be shown to give reliable data, free from bias and suitable for the intended use. Most methods are multi-step procedures, and the process of validation generally involves a stepwise approach in which optimized experimental parameters are tested for **robustness** (ruggedness), that is sensitivity to variations in the conditions, and sources of errors investigated.

A common approach is to start with the final measurement stage, using calibration standards of known high purity for each analyte to establish the performance characteristics of the detection system (i.e. specificity, range, quantitative response (linearity), sensitivity, stability and reproducibility). Robustness in terms of temperature, humidity and pressure variations would be included at this stage, and a statistical assessment made of the reproducibility of repeated identical measurements (replicates). The process is then extended backwards in sequence through the preceding stages of the method, checking that the optimum conditions and performance established for the final measurement on analyte calibration standards remain valid throughout. Where this is not the case, new conditions must be investigated by modification of the procedure and the process repeated. A summary of this approach is shown in *Figure 1* in the form of a flow diagram. At each stage, the results are assessed using appropriate statistical tests (Section B) and compared for consistency with those of the previous stage. Where unacceptable variations arise, changes to the procedure are implemented and the assessment process repeated. The performance and robustness of the overall method are finally tested with field trials in one or more routine analytical laboratories before the method is considered to be fully validated.

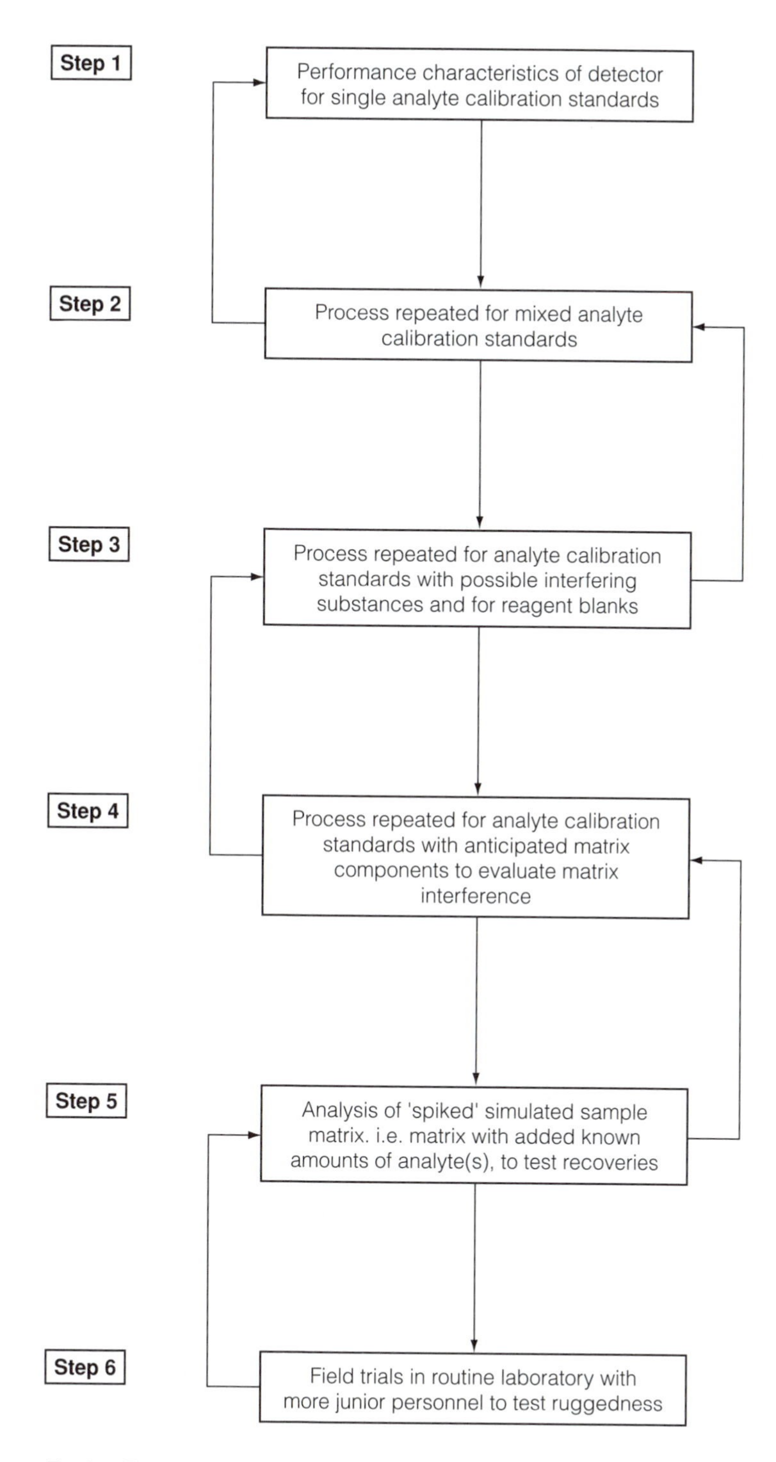

Fig. 1. Flow chart for method validation.

A4 Sampling and sample handling

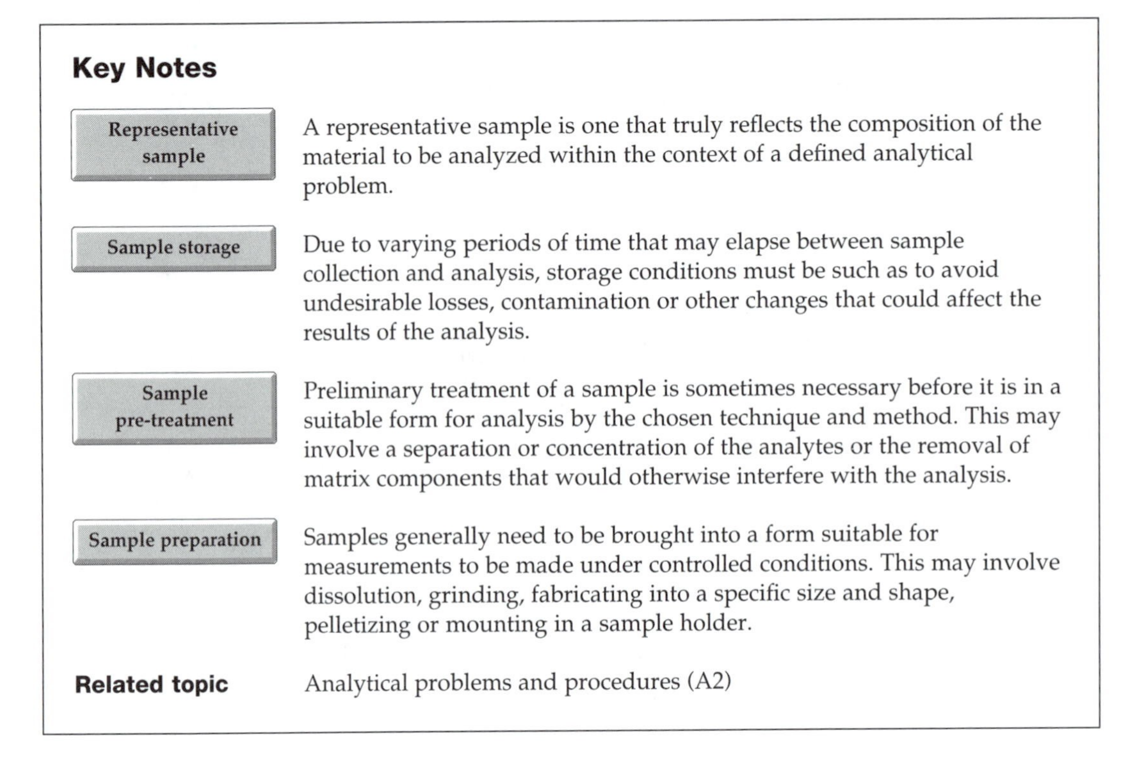

Key Notes

Representative sample	A representative sample is one that truly reflects the composition of the material to be analyzed within the context of a defined analytical problem.
Sample storage	Due to varying periods of time that may elapse between sample collection and analysis, storage conditions must be such as to avoid undesirable losses, contamination or other changes that could affect the results of the analysis.
Sample pre-treatment	Preliminary treatment of a sample is sometimes necessary before it is in a suitable form for analysis by the chosen technique and method. This may involve a separation or concentration of the analytes or the removal of matrix components that would otherwise interfere with the analysis.
Sample preparation	Samples generally need to be brought into a form suitable for measurements to be made under controlled conditions. This may involve dissolution, grinding, fabricating into a specific size and shape, pelletizing or mounting in a sample holder.
Related topic	Analytical problems and procedures (A2)

Representative sample

The importance of obtaining a representative sample for analysis cannot be overemphasized. Without it, results may be meaningless or even grossly misleading. Sampling is particularly crucial where a **heterogeneous** material is to be analyzed. It is vital that the aims of the analysis are understood and an appropriate sampling procedure adopted. In some situations, a **sampling plan** or strategy may need to be devised so as to optimize the value of the analytical information collected. This is necessary particularly where environmental samples of soil, water or the atmosphere are to be collected or a complex industrial process is to be monitored. Legal requirements may also determine a sampling strategy, particularly in the food and drug industries. A small sample taken for analysis is described as a **laboratory sample**. Where duplicate analyses or several different analyses are required, the laboratory sample will be divided into **sub-samples** which should have identical compositions.

Homogeneous materials (e.g., single or mixed solvents or solutions and most gases) generally present no particular sampling problem as the composition of any small laboratory sample taken from a larger volume will be representative of the bulk solution. **Heterogeneous materials** have to be homogenized prior to obtaining a laboratory sample if an average or bulk composition is required. Conversely, where analyte levels in different parts of the material are to be measured, they may need to be physically separated before laboratory samples are taken. This is known as **selective sampling**. Typical examples of heterogeneous materials where selective sampling may be necessary include:

- surface waters such as streams, rivers, reservoirs and seawater, where the concentrations of trace metals or organic compounds in solution and in sediments or suspended particulate matter may each be of importance;
- materials stored in bulk, such as grain, edible oils, or industrial organic chemicals, where physical segregation (stratification) or other effects may lead to variations in chemical composition throughout the bulk;
- ores, minerals and alloys, where information about the distribution of a particular metal or compound is sought;
- laboratory, industrial or urban atmospheres where the concentrations of toxic vapors and fumes may be localized or vary with time.

Obtaining a laboratory sample to establish an average analyte level in a highly heterogeneous material can be a lengthy procedure. For example, sampling a large shipment of an ore or mineral, where the economic cost needs to be determined by a very accurate assay, is typically approached in the following manner.

(i) Relatively large pieces are **randomly** selected from different parts of the shipment.
(ii) The pieces are crushed, ground to coarse granules and thoroughly mixed.
(iii) A repeated **coning and quartering** process, with additional grinding to reduce particle size, is used until a laboratory-sized sample is obtained. This involves creating a conical heap of the material, dividing it into four equal portions, discarding two diagonally opposite portions and forming a new conical heap from the remaining two quarters. The process is then repeated as necessary (*Fig. 1*).

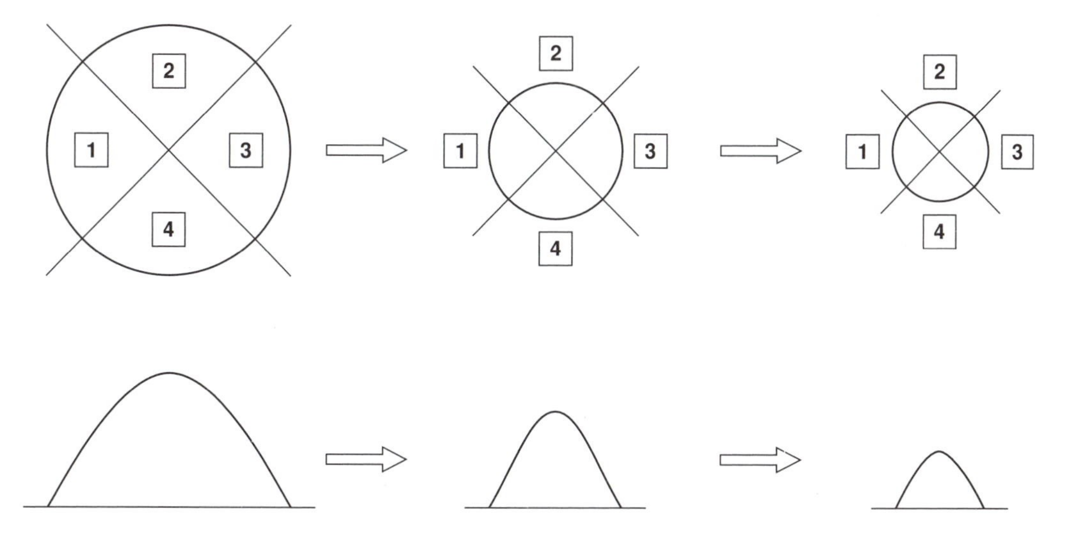

Fig. 1. A diagrammatic representation of coning and quartering (quarters 1 and 3, or 2 and 4 are discarded each time).

The distribution of toxic heavy metals or organic compounds in a land redevelopment site presents a different problem. Here, to economize on the number of analyses, a grid is superimposed on the site dividing it up into approximately one- to five-metre squares. From each of these, samples of soil will be taken at several specified depths. A three-dimensional representation of the distribution of each analyte over the whole site can then be produced, and any localized high concentrations, or **hot spots**, can be investigated by taking further, more closely-spaced, samples. Individual samples may need to be ground, coned and quartered as part of the sampling strategy.

Repeated sampling over a period of time is a common requirement. Examples include the continuous monitoring of a process stream in a manufacturing plant and the frequent sampling of patients' body fluids for changes in the levels of drugs, metabolites, sugars or enzymes, etc., during hospital treatment. Studies of seasonal variations in the levels of pesticide, herbicide and fertilizer residues in soils and surface waters, or the continuous monitoring of drinking water supplies are two further examples.

Having obtained a representative sample, it **must** be **labeled** and stored under appropriate conditions. Sample identification through proper labeling, increasingly done by using bar codes and optical readers under computer control, is an essential feature of sample handling.

Sample storage

Samples often have to be collected from places remote from the analytical laboratory and several days or weeks may elapse before they are received by the laboratory and analyzed. Furthermore, the workload of many laboratories is such that incoming samples are stored for a period of time prior to analysis. In both instances, sample containers and storage conditions (e.g., temperature, humidity, light levels and exposure to the atmosphere) must be controlled such that no significant changes occur that could affect the validity of the analytical data. The following effects during storage should be considered:

- increases in temperature leading to the loss of volatile analytes, thermal or biological degradation, or increased chemical reactivity;
- decreases in temperature that lead to the formation of deposits or the precipitation of analytes with low solubilities;
- changes in humidity that affect the moisture content of hygroscopic solids and liquids or induce hydrolysis reactions;
- UV radiation, particularly from direct sunlight, that induces photochemical reactions, photodecomposition or polymerization;
- air-induced oxidation;
- physical separation of the sample into layers of different density or changes in crystallinity.

In addition, containers may leak or allow contaminants to enter.

A particular problem associated with samples having very low (**trace** and **ultra-trace**) levels of analytes in solution is the possibility of losses by adsorption onto the walls of the container or contamination by substances being leached from the container by the sample solvent. Trace metals may be depleted by adsorption or ion-exchange processes if stored in glass containers, whilst sodium, potassium, boron and silicates can be leached from the glass into the sample solution. Plastic containers should always be used for such samples.

Conversely, sample solutions containing organic solvents and other organic liquids should be stored in glass containers because the base plastic or additives such as plasticizers and antioxidants may be leached from the walls of plastic containers.

Sample pre-treatment

Samples arriving in an analytical laboratory come in a very wide assortment of sizes, conditions and physical forms and can contain analytes from major constituents down to ultra-trace levels. They can have a variable moisture content and the matrix components of samples submitted for determinations of the same analyte(s) may also vary widely. A preliminary, or **pre-treatment**, is often used to **condition** them in readiness for the application of a specific method of analysis or to **pre-concentrate** (enrich) analytes present at very low levels. Examples of pre-treatments are:

- drying at 100°C to 120°C to eliminate the effect of a **variable moisture content**;
- weighing before and after drying enables the water content to be calculated or it can be established by thermogravimetric analysis (Topic G1);
- separating the analytes into groups with common characteristics by distillation, filtration, centrifugation, solvent or solid phase extraction (Topic D1);
- removing or reducing the level of **matrix components** that are known to cause **interference** with measurements of the analytes;
- concentrating the analytes if they are below the concentration range of the analytical method to be used by evaporation, distillation, co-precipitation, ion exchange, solvent or solid phase extraction or electrolysis.

Sample **clean-up** in relation to **matrix interference** and to protect specialized analytical equipment such as chromatographic columns and detection systems from high levels of matrix components is widely practised using **solid phase extraction** (**SPE**) cartridges (Topic D1). Substances such as lipids, fats, proteins, pigments, polymeric and tarry substances are particularly detrimental.

Sample preparation

A laboratory sample generally needs to be prepared for analytical measurement by treatment with reagents that convert the analyte(s) into an appropriate chemical form for the selected technique and method, although in some instances it is examined directly **as received** or mounted in a sample holder for surface analysis. If the material is readily soluble in aqueous or organic solvents, a simple dissolution step may suffice. However, many samples need first to be decomposed to release the analyte(s) and facilitate specific reactions in solution. Sample solutions may need to be diluted or concentrated by enrichment so that analytes are in an optimum concentration range for the method. The stabilization of solutions with respect to pH, ionic strength and solvent composition, and the removal or **masking** of interfering matrix components not accounted for in any pre-treatment may also be necessary. An **internal standard** for reference purposes in quantitative analysis (Topic A5 and Section B) is sometimes added before adjustment to the final prescribed volume. Some common methods of decomposition and dissolution are given in *Table 1*.

Table 1. Some methods for sample decomposition and dissolution

Method of attack	Type of sample
Heated with concentrated mineral acids (HCl, HNO_3, aqua regia) or strong alkali, including microwave digestion	Geological, metallurgical
Fusion with flux (Na_2O_2, Na_2CO_3, $LiBO_2$, $KHSO_4$, KOH)	Geological, refractory materials
Heated with HF and H_2SO_4 or $HClO_4$	Silicates where SiO_2 is not the analyte
Acid leaching with HNO_3	Soils and sediments
Dry oxidation by heating in a furnace or wet oxidation by boiling with concentrated H_2SO_4 and HNO_3 or $HClO_4$	Organic materials with inorganic analytes

A5 CALIBRATION AND STANDARDS

Key Notes

Calibration Calibration or standardization is the process of establishing the response of a detection or measurement system to known amounts or concentrations of an analyte under specified conditions, or the comparison of a measured quantity with a reference value.

Chemical standard A chemical standard is a material or substance of very high purity and/or known composition that is used to standardize a reagent or calibrate an instrument.

Reference material A reference material is a material or substance, one or more properties of which are sufficiently homogeneous and well established for it to be used for the calibration of apparatus, the assessment of a measurement method or for assigning values to materials.

Related topic Calibration and linear regression (B4)

Calibration

With the exception of absolute methods of analysis that involve chemical reactions of known stoichiometry (e.g., gravimetric and titrimetric determinations), a **calibration** or **standardization procedure** is required to establish the relation between a measured physico-chemical response to an analyte and the amount or concentration of the analyte producing the response. Techniques and methods where calibration is necessary are frequently instrumental, and the detector response is in the form of an electrical signal. An important consideration is the effect of matrix components on the analyte detector signal, which may be supressed or enhanced, this being known as the **matrix effect**. When this is known to occur, **matrix matching** of the calibration standards to simulate the gross composition expected in the samples is essential (i.e. matrix components are added to all the analyte standards in the same amounts as are expected in the samples).

There are several methods of calibration, the choice of the most suitable depending on the characteristics of the analytical technique to be employed, the nature of the sample and the level of analyte(s) expected. These include:

- **External standardization**. A series of at least four **calibration standards** containing known amounts or concentrations of the analyte and matrix components, if required, is either prepared from laboratory chemicals of guaranteed purity (AnalaR or an equivalent grade) or purchased as a concentrated standard ready to use. The response of the detection system is recorded for each standard under specified and stable conditions and additionally for a **blank**, sometimes called a **reagent blank** (a standard prepared in an identical fashion to the other standards but omitting the analyte). The data is either plotted as a **calibration graph** or used to calculate a **factor** to convert detector responses measured for the analyte in samples into corresponding masses or concentrations (Topic B4).

- Standard addition.
- Internal standardization.

The last two methods of calibration are described in Topic B4.

Instruments and apparatus used for analytical work must be correctly maintained and calibrated against reference values to ensure that measurements are accurate and reliable. Performance should be checked regularly and records kept so that any deterioration can be quickly detected and remedied. Microcomputer and microprocessor controlled instrumentation often has built-in performance checks that are automatically initiated each time an instrument is turned on. Some examples of instrument or apparatus calibration are

- manual calibration of an electronic balance with certified weights;
- calibration of volumetric glassware by weighing volumes of pure water;
- calibration of the wavelength and absorbance scales of spectrophotometers with certified emission or absorption characteristics;
- calibration of temperature scales and electrical voltage or current readouts with certified measurement equipment.

Chemical standard

Materials or substances suitable for use as chemical standards are generally single compounds or elements. They must be of known composition, and high purity and stability. Many are available commercially under the name **AnalaR**. **Primary standards**, which are used principally in titrimetry (Section C) to standardize a reagent (titrant) (i.e. to establish its exact concentration) must be internationally recognized and should fulfil the following requirements:

- be easy to obtain and preserve in a high state of purity and of known chemical composition;
- be non-hygroscopic and stable in air allowing accurate weighing;
- have impurities not normally exceeding 0.02% by weight;
- be readily soluble in water or another suitable solvent;
- react rapidly with an analyte in solution;
- other than pure elements, to have a high relative molar mass to minimize weighing errors.

Primary standards are used directly in titrimetric methods or to standardize solutions of **secondary** or **working standards** (i.e. materials or substances that do not fulfill all of the above criteria, that are to be used subsequently as the titrant in a particular method). Chemical standards are also used as reagents to effect reactions with analytes before completing the analysis by techniques other than titrimetry.

Some approved primary standards for titrimetric analysis are given in *Table 1*.

Reference material

Reference materials are used to demonstrate the accuracy, reliability and comparability of analytical results. A **certified** or **standard reference material** (**CRM** or **SRM**) is a reference material, the values of one or more properties of which have been certified by a technically valid procedure and accompanied by a traceable certificate or other documentation issued by a certifying body such as the

Table 1. Some primary standards used in titrimetric analysis

Type of titration	Primary standard
Acid-base	Sodium carbonate, Na_2CO_3
	Sodium tetraborate, $Na_2B_4O_7.10H_2O$
	Potassium hydrogen phthalate, $KH(C_8H_4O_4)$
	Benzoic acid, C_6H_5COOH
Redox	Potassium dichromate, $K_2Cr_2O_7$
	Potassium iodate, KIO_3
	Sodium oxalate, $Na_2C_2O_4$
Precipitation (silver halide)	Silver nitrate, $AgNO_3$
	Sodium chloride, NaCl
Complexometric (EDTA)	Zinc, Zn
	Magnesium, Mg
	EDTA (disodium salt), $C_{10}H_{14}N_2O_8Na_2$

Bureau of Analytical Standards. CRMs or SRMs are produced in various forms and for different purposes and they may contain one or more certified components, such as

- pure substances or solutions for calibration or identification;
- materials of known matrix composition to facilitate comparisons of analytical data;
- materials with approximately known matrix composition and specified components.

They have a number of principal uses, including

- validation of new methods of analysis;
- standardization/calibration of other reference materials;
- confirmation of the validity of standardized methods;
- support of quality control and quality assurance schemes.

A6 QUALITY IN ANALYTICAL LABORATORIES

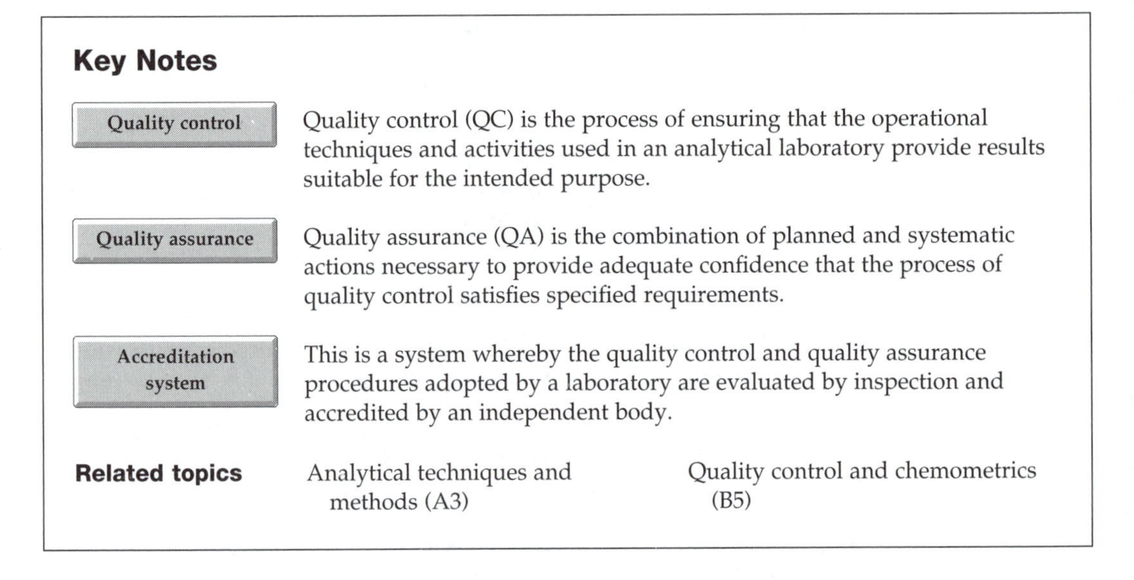

Key Notes

Quality control	Quality control (QC) is the process of ensuring that the operational techniques and activities used in an analytical laboratory provide results suitable for the intended purpose.
Quality assurance	Quality assurance (QA) is the combination of planned and systematic actions necessary to provide adequate confidence that the process of quality control satisfies specified requirements.
Accreditation system	This is a system whereby the quality control and quality assurance procedures adopted by a laboratory are evaluated by inspection and accredited by an independent body.

Related topics Analytical techniques and methods (A3) Quality control and chemometrics (B5)

Quality control

Analytical data must be of demonstrably high quality to ensure confidence in the results. **Quality control (QC)** comprises a system of planned activities in an analytical laboratory whereby analytical methods are monitored at every stage to verify compliance with validated procedures and to take steps to eliminate the causes of unsatisfactory performance. Results are considered to be of sufficiently high quality if

- they meet the specific requirements of the requested analytical work within the context of a defined problem;
- there is confidence in their validity;
- the work is cost effective.

To implement a QC system, a complete understanding of the chemistry and operations of the analytical method and the likely sources and magnitudes of errors at each stage is essential. The use of reference materials (Topic A5) during method validation (Topic A3) ensures that results are **traceable** to certified sources. QC processes should include:

- checks on the **accuracy** and **precision** of the data using statistical tests (Section B);
- detailed records of calibration, raw data, results and instrument performance;
- observations on the nature and behavior of the sample and unsatisfactory aspects of the methodology;
- **control charts** to determine system control for instrumentation and repeat analyses (Topic B5);

- provision of full documentation and **traceability** of results to recognized reference materials through recorded identification;
- maintenance and calibration of instrumentation to manufacturers' specifications;
- management and control of laboratory chemicals and other materials including checks on quality;
- adequate training of laboratory personnel to ensure understanding and competence;
- external verification of results wherever possible;
- accreditation of the laboratory by an independent organization.

Quality assurance The overall management of an analytical laboratory should include the provision of evidence and assurances that appropriate QC procedures for laboratory activities are being correctly implemented. **Quality assurance (QA)** is a managerial responsibility that is designed to ensure that this is the case and to generate confidence in the analytical results. Part of QA is to build confidence through the laboratory participating in **interlaboratory studies** where several laboratories analyze one or more identical homogeneous materials under specified conditions. **Proficiency testing** is a particular type of study to assess the performance of a laboratory or analyst relative to others, whilst **method performance studies** and **certification studies** are undertaken to check a particular analytical method or reference material respectively. The results of such studies and their statistical assessment enable the performances of individual participating laboratories to be demonstrated and any deficiencies in methodology and the training of personnel to be addressed.

Accreditation system Because of differences in the interpretation of the term **quality**, which can be defined as **fitness for purpose**, QC and QA systems adopted by analyical laboratories in different industries and fields of activity can vary widely. For this reason, defined **quality standards** have been introduced by a number of organizations throughout the world. Laboratories can design and implement their own quality systems and apply to be inspected and accredited by the organization for the standard most appropriate to their activity. A number of organizations that offer accreditation suitable for analytical laboratories and their corresponding quality standards are given in *Table 1*.

Table 1. Accreditation organizations and their quality standards

Name of accreditation organization	Quality standard
Organization for Economic Co-operation and Development (OECD)	Good Laboratory Practice (GLP)
The International Organization for Standardization (ISO)	ISO 9000 series of quality standards ISO Guide 25 general requirements for competence of calibration and testing laboratories
European Committee for Standardization (CEN)	EN 29000 series EN 45000 series
British Standards Institution (BSI)	BS 5750 quality standard BS 7500 series
National Measurement Accreditation Service (NAMAS)	NAMAS

B1 Errors in analytical measurements

Key Notes

Measurement errors All measurement processes are subject to measurement errors that affect numerical data and which arise from a variety of sources.

Absolute and relative errors An absolute error is the numerical difference between a measured value and a true or accepted value. A relative error is the absolute error divided by the true or accepted value.

Determinate errors Also known as systematic errors, or bias, these generally arise from determinate or identifiable sources causing measured values to differ from a true or accepted value.

Indeterminate errors Also known as random errors, these arise from a variety of uncontrolled sources and cause small random variations in a measured quantity when the measurement is repeated a number of times.

Accumulated errors Where several different measurements are combined to compute an overall analytical result, the errors associated with each individual measurement contribute to a total or accumulated error.

Related topic Assessment of accuracy and precision (B2)

Measurement errors

The causes of measurement errors are numerous and their magnitudes are variable. This leads to **uncertainties** in reported results. However, measurement errors can be minimized and some types eliminated altogether by careful experimental design and control. Their effects can be assessed by the application of **statistical methods** of data analysis and **chemometrics** (Topic B5). **Gross errors** may arise from faulty equipment or bad laboratory practice; proper equipment maintenance and appropriate training and supervision of personnel should eliminate these.

Nevertheless, whether it is reading a burette or thermometer, weighing a sample or timing events, or monitoring an electrical signal or liquid flow, there will always be inherent variations in the measured parameter if readings are repeated a number of times under the same conditions. In addition, errors may go undetected if the **true** or **accepted** value is not known for comparison purposes.

Errors must be controlled and assessed so that valid analytical measurements can be made and reported. The **reliability** of such data must be demonstrated so that an end-user can have an acceptable **degree of confidence** in the results of an analysis.

Absolute and relative errors

The **absolute error**, E_A, in a measurement or result, x_M, is given by the equation

$$E_A = x_M - x_T$$

where x_T is the true or accepted value. Examples are shown in *Figure 1* where a 200 mg aspirin standard has been analyzed a number of times. The **absolute errors** range from –4 mg to +10 mg.

The **relative error**, E_R, in a measurement or result, x_M, is given by the equation

$$E_R = (x_M - x_T)/x_T$$

Often, E_R is expressed as a percentage relative error, $100E_R$. Thus, for the aspirin results shown in *Figure 1*, the relative error ranges from –2% to +5%. Relative errors are particularly useful for comparing results of differing magnitude.

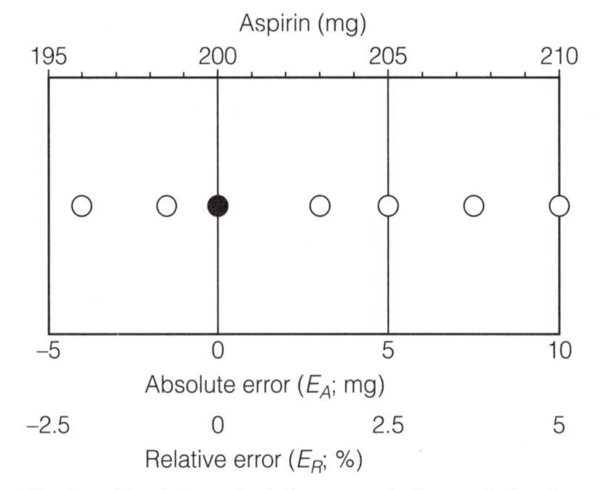

Fig. 1. Absolute and relative errors in the analysis of an aspirin standard.

Determinate errors

There are three basic sources of **determinate** or **systematic errors** that lead to a **bias** in measured values or results:

- the analyst or operator;
- the equipment (apparatus and instrumentation) and the laboratory environment;
- the method or procedure.

It should be possible to eliminate errors of this type by careful observation and record keeping, equipment maintenance and training of laboratory personnel.

Operator errors can arise through carelessness, insufficient training, illness or disability. **Equipment errors** include substandard volumetric glassware, faulty or worn mechanical components, incorrect electrical signals and a poor or insufficiently controlled laboratory environment. **Method** or **procedural errors** are caused by inadequate method validation, the application of a method to samples or concentration levels for which it is not suitable or unexpected variations in sample characteristics that affect measurements. Determinate errors that lead to a higher value or result than a true or accepted one are said to show a **positive bias**; those leading to a lower value or result are said to show a **negative bias**. Particularly large errors are described as **gross errors**; these should be easily apparent and readily eliminated.

Determinate errors can be **proportional** to the size of sample taken for analysis. If so, they will have the same effect on the magnitude of a result regardless of the size of the sample, and their presence can thus be difficult to detect. For example, copper(II) can be determined by titration after reaction with potassium iodide to release iodine according to the equation

$$2Cu^{2+} + 4I^{-} \rightarrow 2CuI + I_2$$

However, the reaction is not specific to copper(II), and any iron(III) present in the sample will react in the same way. Results for the determination of copper in an alloy containing 20%, but which also contained 0.2% of iron are shown in *Figure 2* for a range of sample sizes. The same absolute error of +0.2% or relative error of 1% (i.e. a **positive bias**) occurs regardless of sample size, due to the presence of the iron. This type of error may go undetected unless the constituents of the sample and the chemistry of the method are known.

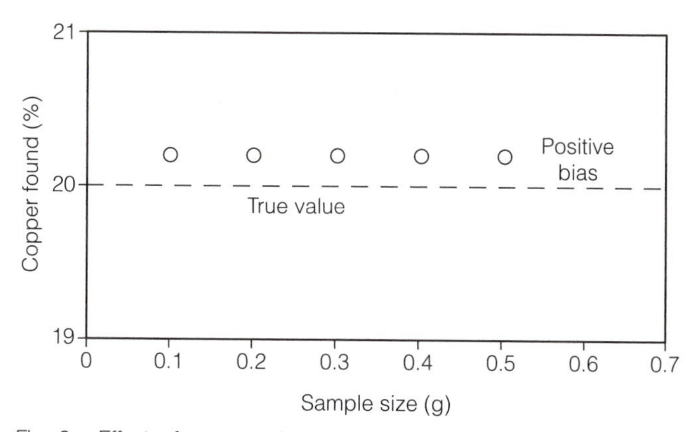

Fig. 2. Effect of a proportional error on the determination of copper by titration in the presence of iron.

Constant determinate errors are independent of sample size, and therefore become less significant as the sample size is increased. For example, where a visual indicator is employed in a volumetric procedure, a small amount of titrant is required to change the color at the end-point, even in a **blank** solution (i.e. when the solution contains none of the species to be determined). This **indicator blank** (Topic C5) is the same regardless of the size of the titer when the species being determined is present. The **relative error**, therefore, decreases with the magnitude of the titer, as shown graphically in *Figure 3*. Thus, for an indicator blank of 0.02 cm^3, the relative error for a 1 cm^3 titer is 2%, but this falls to only 0.08% for a 25 cm^3 titer.

Indeterminate errors

Known also as **random errors**, these arise from random fluctuations in measured quantities, which always occur even under closely controlled conditions. It is impossible to eliminate them entirely, but they can be minimized by careful experimental design and control. Environmental factors such as temperature, pressure and humidity, and electrical properties such as current, voltage and resistance are all susceptible to small continuous and random variations described as **noise**. These contribute to the overall indeterminate error in any

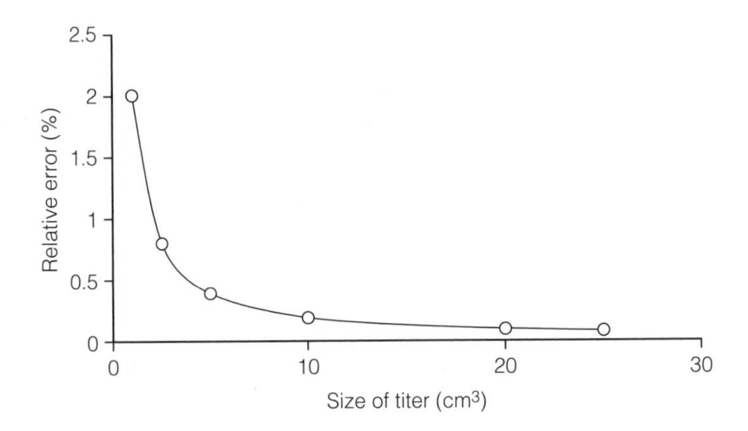

Fig. 3. Effect of a constant error on titers of differing magnitudes.

physical or physico-chemical measurement, but no one specific source can be identified.

A series of measurements made under the same prescribed conditions and represented graphically is known as a **frequency distribution**. The frequency of occurrence of each experimental value is plotted as a function of the magnitude of the error or **deviation** from the **average** or **mean** value. For analytical data, the values are often distributed symmetrically about the mean value, the most common being the **normal error** or **Gaussian distribution curve**. The curve (*Fig. 4*) shows that

- small errors are more probable than large ones,
- positive and negative errors are equally probable, and
- the maximum of the curve corresponds to the mean value.

The normal error curve is the basis of a number of statistical tests that can be applied to analytical data to assess the effects of indeterminate errors, to compare values and to establish levels of confidence in results (Topics B2 and B3).

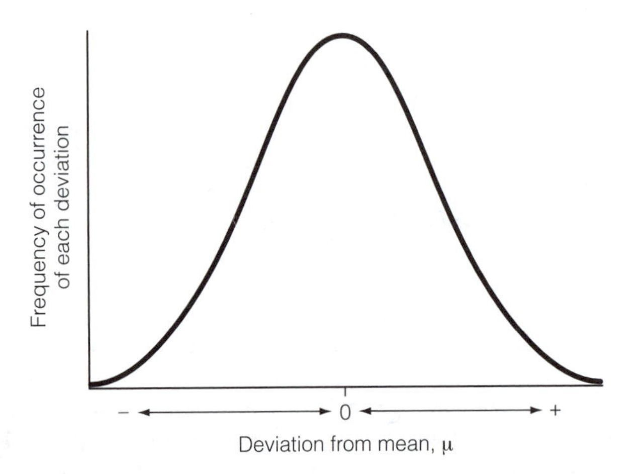

Fig. 4. The normal error or Gaussian distribution curve.

Accumulated errors

Errors are associated with every measurement made in an analytical procedure, and these will be aggregated in the final calculated result. The **accumulation** or **propagation** of errors is treated similarly for both determinate (systematic) and indeterminate (random) errors.

Determinate (systematic) errors can be either positive or negative, hence some cancellation of errors is likely in computing an overall determinate error, and in some instances this may be zero. The overall error is calculated using one of two alternative expressions, that is

- where only a linear combination of individual measurements is required to compute the result, the overall **absolute** determinate error, E_T, is given by

  $$E_T = E_1 + E_2 + E_3 + \ldots\ldots$$

  E_1 and E_2 etc., being the **absolute** determinate errors in the individual measurements taking sign into account
- where a multiplicative expression is required to compute the result, the overall **relative** determinate error, E_{TR}, is given by

$$E_{TR} = E_{1R} + E_{2R} + E_{3R} + \ldots\ldots$$

E_{1R} and E_{2R} etc., being the **relative** determinate errors in the individual measurements taking sign into account.

The accumulated effect of indeterminate (random) errors is computed by combining statistical parameters for each measurement (Topic B2).

B2 ASSESSMENT OF ACCURACY AND PRECISION

Key Notes

Accuracy and precision — Accuracy is the closeness of an experimental measurement or result to the true or accepted value. Precision is the closeness of agreement between replicated measurements or results obtained under the same prescribed conditions.

Standard deviation — The standard deviation of a set of values is a statistic based on the normal error (Gaussian) curve and used as a measure of precision.

Relative standard deviation — Relative standard deviation (coefficient of variation) is the standard deviation expressed as a percentage of the measured value.

Pooled standard deviation — A standard deviation can be calculated for two or more sets of data by pooling the values to give a more reliable measure of precision.

Variance — This is the square of the standard deviation, which is used in some statistical tests.

Overall precision — An estimate of the overall precision of an analytical procedure can be made by combining the precisions of individual measurements.

Confidence interval — This is the range of values around an experimental result within which the true or accepted value is expected to lie with a defined level of probability.

Related topic — Errors in analytical measurements (B1)

Accuracy and precision

These two characteristics of numerical data are the most important and the most frequently confused. It is vital to understand the difference between them, and this is best illustrated diagrammatically as in *Figure 1*. Four analysts have each performed a set of five titrations for which the correct titer is known to be 20.00 cm^3. The titers have been plotted on a linear scale, and inspection reveals the following:

- the average titers for analysts B and D are very close to 20.00 cm^3 – these two sets are therefore said to have **good accuracy**;
- the average titers for analysts A and C are well above and below 20.00 cm^3 respectively – these are therefore said to have **poor accuracy**;
- the five titers for analyst A and the five for analyst D are very close to one another within each set – these two sets therefore both show **good precision**;
- the five titers for analyst B and the five for analyst C are spread widely within each set – these two sets therefore both show **poor precision**.

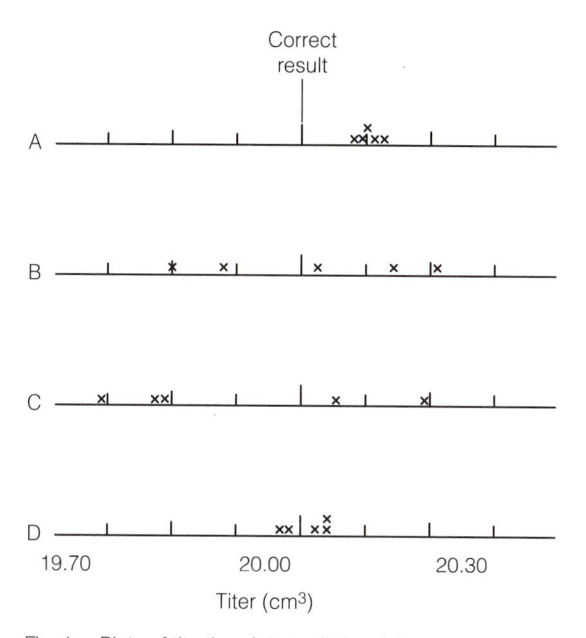

Fig. 1. Plots of titration data to distinguish accuracy and precision.

It should be noted that good precision does not necessarily produce good accuracy (analyst A) and poor precision does not necessarily produce poor accuracy (analyst B). However, confidence in the analytical procedure and the results is greater when good precision can be demonstrated (analyst D).

Accuracy is generally the more important characteristic of quantitative data to be assessed, although consistency, as measured by precision, is of particular concern in some circumstances. **Trueness** is a term associated with accuracy, which describes the closeness of agreement between the average of a large number of results and a true or accepted reference value. The degree of accuracy required depends on the context of the analytical problem; results must be shown to be fit for the purpose for which they are intended. For example, one result may be satisfactory if it is within 10% of a true or accepted value whilst it may be necessary for another to be within 0.5%. By repeating an analysis a number of times and computing an average value for the result, the level of accuracy will be improved, provided that no systematic error (bias) has occurred. Accuracy cannot be established with certainty where a true or accepted value is not known, as is often the case. However, statistical tests indicating the accuracy of a result with a given **probability** are widely used (*vide infra*).

Precision, which is a measure of the **variability** or **dispersion** within a set of **replicated** values or results obtained under **the same prescribed conditions**, can be assessed in several ways. The **spread** or **range** (i.e. the difference between the highest and lowest value) is sometimes used, but the most popular method is to **estimate** the **standard deviation** of the data (*vide infra*). The precision of results obtained within one working session is known as **repeatability** or **within-run precision**. The precision of results obtained over a series of working sessions is known as **reproducibility** or **between-runs precision**. It is sometimes necessary to separate the contributions made to the overall precision by **within-run** and

between-runs variability. It may also be important to establish the precision of individual steps in an analysis.

Standard deviation

This is the most widely used measure of **precision** and is a parameter of the **normal error** or **Gaussian curve** (Topic B1, *Fig. 4*). *Figure 2* shows two curves for the frequency distribution of two theoretical sets of data, each having an infinite number of values and known as a **statistical population**.

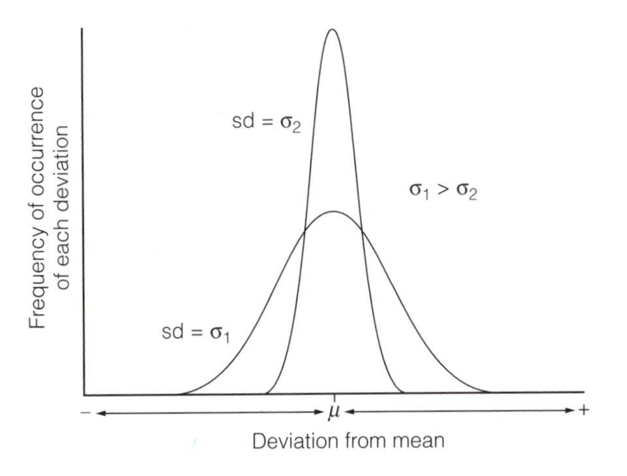

Fig. 2. Normal error or Gaussian curves for the frequency distributions of two statistical populations with differing spreads.

The maximum in each curve corresponds to the **population mean**, which for these examples has the same value, μ. However, the spread of values for the two sets is quite different, and this is reflected in the half-widths of the two curves at the points of inflection, which, by definition, is the **population standard deviation**, σ. As σ_2 is much less than σ_1, the **precision** of the second set is much better than that of the first. The abscissa scale can be calibrated in absolute units or, more commonly, as positive and negative deviations from the mean, μ.

In general, the smaller the spread of values or deviations, the smaller the value of σ and hence the better the precision. In practice, the true values of μ and σ can never be known because they relate to a population of infinite size. However, an assumption is made that a small number of experimental values or a **statistical sample** drawn from a statistical population is also distributed normally or approximately so. The **experimental mean**, $\bar{x}$, of a set of values x_1, x_2, x_3,.......x_n is therefore considered to be an **estimate** of the true or population mean, μ, and the **experimental standard deviation**, s, is an **estimate** of the true or population standard deviation, σ.

A useful property of the normal error curve is that, regardless of the magnitude of μ and σ, the area under the curve within defined limits on either side of μ (usually expressed in multiples of $\pm\sigma$) is a constant proportion of the total area. Expressed as a percentage of the total area, this indicates that a particular percentage of the population will be found between those limits.

Thus, approximately 68% of the area, and therefore of the population, will be found within ±1σ of the mean, approximately 95% will be found within ±2σ and approximately 99.7% within ±3σ. More practically convenient levels, as shown in *Figure 3*, are those corresponding to 90%, 95% and 99% of the population, which are defined by ±1.64σ, ±1.96σ and ±2.58σ respectively. Many statistical tests are based on these **probability levels**.

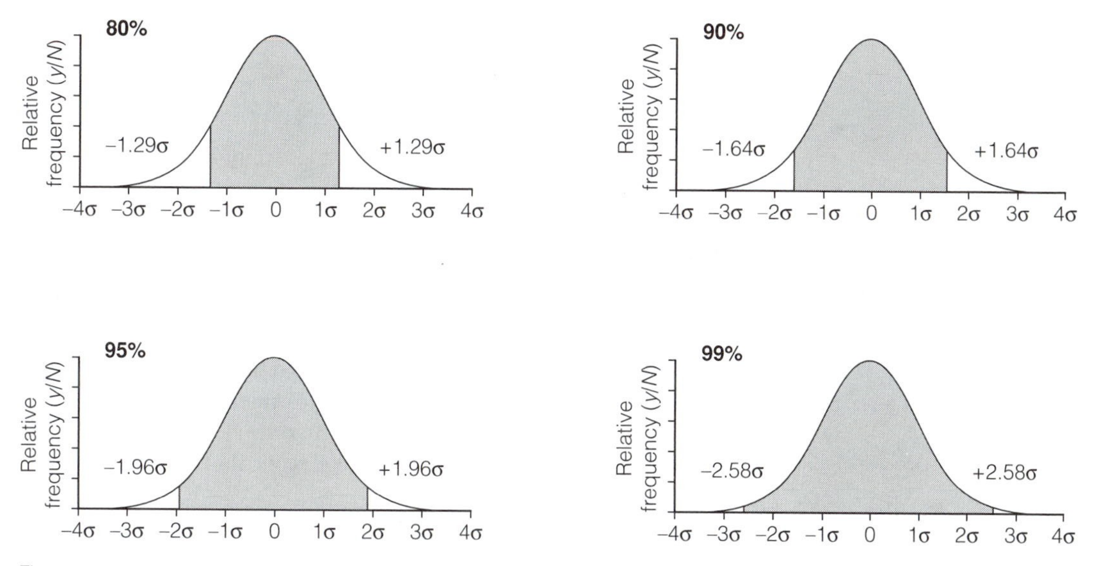

Fig. 3. Proportions of a population within defined limits of the mean.

The value of the population standard deviation, σ, is given by the formula

$$\sigma = \sqrt{\frac{\sum_{i=1}^{i=N}(x_i - \mu)^2}{N}} \tag{1}$$

where x_i represents any individual value in the population and N is the total number of values, strictly infinite. The summation symbol, Σ, is used to show that the numerator of the equation is the sum for $i = 1$ *to* $i = N$ of the **squares** of the **deviations** of the individual x values from the population mean, μ. For very large sets of data (e.g., when $N > 50$), it may be justifiable to use this formula as the difference between σ and s will then be negligible. However, most analytical data consists of sets of values of less than ten and often as small as three. Therefore, a modified formula is used to calculate an **estimated standard deviation**, s, to replace σ, and using an experimental mean, $\bar{x}$, to replace the population mean, μ:

$$s = \sqrt{\frac{\sum_{i=1}^{i=N}(x_i - \bar{x})^2}{N-1}} \tag{2}$$

Note that N in the denominator is replaced by $N-1$, which is known as the **number of degrees of freedom** and is defined as the number of **independent deviations** $(x_i - \bar{x})$ used to calculate s. For single sets of data, this is always one less than the number in the set because when $N-1$ deviations are known the last one can be deduced as, taking sign into account, $\sum_{i=1}^{i=N}(x_i - \bar{x})$ must be zero (see *Example 1* below).

In summary, the calculation of an estimated standard deviation, s, for a small number of values involves the following steps:

- calculation of an experimental mean;
- calculation of the deviations of individual x_i values from the mean;
- squaring the deviations and summing them;
- dividing by the number of degrees of freedom, $N-1$, and
- taking the square root of the result.

Note that if N were used in the denominator, the calculated value of s would be an underestimate of σ.

Estimated standard deviations are easily obtained using a calculator that incorporates statistical function keys or with one of the many computer software packages. It is, however, useful to be able to perform a stepwise arithmetic calculation, and an example using the set of five replicate titers by analyst A (*Fig. 1*) is shown below.

Example 1

	x_i/cm^3	$(x_i - x)$	$(x_i - x)^2$
	20.16	−0.04	1.6×10^{-3}
	20.22	+0.02	4×10^{-4}
	20.18	−0.02	4×10^{-4}
	20.20	0.00	0
	20.24	+0.04	1.6×10^{-3}
Σ	101.00		$\mathbf{4 \times 10^{-3}}$
$\bar{x}$	**20.20**		

$$s = \sqrt{\frac{4 \times 10^{-3}}{4}} = 0.032\ \text{cm}^3$$

Relative standard deviation

The **relative standard deviation, RSD** or s_r, is also known as the **coefficient of variation**, CV. It is a measure of **relative precision** and is normally expressed as a percentage of the mean value or result

$$s_r = (s/\bar{x}) \times 100 \tag{3}$$

It is an example of a **relative error** (Topic B1) and is particularly useful for comparisons between sets of data of differing magnitude or units, and in calculating accumulated (propagated) errors. The **RSD** for the data in *Example 1* is given below.

Example 2

$$s_r = \frac{0.032}{20.20} \times 100 = 0.16\%$$

Pooled standard deviation

Where replicate samples are analyzed on a number of occasions under the same prescribed conditions, an improved estimate of the standard deviation can be obtained by pooling the data from the individual sets. A general formula for the **pooled standard deviation,** s_{pooled}, is given by the expression

$$s_{pooled} = \sqrt{\frac{\sum_{i=1}^{i=N_1}(x_i - \bar{x}_1)^2 + \sum_{i=1}^{i=N_2}(x_i - \bar{x}_2)^2 + \sum_{i=1}^{i=N_3}(x_i - \bar{x}_3)^2 + \ldots + \sum_{i=1}^{i=N_k}(x_i - \bar{x}_k)^2}{\sum_{i=1}^{i=k} N_i = k}} \quad (4)$$

where N_1, N_2, $N_3 \ldots N_k$ are the numbers of results in each of the k sets, and $\bar{x}_1$, $\bar{x}_2$, $\bar{x}_3, \ldots \bar{x}_k$, are the means for each of the k sets.

Variance

The **square** of the standard deviation, σ^2, or estimated standard deviation, s^2, is used in a number of statistical computations and tests, such as for calculating accumulated (propagated) errors (Topic B1 and below) or when comparing the precisions of two sets of data (Topic B3).

Overall precision

Random errors accumulated within an analytical procedure contribute to the overall precision. Where the calculated result is derived by the addition or subtraction of the individual values, the overall precision can be found by summing the **variances** of all the measurements so as to provide an estimate of the overall standard deviation, i.e.

$$s_{overall} = \sqrt{\left(s_1^{\,2} + s_2^{\,2} + s_3^{\,2} + \ldots\right)}$$

Example

In a titrimetric procedure, the buret must be read twice, and the error associated with each reading must be taken into account in estimating the overall precision. If the reading error has an estimated standard deviation of 0.02 cm^3, then the overall estimated standard deviation of the titration is given by

$$s_{overall} = \sqrt{\left(0.02^2 + 0.02^2\right)} = 0.028 \text{ cm}^3$$

Note that this is less than twice the estimated standard deviation of a single reading. The overall standard deviation of weighing by difference is estimated in the same way.

If the calculated result is derived from a multiplicative expression, the overall *relative* precision is found by summing the squares of the *relative* standard deviations of all the measurements, i.e.

$$s_{r(overall)} = \sqrt{\left(s_{r1}^2 + s_{r2}^2 + s_{r3}^2 + \ldots\right)}$$

Confidence interval

The true or accepted mean of a set of experimental results is generally unknown except where a certified reference material is being checked or analyzed for calibration purposes. In all other cases, an **estimate of the accuracy** of the experimental mean, $\bar{x}$, must be made. This can be done by defining a range of values on either side of $\bar{x}$ within which the true mean, μ, is expected to lie with a defined level of **probability**. This range, which ideally should be as narrow as possible, is based on the standard deviation and is known as the

confidence interval, CI, and the upper and lower limits of the range as **confidence limits**, CL. Confidence limits can be calculated using the standard deviation, σ, if it is known, or the estimated standard deviation, s, for the data. In either case, a **probability level** must be defined, otherwise the test is of no value.

When the standard deviation is already known from past history, the confidence limits are given by the equation

$$CL(\mu) = \bar{x} \pm \frac{z\sigma}{\sqrt{N}} \tag{5}$$

where z is a **statistical factor** related to the probability level required, usually 90%, 95% or 99%. The values of z for these levels are 1.64, 1.96 and 2.58, respectively, and correspond to the multiples of the standard deviation shown in *Figure 3*.

Where an estimated standard deviation is to be used, σ is replaced by s, which must first be calculated from the current data. The confidence limits are then given by the equation

$$CL(\mu) = \bar{x} \pm \frac{ts}{\sqrt{N}} \tag{6}$$

where z is replaced by an alternative **statistical factor**, t, also related to the probability level but in addition determined by the **number of degrees of freedom** for the set of data, i.e. **one less** than the number of results. It should be noted that (i) the confidence interval is inversely proportional to $\sqrt{N}$, and (ii) the higher the selected probability level, the greater the confidence interval becomes as both z and t increase. A probability level of 100 percent is meaningless, as the confidence limits would then have to be $\pm\infty$.

The following examples demonstrate the calculation of confidence limits using each of the two formulae.

Example 3

The chloride content of water samples has been determined a very large number of times using a particular method, and the standard deviation found to be 7 ppm. Further analysis of a particular sample gave experimental values of 350 ppm for a single determination, for the mean of two replicates and for the mean of four replicates. Using equation (5), and at the 95% probability level, $z = 1.96$ and the confidence limits are:

1 determination $\quad CL(\mu) = 350 \pm \frac{1.96 \times 7}{\sqrt{1}} = \mathbf{350 \pm 14\ ppm}$

2 determinations $\quad CL(\mu) = 350 \pm \frac{1.96 \times 7}{\sqrt{2}} = \mathbf{350 \pm 10\ ppm}$

4 determinations $\quad CL(\mu) = 350 \pm \frac{1.96 \times 7}{\sqrt{4}} = \mathbf{350 \pm 7\ ppm}$

Example 4

The same chloride analysis as in *Example 3*, but using a new method for which the standard deviation was not known, gave the following replicate results, mean and estimated standard deviation:

Chloride/ppm	Mean	Estimated standard deviation
346	351.67 ppm	6.66 ppm
359		
350		

Using equation (6), and at the 95% probability level, $t = 4.3$ for **two degrees of freedom**, and the confidence limits are:

$$\text{3 determinations} \quad CL(\mu) = 352 \pm \frac{4.3 \times 6.66}{\sqrt{3}} = \mathbf{352 \pm 17\ ppm}$$

The wider limits given by equation (6) when the standard deviation is estimated with only three results reflects the much greater uncertainty associated with this value, which in turn affects the confidence in the degree of accuracy. To demonstrate good accuracy, the confidence interval, CI, should be as small as possible and increasing the number of replicates will clearly achieve this. However, due to the $\sqrt{N}$ term in the denominator, to reduce the interval by, say, a factor of two requires an increase in the number of replicates by a factor of four as shown by *Example 3*. Unfortunately, the law of diminishing returns applies here, so if the CI is to be halved again, the number of replicates must be increased from *four* to *sixteen*. Similarly, in *Example 4*, the number of replicates would have to be increased from three to twelve to halve the CI, which would represent an unacceptable amount of time and money for most analytical laboratories.

B3 SIGNIFICANCE TESTING

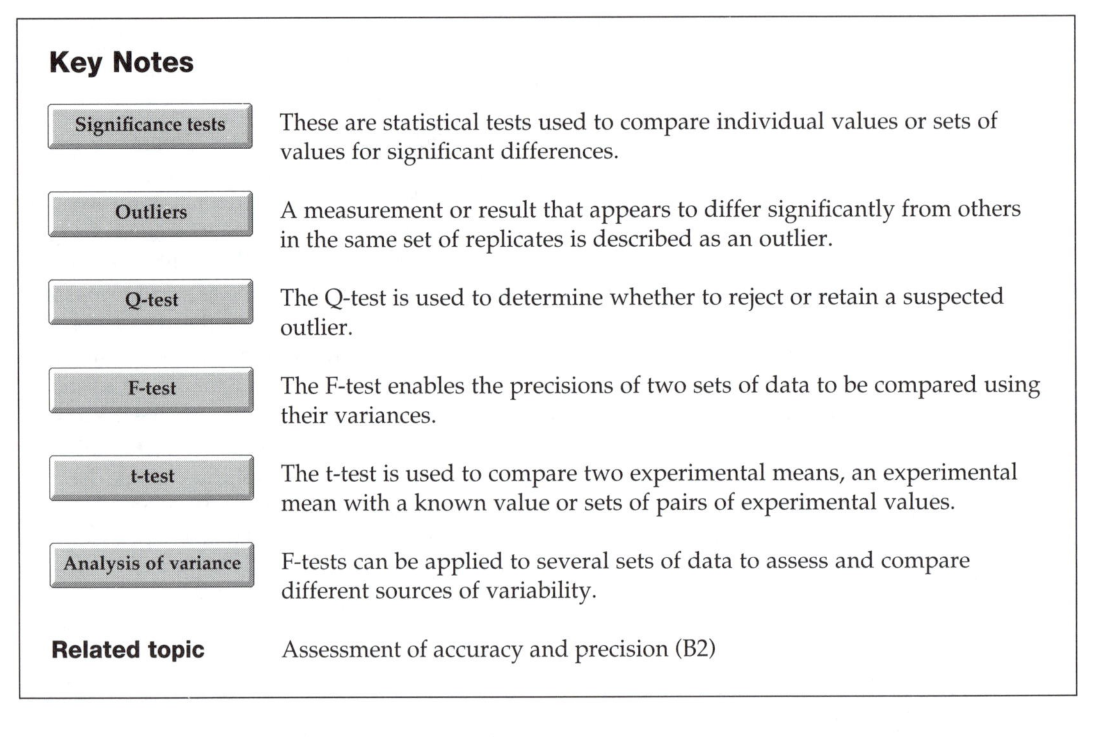

Key Notes

Significance tests	These are statistical tests used to compare individual values or sets of values for significant differences.
Outliers	A measurement or result that appears to differ significantly from others in the same set of replicates is described as an outlier.
Q-test	The Q-test is used to determine whether to reject or retain a suspected outlier.
F-test	The F-test enables the precisions of two sets of data to be compared using their variances.
t-test	The t-test is used to compare two experimental means, an experimental mean with a known value or sets of pairs of experimental values.
Analysis of variance	F-tests can be applied to several sets of data to assess and compare different sources of variability.
Related topic	Assessment of accuracy and precision (B2)

Significance tests

Significance tests involve a comparison between a calculated experimental factor and a tabulated factor determined by the number of values in the set(s) of experimental data and a selected probability level that the conclusion is correct. They are used for several purposes, such as:

- to check individual values in a set of data for the presence of determinate errors (bias);
- to compare the precision of two or more sets of data using their variances;
- to compare the means of two or more sets of data with one another or with known values to establish levels of accuracy.

Tests are based on a **null hypothesis** – an assumption that there is **no significant difference** between the values being compared. The hypothesis is accepted if the calculated experimental factor is less than the corresponding tabulated factor, otherwise it is rejected and there is said to be a **significant difference** between the values at the selected probability level. The conclusion should always be stated clearly and unambiguously.

Probability levels of 90%, 95% and 99% are generally considered appropriate for most purposes, but it should be remembered that there are also corresponding 10%, 5% or 1% probabilities, respectively, of the opposite conclusion being valid. For example, if a test indicates that the null hypothesis is correct and that there is no significant difference between two values at the 95% probability level, it also allows the possibility that there is a significant difference at the 5% level.

Separate tabular values for some significance test factors have been compiled for what are described as **one-tailed** and **two-tailed** tests. The exact purpose of the comparison that is to be made determines which table to use.

- The **one-tailed test** is used EITHER to establish whether one experimental value is significantly greater than the other OR the other way around.
- The **two-tailed test** is used to establish whether there is a **significant difference** between the two values being compared, whether one is higher or lower than the other not being specified.

The two-tailed test is by far the most widely used. Examples are given below.

Outliers

Inspection of a set of replicate measurements or results may reveal that one or more is considerably higher or lower than the remainder and appears to be outside the range expected from the inherent effects of indeterminate (random) errors alone. Such values are termed **outliers,** or **suspect values**, because it is possible that they may have a bias due to a determinate error. On occasions, the source of error may already be known or it is discovered on investigation, and the outlier(s) can be rejected without recourse to a statistical test. Frequently, however, this is not the case, and a test of significance such as the Q-test should be applied to a suspect value to determine whether it should be rejected and therefore not included in any further computations and statistical assessments of the data.

Q-test

Also known as **Dixon's Q-test**, this is one of several that have been devised to test suspected **outliers** in a set of replicates. It involves the calculation of a ratio, Q_{exptl}, defined as the absolute difference between a suspect value and the value closest to it divided by the spread of all the values in the set:

$$Q_{exptl} = |\,suspect\ value - nearest\ value\,| / (largest\ value - smallest\ value)$$

Q_{exptl} is then compared with a tabulated value, Q_{tab}, at a selected level of probability, usually 90% or 95%, for a set of n values (*Table 1*). If Q_{exptl} is **less** than Q_{tab}, then the null hypothesis that there is **no significant difference** between the suspect value and the other values in the set is accepted, and the suspect value is retained for further data processing. However, if Q_{exptl} is **greater** than Q_{tab}, then the suspect value is regarded as an outlier and is rejected. A rejected value should NOT be used in the remaining calculations.

Table 1. Critical values of Q at the 95% (P = 0.05) level for a two-tailed test

Sample size	Critical value
4	0.831
5	0.717
6	0.621
7	0.570
8	0.524

Example 1

Four replicate values were obtained for the determination of a pesticide in river water

$$0.403,\ 0.410,\ 0.401,\ 0.380\ \mu g\ dm^{-3}$$

Inspection of the data suggests that 0.380 μg dm^{-3} is a possible outlier.

$$Q_{exptl} = |0.380 - 0.401| / (0.410 - 0.380) = 0.021/0.03 = 0.70$$

$$Q_{tab} = 0.83 \text{ for four values at the 95\% probability level}$$

As Q_{exptl} is **less** than Q_{tab}, 0.380 μg dm^{-3} is not an outlier at the 95% level and should be retained.

Example 2

If, in Example 1, three additional values of 0.400, 0.413 and 0.411 μg dm^{-3} were included, 0.380 μg dm^{-3} is still a possible outlier.

$$Q_{exptl} = |0.380 - 0.400| / (0.413 - 0.380) = 0.020/0.033 = 0.61$$

$$Q_{tab} = 0.57 \text{ for seven values at the 95\% probability level}$$

Now, as Q_{exptl} is **greater** than Q_{tab}, 0.380 μg dm^{-3} is an outlier at the 95% level and should be rejected. Note that because the three additional values are all around 0.4 μg dm^{-3}, the suspect value of 0.380 μg dm^{-3} appears even more anomalous.

F-test

This test is used to compare the precisions of two sets of data which may originate from two analysts in the same laboratory, two different methods of analysis for the same analyte or results from two different laboratories. A statistic, F, is defined as the ratio of the population variances, σ_1^2/σ_2^2, or the sample variances, s_1^2/s_2^2, of the two sets of data where the larger variance is always placed in the numerator so that $F \geq 1$.

If the null hypothesis is true, the variances are equal and the value of F will be one or very close to it. As for the Q-test, an experimental value, F_{exptl}, is calculated and compared with a tabulated value, F_{tab}, at a defined probability level, usually 90% or 95%, and for the number of degrees of freedom, $N-1$, for each set of data. If F_{exptl} is **less** than F_{tab}, then the null hypothesis that there is **no significant difference** between the two variances and hence between the precision of the two sets of data, is accepted. However, if F_{exptl} is **greater** than F_{tab}, there **is a significant difference** between the two variances and hence between the precisions of the two sets of data.

Some values of F_{tab} at the 95% probability level are given in *Table 2*. The columns in the table correspond to the numbers of degrees of freedom for the numerator set of data, while the rows correspond to the number of degrees of freedom for the denominator set. Two versions of the table are available, depending on the exact purpose of the comparison to be made: a **one-tailed F-test** will show whether the precision of one set of data is significantly better than the other, while a **two-tailed F-test** will show whether the two precisions are significantly different.

Table 2. Critical values of F at the 95% (P = 0.05) level for a two-tailed test

v_1	5	7	9
v_2			
5	7.146	6.853	6.681
7	5.285	4.995	4.823
9	4.484	4.197	4.026

v_1 = number of degrees of freedom of the numerator. v_2 = number of degrees of freedom of the denominator

The application of a two-tailed F-test is demonstrated by the following example.

Example 3

A proposed new method for the determination of sulfate in an industrial waste effluent is compared with an existing method, giving the following results:

Method	Mean/g dm^{-3}	No. of replicates	No. of degrees of freedom	s/mg dm^{-3}
Existing	72	8	7	3.38
New	72	8	7	1.50

Is there a significant difference between the precisions of the two methods?

$$F_{exptl} = s^2_{existing} / s^2_{new} = \frac{(3.38)^2}{(1.50)^2} = 5.08$$

The two-tailed tabular value for F with 7 degrees of freedom for both the numerator and the denominator is

$$F_{7,7} = 5.00 \text{ at the 95\% probability level}$$

As F_{exptl} is **greater** than F_{tab}, the null hypothesis is rejected; the two methods are giving significantly different precisions.

t-test

This test is used to compare the experimental means of two sets of data or to compare the experimental mean of one set of data with a known or reference value. A statistic, t, is defined, depending on the circumstances, by one of three alternative equations.

Comparison of two experimental means, $\bar{x}_A$ and $\bar{x}_B$

$$t = \frac{(\bar{x}_A - \bar{x}_B)}{s_{pooled}} \times \left(\frac{NM}{N+M}\right)^{1/2} \tag{1}$$

where s_{pooled} is the pooled estimated standard deviation (Topic B2) for sets A and B, and N and M are the numbers of values in sets A and B respectively. If $N = M$, then the second term reduces to $(N/2)^{1/2}$. A simplified version of equation (4), Topic B2, can be used to calculate s_{pooled} as there are only two sets of data.

$$s_{pooled} = \left\{\left[(N-1)s_A^2 + (M-1)s_B^2\right] \Big/ \left[N + M - 2\right]\right\}^{1/2} \tag{2}$$

In some circumstances, the use of equation (1) may not be appropriate for the comparison of two experimental means. Examples of when this may be the case are if

- the amount of sample is so restricted as to allow only one determination by each of the two methods;
- the methods are to be compared for a series of samples containing different levels of analyte rather than replicating the analysis at one level only;
- samples are to be analyzed over a long period of time when the same experimental conditions cannot be guaranteed.

It may therefore be essential or convenient to pair the results (one from each method) and use a **paired t-test** where t is defined by

$$t = \frac{\bar{x}_d}{s_d} \times N^{1/2} \quad (3)$$

$\bar{x}_d$ being the mean difference between paired values and s_d the estimated standard deviation of the differences.

Comparison of one experimental mean with a known value, μ

$$t = \frac{(\bar{x} - \mu)}{s} \times N^{1/2} \quad (4)$$

Using the appropriate equation, an experimental value, t_{exptl}, is calculated and compared with a tabulated value, t_{tab}, at a defined probability level, usually between 90 and 99%, and for $N-1$ degrees of freedom (equations (3) and (4)) or $(N+M-2)$ degrees of freedom (equation (1)). If t_{exptl} is less than t_{tab}, then the null hypothesis that there is no significant difference between the two experimental means or between the experimental mean and a known value is accepted, i.e. there is no evidence of a bias. However, if t_{exptl} is greater than t_{tab}, there is a significant difference indicating a bias.

Both one-tailed and two-tailed t-tests can be used, depending on circumstances, but two-tailed are often preferred (*Table* 3). The application of all three t-test equations is demonstrated by the following examples.

Table 3. Critical values of t at the 95% and 99% (P = 0.05 and 0.01) levels for a two-tailed test

Number of degrees of freedom	95 percent level	99 percent level
2	4.30	9.92
5	2.57	4.03
10	2.23	3.10
18	2.10	2.88

Example 1

Two methods for the determination of polyaromatic hydrocarbons in soils were compared by analyzing a standard with the following results:

No. of determinations by each method:	10	
No. of degrees of freedom:	18	
UV spectrophotometry:	$\bar{x} = 28.00$ mg kg^{-1}	$s = 0.30$ mg kg^{-1}
Fluorimetry:	$\bar{x} = 26.25$ mg kg^{-1}	$s = 0.23$ mg kg^{-1}

Do the mean results for the two methods differ significantly?

Equation (2) is first used to calculate a pooled standard deviation:

$$s_{pooled} = \left\{\left[(N-1)s_A^2 + (M-1)s_B^2\right] / \left[N+M-2\right]\right\}^{1/2} = \{(9 \times 0.3^2 + 9 \times 0.23^2)/18\}^{1/2}$$

$$s_{pooled} = 0.267 \text{ mg kg}^{-1}$$

Then equation (1) is used to evaluate t_{exptl}

$$t_{exptl} = \frac{(\bar{x}_A - \bar{x}_B)}{s_{pooled}} \times \left(\frac{NM}{N+M}\right)^{1/2} = \{(28.0 - 26.25)/0.267\} \times 5^{1/2} = \mathbf{14.7}$$

For 18 degrees of freedom, the two-tailed value of t_{tab} at the 95% probability level is 2.10, and at the 99% level it is 2.88.

As t_{exptl} is **greater** than t_{tab} at both the 95 and 99% probability levels, there is a significant difference between the means of the two methods.

Example 2

A new high performance liquid chromatographic method for the determination of pseudoephedrine in a pharmaceutical product at two different levels was compared with an established method with the following results:

Pseudoephedrine per dose (mg)	
Method 1	Method 2
59.9	58.6
59.3	58.3
60.4	60.5
30.7	29.4
30.2	30.4
30.1	28.9

Do the means of the two methods differ significantly?

Because the two levels of pseudoephedrine differ considerably, equation (3) for a paired t-test is used to calculate t_{exptl}. The differences between the pairs of values are 1.3, 1.0, –0.1, 1.3, –0.2 and 1.2 mg per dose, and the estimated standard deviation of the differences from their mean of 0.750 mg per dose is 0.706 mg per dose. Substitution of these values into the equation gives

$$t_{exptl} = \frac{\bar{x}_d}{s_d} \times N^{1/2} = (0.750/0.706) \times 6^{1/2} = \mathbf{2.60}$$

For 5 degrees of freedom, the two-tailed value of t_{tab} at the 95% probability level is 2.57. As t_{exptl} is **greater** than t_{tab}, there is a significant difference between the means of the two methods. (Note: using equation (1) would give a t_{exptl} value of 0.08 and an incorrect conclusion.)

Example 3

A method for the determination of mercury by atomic absorption spectrometry gave values of 400, 385 and 382 ppm for a standard known to contain 400 ppm. Does the mean value differ significantly from the true value, or is there any evidence of systematic error (bias)?

$$\bar{x} = 389 \text{ ppm} \qquad s = 9.64 \text{ ppm} \qquad \mu = 400 \text{ ppm}$$

Using equation (4) to evaluate $\boldsymbol{t_{exptl}}$

$$t_{exptl} = \frac{(\bar{x} - \mu)}{s} \times N^{1/2} = \frac{(389 - 400)}{9.64} \times 3^{1/2} = \mathbf{1.98}$$

For 2 degrees of freedom, the two-tailed t_{tab} value is 4.30 at the 95% probability level. As t_{exptl} is **less** than the two-tailed value of t_{tab}, the mean is not significantly different from the true value. There is, therefore, no evidence of a systematic error, or bias.

Analysis of variance

Analysis of variance, also known as **ANOVA**, is a statistical technique for investigating different sources of variability associated with a series of results. It enables the effect of each source to be assessed separately and compared with the other(s) using F-tests. Indeterminate or random errors affect all measurements, but additional sources of variability may also arise. The additional sources can be divided into two types:

- additional random effects, described as **random-effect factors**;
- specific effects from determinate sources, described as **controlled** or **fixed-effect factors**.

Where one additional effect may be present, a **one-way ANOVA** is used, whilst for two additional effects, **two-way ANOVA** is appropriate. Both involve much lengthier calculations than the simpler tests of significance, but facilities for these are available with computer packages such as Microsoft Excel and Minitab.

Typical examples of the use of ANOVA are:

- analysis of a heterogeneous material where variation in composition is an additional random factor;
- analysis of samples by several laboratories, methods or analysts where the laboratories, methods or analysts are additional fixed-effect factors;
- analysis of a material stored under different conditions to investigate stability where the storage conditions provide an additional fixed-effect factor.

B4 Calibration and linear regression

Key Notes

Calibration

Calibration is the process of establishing a relation between a detection or measurement system and known amounts or concentrations of an analyte under specified conditions.

Correlation coefficient

The coefficient is used to assess the degree of linearity between two variables, e.g. an instrument response and an analyte mass or concentration.

Linear regression

Calculations to define the best straight line through a series of calibration points represented graphically are described as linear regression.

Limit of detection

The smallest mass or concentration of an analyte that can be measured quantitatively at a defined level of probability defines a limit of detection.

Standard addition

This is a calibration procedure that avoids matrix interference by measuring instrument response for an analyte in both the sample and a sample to which known amounts of an analyte standard have been added.

Internal standardization

This is a calibration procedure where the ratio of the instrument response for an analyte to that of an added standard is measured for a series of analyte standards and samples.

Internal normalization

Internally normalized results give the relative composition of a mixture by expressing the instrument response for each analyte as a fraction or percentage of the sum of the responses for all of the analytes.

Related topic

Calibration and standards (A5)

Calibration

Many quantitative analytical procedures rely on instrumental measurements where a property of the analyte(s) is monitored by a suitable detection system. The detector generates an electrical signal, the magnitude of which is determined by the mass or concentration of the analyte. Before using a particular analytical procedure to analyze samples, it is first necessary to establish the **detector responses** to known amounts of the analyte (**calibration standards**) over a selected mass or concentration range, a process known as **calibration**. The relation between the two variables is often **linear** (directly proportional), but there is generally an upper limit to the range of values beyond which a curved or **curvilinear** relation is observed. In some instances, there may be no direct linear relation at all, or a logarithmic or more complex mathematical correlation may be found.

Calibration data are generally used to construct a **calibration graph**, where detector response is plotted on the ordinate axis (y-values) and **mass** or **concentration** of the analyte on the abscissa axis (x-values) as shown in *Figure 1*.

The graphs are often linear, being defined by the equation

$$y = bx + a \tag{1}$$

where b is the slope and a the intercept on the y-axis. In some cases, it is preferable to plot a logarithmic function of the detector response or analyte concentration to obtain a linear calibration curve.

Unknown levels of the analyte are determined from the graph by **interpolation**. Where a linear relation has been established, a calibration factor can be used to convert detector response to mass or concentration of analyte when analyzing samples.

Theoretically, the graph should pass through the origin, but frequently in practice there is a small positive intercept due to traces of analyte in the reagent blank or contributions to the detector signal by other components in the standards. Calibration points also show a degree of scatter due to the effects of experimental errors in preparing the standards, or noise in the measuring circuitry. **A line of best fit** through the points, known as a **regression line**, is therefore drawn or computed.

Calibration graphs may show curvature, particularly at higher mass or concentration levels, but this does not invalidate their use if the data are reproducible. However, it is advisable to prepare additional standards to define the curve more closely, and the use of a factor to compute analyte levels in samples is precluded.

Statistical methods are used to assess calibration data

- for linearity or otherwise;
- to calculate the parameters defining a calibration curve;
- to assess the effects of determinate and indeterminate errors on standards and samples.

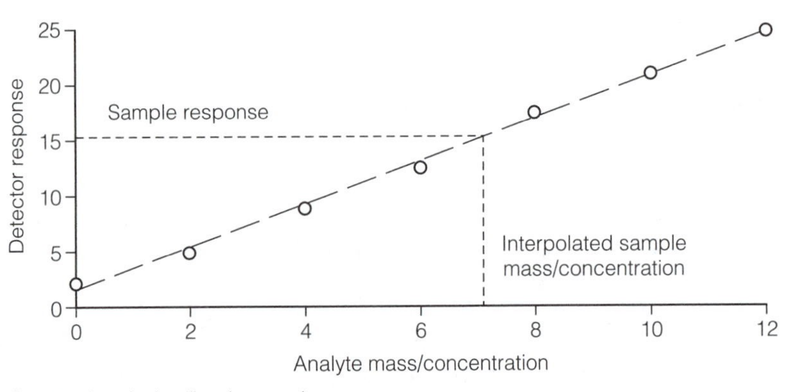

Fig. 1. A typical calibration graph.

Correlation coefficient

The **correlation coefficient**, r, indicates the **degree of linearity** between x and y and is given by the expression

$$r = \frac{\sum_{i=1}^{i=N}\{(x_i - \bar{x})(y_i - \bar{y})\}}{\left\{\left[\sum_{i=1}^{i=N}(x_i - \bar{x})^2\right]\left[\sum_{i=1}^{i=N}(y_i - \bar{y})^2\right]\right\}^{1/2}} \tag{2}$$

where x_1y_1; x_2y_2; x_3y_3;....x_n,y_n are the co-ordinates of the plotted points, $\bar{x}$ and $\bar{y}$ are the means of the x and y values respectively, and Σ indicates sums of terms (see standard deviation equations (1), (2) and (4), Topic B2).

The range of possible values for r is $-1 \leq r \leq +1$. A value of **unity** indicates a **perfect linear correlation** between x and y, all the points lying exactly on a straight line, whilst a value of **zero** indicates **no linear correlation**. Values may be positive or negative depending on the slope of the calibration graph. These alternatives are illustrated in *Figure 2 (a)* to *(c)*.

Most calibration graphs have a positive slope, and correlation coefficients frequently exceed 0.99. They are normally quoted to four decimal places. (Note that graphs with a slight curvature may still have correlation coefficients exceeding about 0.98 (*Fig. 2(d)*), hence great care must be taken before concluding that the data shows a linear relation. Visual inspection of the plotted points is the only way of avoiding mistakes.)

Linear regression

When inspection of the calibration data and the value of the correlation coefficient show that there is a linear relation between the detector response and the

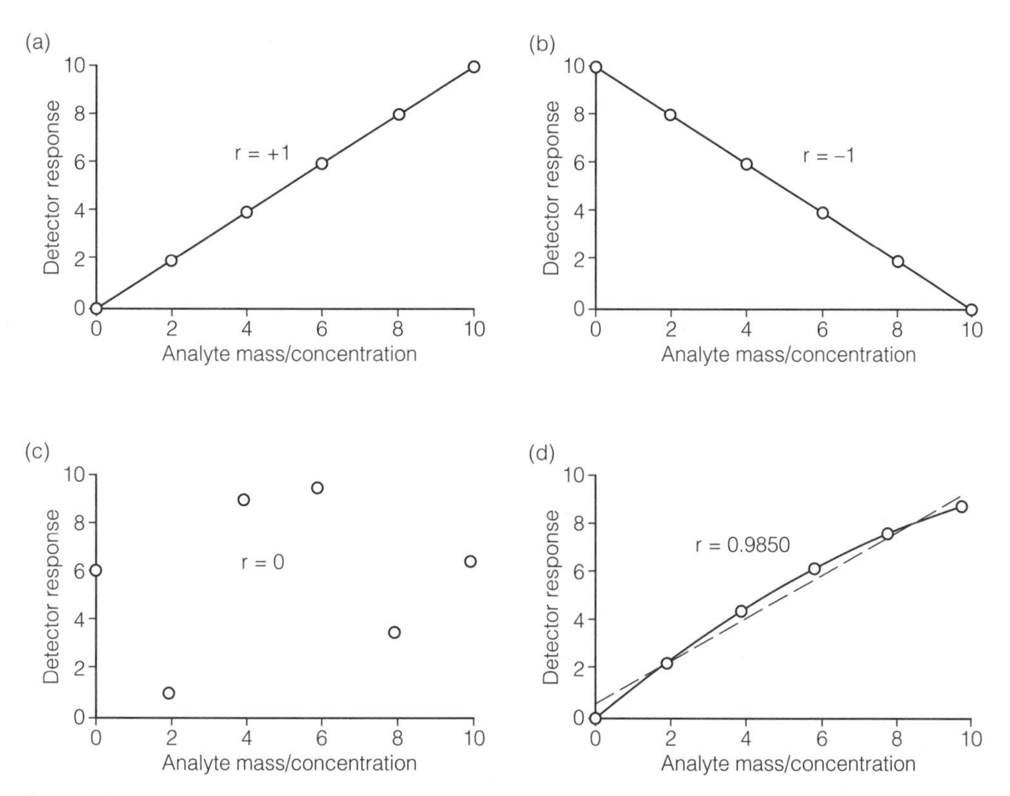

Fig. 2. Examples of correlation coefficients. (a) Perfect positive correlation; (b) perfect negative correlation; (c) no correlation, and (d) curved correlation.

mass or concentration of the analyte, it is necessary to draw a **line of best fit** through the plotted points before they can be used as a working curve. Although this can be done by eye, a more accurate method is to employ **linear regression**. It is invariably the case that, due to the effects of indeterminate errors on the data, most of the points do not lie exactly on the line, as shown in *Figure 1*. Linear regression enables a line of best fit through the points to be defined by calculating values for the slope and y-axis intercept (b and a respectively in equation (1)), and the **method of least squares** is commonly used for this purpose. An assumption is made that only errors in the detector responses (y-values) are significant, any errors in the values for the mass or concentration of the analyte being neglected. The deviations in the y-direction of the individual plotted points from the calculated regression line are known as ***y*-residuals** (*Fig. 3*) and the line represents the regression of y upon x. The method of least squares minimizes the sum of the squares of the y-residuals by equating them to zero in defining equations for the slope and intercept of the regression line.

For the slope, b

$$b = \frac{\sum_{i=1}^{i=N} \{(x_i - \bar{x})(y_i - \bar{y})\}}{\sum_{i=1}^{i=N} (x_i - \bar{x})^2} \quad (3)$$

For the y-axis intercept, a

$$a = \bar{y} - b.\bar{x} \quad (4)$$

N.B. As equation (4) is a re-arrangement of equation (1), it follows that the point $\bar{x}, \bar{y}$, known as the **centroid**, must lie on the regression line.

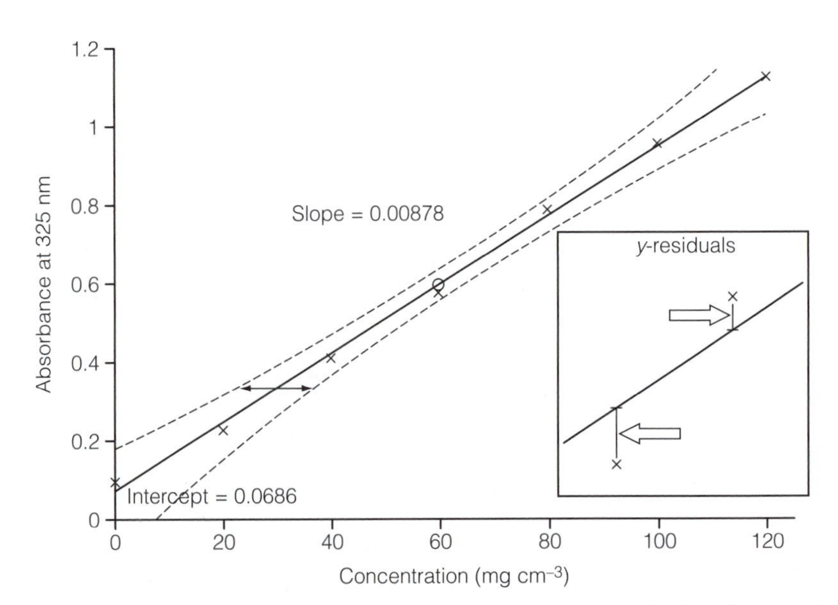

Fig. 3. Calibration graph, regression line, slope and intercept values for the UV spectrophotometric determination of an active ingredient in a sun cream. ——— = regression line; ○ = centroid, $\bar{x}$, $\bar{y}$; ---------- = confidence limits lines at the 99 percent level; ◄──► = confidence limits for sample concentration of 30 mg cm^{-3}. Inset: Illustration of y-residuals.

Example

A calibration graph was prepared as part of a validation procedure for a new method to determine an active constituent of a sun cream by UV spectrophotometry. The following data were obtained:

Analyte conc. (mg cm^{-3})	0	20	40	60	80	100	120
UV absorbance at 325 nm	0.095	0.227	0.409	0.573	0.786	0.955	1.123

The data is first checked for linearity by calculation of the correlation coefficient, r, and visual inspection of a plotted curve. Some calculators and computer software can perform the computation from the raw data, but it is instructive to show the full working, for which tabulation is preferable.

x_i	y_i	$(x_i - \bar{x})$	$(x_i - \bar{x})^2$	$(y_i - \bar{y})$	$(y_i - \bar{y})^2$	$(x_i - \bar{x})(y_i - \bar{y})$
0	0.095	–60	3600	–0.5004	0.2504	30.024
20	0.227	–40	1600	–0.3684	0.1357	14.736
40	0.409	–20	400	–0.1864	0.0347	3.728
60	0.573	0	0	–0.0224	0.0005	0
80	0.786	20	400	0.1906	0.0363	3.812
100	0.955	40	1600	0.3596	0.1293	14.384
120	1.123	60	3600	0.5276	0.2784	31.656
Σ 420	**4.168**	**0**	**11200**	**0**	**0.8653**	**98.340**
$\bar{x}$ **= 60**	$\bar{y}$ **= 0.59543**					

Substitution of the totals in columns 4, 6 and 7 in equation (2) gives

$$r = 98.340/(11200 \times 0.8653)^{1/2} = 98.340/98.445 = 0.9989$$

Figure 3 and the correlation coefficient of 0.9989 show that there is a good linear relation between the measured UV absorbance and the analyte concentration.

The slope and y-axis intercept of the regression line, given by equations (3) and (4) respectively are

$$b = 98.340/11200 = 0.00878 \quad a = 0.59543 - (0.00878 \times 60) = 0.0686$$

The y-axis intercept, slope and analyte masses or concentrations calculated by interpolation from the regression line are all affected by errors. Additional equations can be used to obtain the following statistics:

- estimated standard deviations for the slope and intercept;
- estimated standard deviations for analyte masses or concentrations determined from the calibration graph;
- confidence limits for analyte masses and concentrations at selected probability levels;
- limit of detection of the analyte (*vide infra*).

Confidence limits (Topic B2) over the entire range of the calibration graph at selected probability levels, e.g. 95 or 99 percent, can be displayed (dashed curves, *Fig. 3*). A horizontal line drawn through a given experimental point on the regression line and intersecting the confidence limits lines on either side gives the upper and lower limits for that particular mass or concentration. *Figure 3* shows the 99% limits, the narrowest interval being at the centroid, $\bar{x}, \bar{y}$, of the graph, and widening steadily towards each end.

Some calculators and computer packages have the ability to perform the regression calculations described. Where there is a nonlinear relation between the detector response and the mass or concentration of the analyte more complex **curvilinear** or **logarithmic regression** calculations are required.

Limit of detection

For any analytical procedure, it is important to establish the smallest amount of an analyte that can be detected and/or measured quantitatively. In statistical terms, and for instrumental data, this is defined as the smallest amount of an analyte giving a detector response significantly different from a blank or background response (i.e. the response from standards containing the same reagents and having the same overall composition (matrix) as the samples, where this is known, but containing no analyte). Detection limits are usually based on estimates of the standard deviation of replicate measurements of prepared blanks. A detection limit of two or three times the estimated standard deviation of the blanks above their mean, $\bar{x}_B$, is often quoted, where as many blanks as possible (at least 5 to 10) have been prepared and measured.

This is somewhat arbitrary, and it is perfectly acceptable to define alternatives provided that the basis is clear and comparisons are made at the same probability level.

Standard addition

Where components of a sample other than the analyte(s) (the **matrix**) interfere with the instrument response for the analyte, the use of a calibration curve based on standards of pure analyte may lead to erroneous results. Such **matrix interference effects** can be largely if not entirely avoided by preparing calibration standards where known amounts of pure analyte are added to a series of equal sized portions of the sample, a procedure known as **spiking**. In addition, one portion of sample is not spiked with analyte. (Note: if spiking sample solutions with analyte changes the volume significantly, volume corrections must be applied.)

The effects of the matrix on measurements of the analyte in both the spiked and unspiked samples should be identical. The instrument responses are then used to construct a calibration graph where the x-axis values are the added amounts of analyte and the response for the unspiked sample is at $x = 0$ (i.e., the curve does NOT pass through the origin). The regression line is calculated and extrapolated back to give a negative intercept on the x-axis at $y = 0$, which corresponds to the amount of analyte in the sample (*Fig. 4*).

The less reliable procedure of extrapolation rather than interpolation is outweighed by the advantage of eliminating or minimizing matrix interference.

The method of standard addition is widely used, particularly when the composition of the sample matrix is variable or unknown so that the response of a reagent/matrix blank would be unreliable. At least three and preferably more spiked samples should be prepared, but if the amount of sample is limited, as few as one sample must suffice. It is especially useful with such analytical techniques as **flame** and **plasma emission spectrometry** and **potentiometry** (Topics E4, E5 and C8).

Example

The calcium level in a clinical sample was determined by flame emission spectrometry using a standard addition method, which gave the following data:

Spiked calcium (ppm)	0	10	20	30	40	50
Emission intensity at 423 nm	0.257	0.314	0.364	0.413	0.468	0.528

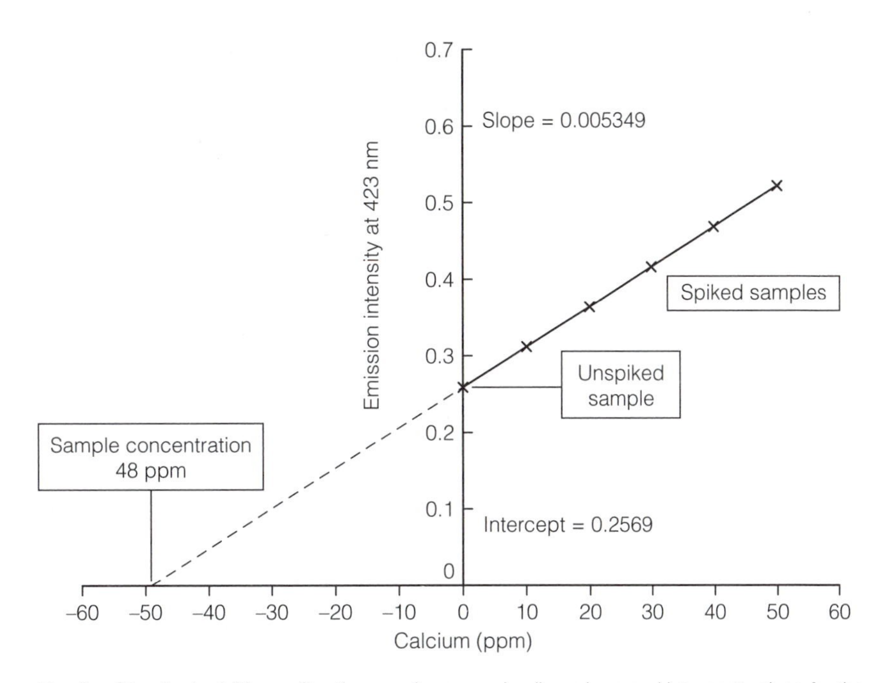

Fig. 4. Standard addition calibration graph, regression line, slope and intercept values for the flame emission determination of calcium in a clinical sample.

Detailed calculations of the correlation coefficient, r, and the slope and intercept values have not been given, but should be set out as in the previous example if a suitable calculator or computer program is not available.

The amount of calcium in the sample can be read from the extrapolated graph or calculated from the slope, b, and the intercept, a

$$\text{Calcium concentration in sample} = a/b = 0.2569/0.005349 = 48 \text{ ppm}$$

Internal standardization

For some analytical techniques, particularly chromatography, variations in experimental conditions can adversely affect the precision of the data. A calibration procedure whereby a constant amount of a selected substance, the **internal standard** is added to all samples and analyte standards alike compensates for variations in sample size and other parameters. The ratio of the detector response for the analyte in each standard to the corresponding response for the added internal standard is plotted on the y-axis of a calibration graph against the mass or concentration of the analyte on the x-axis. The correlation coefficient, slope and intercept values can be computed as shown previously, and **response ratios** for the analyte and added internal standard in the samples can then be used to determine the amount of analyte in the samples by interpolation on the graph. If only one or two analyte standards are prepared, the amount of analyte in a sample can be calculated by simple proportion, i.e.

$$\frac{\textit{analyte in sample}}{\textit{analyte in standard}} = \frac{\textit{response ratio for sample}}{\textit{response ratio for standard}}$$

Internal normalization

For some purposes, only the relative amounts of the analytes in a multicomponent mixture are required. These are normalized to 100 or 1 by expressing each as a percentage or fraction of the total. **Internal normalization** is of particular value in quantitative chromatography where several components of a sample can be determined simultaneously, and absolute levels are not of interest. The relative composition is calculated from the instrument response, peak area in the case of a chromatographic analysis, for each component in the mixture using the formula

$$\%x_i = \frac{A_x}{\sum_{i=1}^{i=n} A_i} \times 100$$

where x_i is one of n components and A is the measured area or response.

Example

Figure 5 is a chromatographic record (chromatogram) of the separation of a 5-component mixture. The measured peak areas (using electronic integration with a computing-integrator, computer and chromatography data processing software or geometric construction such as triangulation, ½ × base × height) and percentages by internal normalization, which must total 100 percent, are given in *Table 1* (e.g., for component 1, relative percent = (167.8/466.94) × 100 = 35.9 percent).

Table 1. Peak areas and percentage composition by internal normalization for a 5-component mixture

Component	Measured peak area (arbitrary units)	Relative percent
1	167.8	35.9
2	31.63	6.8
3	108.5	23.2
4	80.63	17.3
5	78.38	16.8
Totals	466.94	100.0

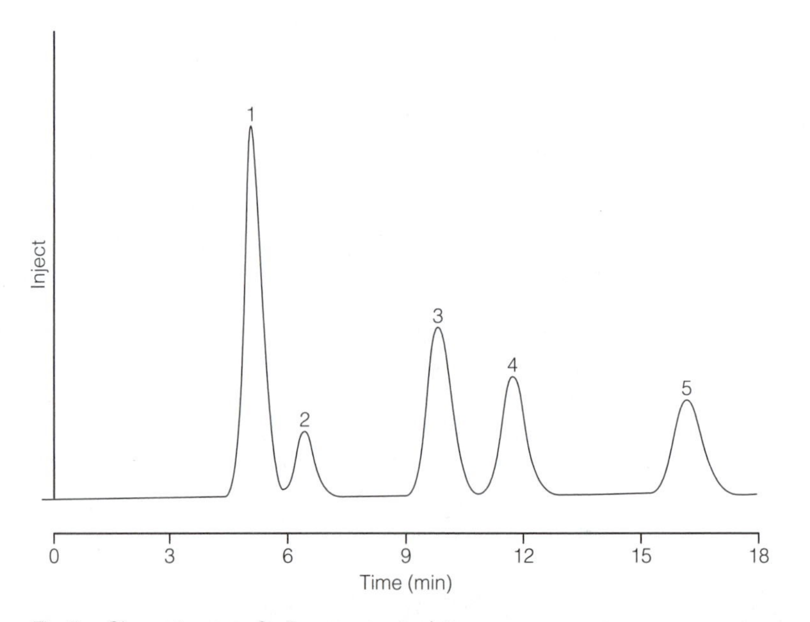

Fig. 5. Chromatogram of a 5-component mixture.

B5 QUALITY CONTROL AND CHEMOMETRICS

Key Notes

Control charts — Graphical representations of quantitative data from an ongoing series of measurements can be used to monitor the stability of a system for quality control (QC) purposes.

Collaborative testing — Schemes have been devised to compare results for the analysis of homogeneous samples or standard materials from groups of analytical laboratories to test specified methods of analysis.

Multivariate statistics — Advanced chemometric procedures can be applied to the statistical analysis of large amounts of data and numerous experimental variables so as to extract the maximum amount of useful information.

Related topic — Quality in analytical laboratories (A6)

Control charts

The purpose of a **control chart** is to monitor data from an ongoing series of quantitative measurements so that the occurrence of determinate (systematic) errors (bias), or any changes in the indeterminate (random) errors affecting the precision of replicates can be detected and remedial action taken. The predominant use of control charts is for quality control (QC) in manufacturing industries where a product or intermediate is sampled and analyzed continually in a process stream or periodically from batches. They may also be used in analytical laboratories, such as those involved in clinical or environmental work, to monitor the condition of reagents, standards and instrument components, which may deteriorate over time.

Shewart charts consist of a y-axis calibrated either in the mass, concentration or range of replicated results of an analyte or in multiples of the estimated standard deviation, s, of the analytical method employed, and the sample number along the x-axis. An **averages** or **X-chart,** the most common type, is pre-prepared with a series of five parallel horizontal lines, the centre one being positioned along the y-axis to correspond to the true, accepted or **target value** for the analyte (*Fig. 1*). The other four lines are positioned in pairs on either side of the target value line and act as critical levels that, when exceeded, indicate probable instability in the system. The inner pair are defined as **warning levels** and the outer pair as **action levels**.

An averages chart is used to monitor the level of an analyte either as single values or as means of N replicates to check for determinate errors (bias) in the results. Chart criteria and decisions are based on the following:

- plotted values have a Gaussian or normal distribution (Topic B2);
- **warning lines** are positioned to correspond to a selected **probability level**

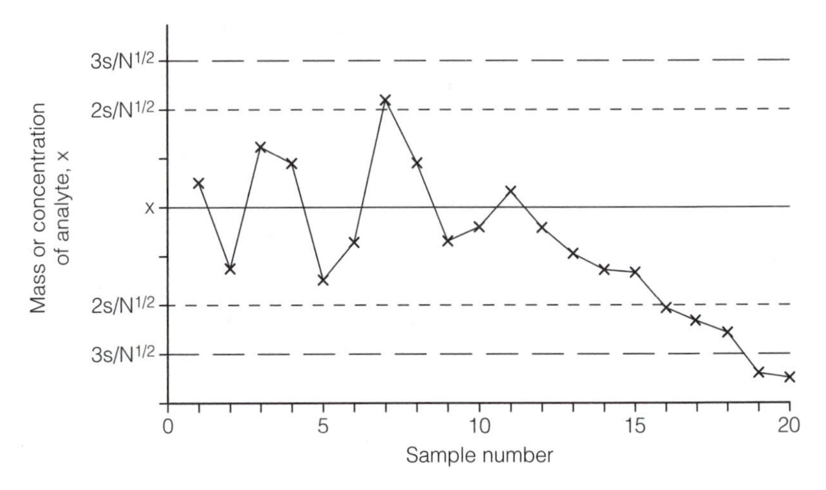

Fig. 1. A Shewart averages control chart.

multiple of the estimated standard deviation of the method, usually 95% or $\pm1.96s/N^{1/2}$;

- **action lines** are positioned to correspond to a selected probability level or multiple of the estimated standard deviation of the method usually 99.7% or $\pm3.01s/N^{1/2}$;
- the pairs of warning lines will move inwards towards the target line as the number of results per plotted point, N, increases, so the value of N must be fixed before the chart is used;
- in the absence of determinate errors (bias), 95% of values should fall within the upper and lower warning lines. The system is then considered to be stable or under control;
- two or more consecutive values falling outside the warning lines but within the action lines indicate a possible loss of control;
- two or more consecutive values falling outside the action lines indicate the occurrence of one or more determinate errors; the system is then considered to be out of control and remedial action should be taken;
- trends in plotted values are an indication of incipient problems, as are ten or more consecutive values on one side of the target line.

The chart shows that the system is in control during the first 6 samples analyzed, but the upper warning level has been breached by result 7. However, the next 8 results are within the warning levels, but results 17 to 20 indicate a downward trend culminating with both lower limits being breached indicating a loss of control due to one or more determinate errors. At this point, the causes should be sought and remedial action taken.

Collaborative testing

The principal purpose of **collaborative testing** is to assess the accuracy of results from a group of laboratories performing one or more specific quantitative determinations. Interlaboratory studies are also used to assess other performance characteristics of a particular method or the competence of individual analysts in relation to a specified technique, analyte or group of analytes. They are frequently used by trade associations, standards organizations or agencies with stringent analytical requirements to develop, test and validate analytical methods.

Proficiency testing schemes are designed to test the competence of individual laboratories to perform specific analyses or a range of determinations of analytes in various matrixes and possibly using alternative techniques.

Some typical examples of collaborative studies and proficiency testing are

- alcohol in beverages;
- metals or volatile organics in soils;
- hazardous airborne substances;
- trace metals or pesticides in domestic water supplies;
- drugs of abuse by chromatographic methods;
- adulterants in foodstuffs;
- additives in polymers and composites.

In statistical terms, collaborative trials are designed to reveal the presence of a bias in the results arising from determinate errors that may occur in addition to inherent indeterminate (random) errors. Statistics that are used in these schemes include **estimated standard deviation** and **coefficient of variation**, **confidence limits** (Topic B2) and **tests of significance** including **ANOVA** (Topic B3). Of particular value is a ***z*-score** to indicate the accuracy of results from the participating laboratories. The ***z*-value** is a statistical factor based on a Gaussian or normal distribution, which is included in the equation for confidence limits of an experimental value in relation to a true or accepted value (Topic B2, equation (5)). Rearrangement of this equation, for a single determination, x_i, (N=1) gives

$$z = \left| \frac{x_i - \mu}{\sigma} \right|$$

where μ is the true or accepted mass or concentration of the analyte and σ is a value for the standard deviation for the method selected by the organizers of the study (strictly an estimated value, s). A typical chart of ***z*-scores** for a collaborative study involving 22 laboratories is shown in *Figure 2*.

Results that have ***z*-scores** of between +1.96 and −1.96 are considered to have acceptable accuracy, as these values correspond to confidence limits at the 95% probability level (Topic B2, *Fig. 3*). In the example, only laboratories 12 and 18

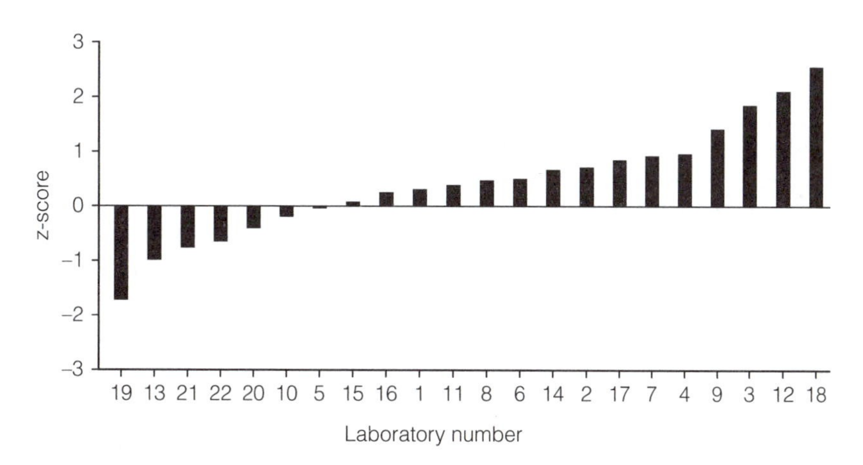

Fig. 2. z-score chart for results from 22 laboratories.

have results with unacceptable accuracy, whilst laboratories 5 and 15 have results closest to the true or accepted value. In practice, the situation is often worse than this, with some laboratories incurring worryingly large determinate errors that need to be identified and rectified.

Although accuracy is of prime importance in collaborative studies and proficiency testing, precision should also be monitored. The level to be expected varies considerably with the concentration of an analyte and type of sample. A useful guide for different levels of an analyte as measured by the coefficient of variation (Topic B2) is exemplified in *Figure 3*.

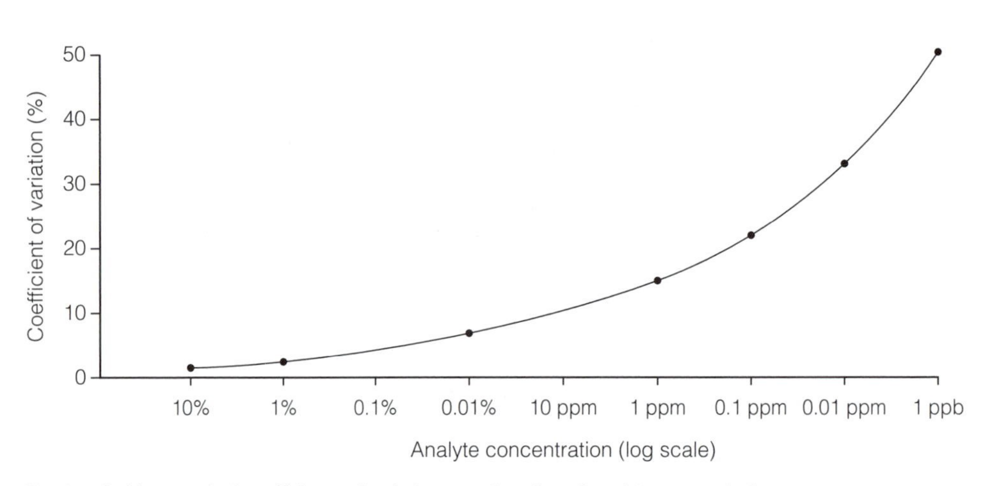

Fig. 3. Guide to typical coefficients of variation as a function of analyte concentration.

Note that values of less than 1% should be attainable for major components of a sample, whilst at ppb (parts per billion) levels, over 50% is acceptable.

Multivariate statistics

Computerized and automated analytical instrumentation facilitates the collection of large amounts of data and the simultaneous monitoring of numerous experimental parameters. To maximize the useful information that can be extracted, sophisticated **multivariate chemometric techniques** are employed. The mathematical computations involve **matrix algebra** and **vectors**, and rely on the availability of specialized computer software. Typical applications include the characterization of substances from profiles based on spectral, chromatographic and other data, and quantitative analysis based on multiple simultaneous measurements. Two important applications of multivariate statistics are pattern recognition and multivariate modeling.

Pattern recognition

Sets of measurements characterizing a sample, e.g. the position of prominent infrared absorption bands (Topic H4), significant mass spectral fragments, levels of particular analytes, and selected physical properties, are described as **patterns**. These can be used to classify substances or to identify unknowns by **pattern matching**. *Figure 4* shows the distribution of trace levels of copper and manganese in twelve geological samples where three clusters are evident. Just

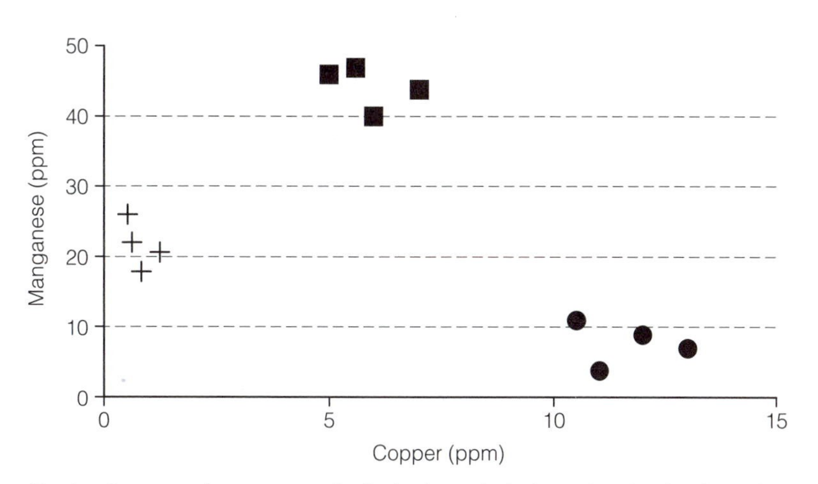

Fig. 4. Copper and manganese distribution in geological samples showing three clusters with differing proportions of each metal.

as two parameters can be plotted as a single point with specified x and y co-ordinates on a two-dimensional graph, the values of n parameters can be represented by a point in **n-dimensional space**. Although n-co-ordinate graphs cannot be visualized, they can be studied through appropriate computer processing and manipulation. Where a number of substances have similar sets of n co-ordinates, and therefore similar characteristics, they produce closely-spaced groups of points described as **clusters**, the interpretation of this data being described as **cluster analysis**. Mathematical procedures to detect clusters include **principal component analysis** (PCA) and **factor analysis** (FA), which seek to simplify the data by projection from n dimensions onto a line, plane or 3-D graph to reduce the number of dimensions without losing information.

Cluster analysis can be used in many ways, e.g. to monitor clinical specimens from hospital patients where, for example, the levels of pH, glucose, potassium, calcium, phosphate and specific enzymes vary according to the absence, presence or severity of a particular disease. It can also be applied to the characterization of glass fragments for forensic purposes through profiling and comparisons of their trace metal contents, for identifying the source of a crude oil spillage on the basis of the proportions of minor organic compounds and metals, and in the classification of organic compounds with similar structural features to facilitate the elucidation of unknown structures.

Multivariate modeling

Quantitative analysis for one or more analytes through the simultaneous measurement of experimental parameters such as molecular UV or infrared absorbance at multiple wavelengths can be achieved even where clearly defined spectral bands are not discernible. Standards of known composition are used to compute and refine quantitative calibration data assuming linear or nonlinear models. **Principal component regression** (PCR) and **partial least squares (PLS) regression** are two multivariate regression techniques developed from linear regression (Topic B4) to optimize the data.

Multivariate modeling is the basis of quantitative analysis by **near infrared spectrometry** (Topic E11) to determine moisture or fats in cereals, meat and other foodstuffs. It is particularly applicable to environmental analysis where complex and inter-related variables affect the distributions of a wide range of organic compounds and elements occurring naturally or as pollutants.

C1 SOLUTION EQUILIBRIA

Key Notes

Solvents

The major component of a solution is referred to as the solvent, and there is a wide range of inorganic and organic solvents used in analytical chemistry. Their properties determine their use.

Solubility

When a substance called the solute is dissolved in a solvent to form a solution, its behavior is often altered. Reactions in solution are faster than in the solid state. The amount of substance that can dissolve in a given amount of solvent at a stated temperature and pressure is called the solubility and is determined by the nature of the materials and the laws governing the solubility equilibrium.

Ions in solution

Some substances form ions, which are species possessing a charge. These behave in a distinct way in solution. They may attract molecules of solvent, may associate together, and may react with other species to form complexes or a precipitate.

The pX notation

Since concentrations vary over a very wide range, they are often represented by the logarithmic pX notation where pX = – log(X), where X is the concentration or activity of an ion, or an equilibrium constant.

Equilibria in solution

The laws of thermodynamics govern the behavior of all species in solution. Every reaction depends upon the thermodynamic properties of the species involved. Where those properties are changed by the solvent by association, by reaction or temperature, the behavior will alter. Physical and chemical equilibria in solution are most important.

Related topics

Other topics in Section C (C2–C10)

Separation techniques (D1–D9)

Solvents

The use of solvents for analytical work is determined by their properties, as shown in *Table 1*.

Solvents with high dielectric constants ($\varepsilon_r > 10$), for example, water and ammonia, are referred to as **polar** and are **ionizing solvents**, promoting the formation and separation of ions in their solutions, whereas those where ε_r is about 2, such as diethyl ether, tetrachloromethane and hexane are **nonpolar** and are **nonionizing solvents**. There are also many solvents whose behavior is intermediate between these extremes.

The solution process in a liquid may be represented by a general equation:

$$\underset{\text{solvent}}{A(l)} + \underset{\text{solute}}{B} = \underset{\text{solution}}{B(sol)}$$

The action of solution changes the properties of both solute and solvent. The solute is made more mobile in solution, and its species may **solvate** by attraction

Table 1. *Properties of some solvents*

Solvent	Boiling point (°C)	Density, (g cm^{-3})	Dielectric constant, ε_r
Water	100	1.00	78.6
Ammonia	–34	0.68	22.0
Ethanol	78	0.79	24.3
n-hexane	69	0.66	1.88
Diethyl ether	34	0.71	4.33

Note: density at 25°C or at BP; dielectric constant = relative permittivity

to the solvent. The solvent structure is also disrupted by the presence of species different in size, shape and polarity from the solvent molecules.

Ideally, the behavior should depend on the concentration m (in molarity, mole fraction or other units), but often this must be modified and the **activity, *a*,** used:

$$a = m\,\gamma = p/p^{\ominus}$$

where γ is called the **activity coefficient**. The vapor pressure of the solution is p, and that in the standard state is $p^{\ominus}$. Activities are dimensionless.

Solvents, such as water, with high dielectric constants (or **relative permittivities**) reduce the force F between ions of charges z_1e and z_2e a distance r apart:

$$F = z_1 z_2\, e^2/\varepsilon_o \varepsilon_r\, r^2$$

where ε_o is the permittivity of free space. Also, they will solvate ions more strongly and thus assist ion formation and separation.

Hexane, diethyl ether and tetrachloromethane (CCl_4) all have low dielectric constants and are nonpolar. They are very poor at ionizing solutes. However, they are very good solvents for nonpolar substances.

Solubility

The equilibrium amount of solute which will dissolve in a given amount of solvent at a given temperature and pressure is called the **solubility.** The solubility may be quoted in any units of concentration, for example, mol m^{-3}, molarity, mole fraction, mass per unit volume or parts per million (ppm).

There is a general 'rule of thumb' that **'like dissolves like'**. For example, a nonpolar hydrocarbon solvent such as hexane would be a very good solvent for solid hydrocarbons such as dodecane or naphthalene. An ester would be a good solvent for esters, and water or other polar solvents are appropriate for polar and ionic compounds.

- **Gases** dissolve in solvents according to **Henry's Law**, provided they do not react with the solvent:

  $$p_B = x_B\, K$$

  where x_B is the mole fraction of solute gas B which dissolves at a partial pressure p_B of B, and K is a constant at a given temperature. This is analytically important for several reasons. For example, nitrogen is bubbled through solutions to decrease the partial pressure of oxygen in electrochemical experiments. Similarly, air is removed from liquid chromatography solvents by passing helium through them, or by boiling them, since gas solubility decreases as the temperature is increased.

- **Liquids.** When different liquids are mixed, many types of behavior may occur. If the molecules in the liquids are of similar size, shape, polarity and chemical nature they may mix in all proportions. For example, benzene and methylbenzene (toluene) mix completely. In such **ideal solutions,** obeying **Raoult's law**, the activity coefficient is close to 1:

$$a = p/p^{\ominus} = x$$

If the component molecules differ greatly in polarity, size or chemical nature (e.g., water and tetrachloromethane) they may not mix at all. This is an important condition for solvent extraction (Topic D1). The distribution of a solute between a pair of **immiscible liquids** depends primarily on the solubility of the solute in each liquid.

- **Solids** generally follow the 'like dissolves like' rule. Nonpolar, covalent materials dissolve best in nonpolar solvents. Solid triglycerides such as tristearin are extracted by diethyl ether, but are nearly insoluble in water. Salts, such as sodium chloride are highly soluble in water, but virtually insoluble in ether.

Ions in solution

The behavior of ions in solution may be summarized as follows.

(i) Solids whose structure consists of ions held together by electrostatic forces (e.g. NaCl) must be separated into discrete ions when they dissolve. These ions often gain stability by solvation with molecules of the solvent. Such solutions are described as **strong electrolytes**.

(ii) Some covalent molecules, such as ethanoic acid, may form ions by the unequal breaking of a covalent bond, followed by stabilization by solvation. This occurs only partially and these are called **weak electrolytes**.

$$H : OCOCH_3 \rightleftharpoons H^+ + {}^-OCOCH_3$$

(iii) In some cases ions do not separate completely. In concentrated solutions, oppositely charged ions may exist as **ion-pairs**, large ions of surfactants may aggregate into **micelles**, which are used in capillary electrophoresis (Topics D8 and D9), and the dissociation of covalent molecules may be only partial.

(iv) At 'infinite dilution', that is, as the concentration approaches zero, ions are truly separate. However, even at quite low concentrations, the attractions between ions of opposite charge will cause each ion to become surrounded by an irregular cloud, or **ionic atmosphere**. As the solution becomes even more concentrated, the ionic atmosphere becomes more compact around each ion and alters its behavior greatly.

In very dilute solutions, the effects of the ionic atmosphere may be approximated by using the **Debye-Hückel theory**, which predicts that the mean ionic activity coefficient, $\gamma_{\pm}$ is, for an electrolyte with positive ions with charge z_+, negative ions with charge z_-, is given by:

$$\log(\gamma_{\pm}) = -A\,(z_+ \,.\, z_-)\,\sqrt{(I)}$$

where I is the **ionic strength** $= \frac{1}{2}\,\Sigma\,(c_i\,z_i^2)$ for all the ions in the solution.

For more concentrated ionic solutions, above 0.1 M, no general theory exists, but additional terms are often added to the Debye-Hückel equation to compensate for the change in the activity.

The pX notation

The concentration of species in solution may range from very small to large. For example in a saturated aqueous solution of silver chloride, the concentration of silver ions is about 10^{-5} M, while for concentrated hydrochloric acid the concentration of hydrogen and chloride ions is about 10 M. For convenience, a logarithmic scale is often used:

$$pX = -\log (X)$$

where X is the concentration of the species, or a related quantity. Thus, for the examples above, pAg = 5 in saturated aqueous silver chloride and pH = −1 in concentrated HCl.

Since equilibrium constants are derived from activities or concentrations as noted below, this notation is also used for them:

$$pK = -\log (K)$$

Equilibria in solution

Most reactions will eventually reach equilibrium. That is, the concentrations of reactants and products change no further, since the rates of the forward and reverse reactions are the same.

From the above arguments concerning solutions, and from the laws of thermodynamics, any equilibrium in solution involving species D, F, U and V:

$$D + F \rightleftharpoons U + V$$

will have an equilibrium constant , K_T, at a particular temperature T given by:

$$K_T = (a_U \,.\, a_V)/(a_D \,.\, a_F)$$

where the activities are the values at equilibrium. It should be noted that K_T changes with temperature. The larger the equilibrium constant, the greater will be the ratio of products to reactants at equilibrium.

There are many types of equilibria that occur in solution, but for the important analytical conditions of ionic equilibria in aqueous solution, four examples are very important.

(i) **Acid and base dissociation.** In aqueous solution, strong electrolytes (e.g., NaCl, HNO_3, NaOH) exist in their ionic forms all the time. However, weak electrolytes exhibit **dissociation equilibria**. For **ethanoic acid**, for example:

$$HOOCCH_3 + H_2O \rightleftharpoons H_3O^+ + CH_3COO^-$$
$$K_a = (a_H \,.\, a_A)/(a_{HA} \,.\, a_W) = 1.75 \times 10^{-5}$$

where HA, W, H and A represent each of the species in the above equilibrium. In dilute solutions the activity of the water a_W is close to 1.

For **ammonia:**

$$NH_3 + H_2O \rightleftharpoons NH_4^+ + OH^-$$
$$K_b = (a_{NH4^+} \,.\, a_{OH^-})/(a_{NH_3} \,.\, a_W) = 1.76 \times 10^{-5}$$

Water behaves in a similar way:

$$2\,H_2O \rightleftharpoons H_3O^+ + OH^-$$
$$K_W = (a_{H_3O^+} \,.\, a_{OH^-}) = 10^{-14}$$

(ii) **Complexation equilibria**. The reaction between an acceptor metal ion M and a **ligand** L to form a complex ML is characterized by an equilibrium constant. This is discussed further in Topic C7, but a simple example will suffice here:

$$M(aq) + L(aq) \rightleftharpoons ML(aq)$$

$$K_f = (a_{ML})/(a_M \,.\, a_L)$$

For example, for the copper–EDTA complex at 25°C: $K_f = 6.3 \times 10^{18}$

(iii) **Solubility equilibria**. If a compound is practically insoluble in water, this is useful analytically because it provides a means of separating this compound from others that are soluble. The technique of **gravimetric analysis** has been developed to give very accurate analyses of materials by weighing pure precipitates of insoluble compounds to give quantitative measurements of their concentration. For the quantitative determination of sulfate ions, SO_4^{2-}, the solution may be treated with a solution of a soluble barium salt such as barium chloride $BaCl_2$, when the following reaction occurs:

$$Ba^{2+} + SO_4^{2-} \rightleftharpoons BaSO_4\ (s)$$

Conversely, if solid barium sulfate is put into water:

$$BaSO_4\ (s) = Ba^{2+} + SO_4^{2-}$$

The **solubility product, K_{sp},** is an equilibrium constant for this reaction

$$K_{sp} = a(Ba^{2+}) \,.\, a(SO_4^{2-}) = 1.2 \times 10^{-10}$$

bearing in mind that the pure, solid $BaSO_4$ has $a = 1$. This means that a solution of barium sulfate in pure water has a concentration of sulfate ions of only 1.1×10^{-5} M. The concentration of the barium ions is the same.

(iv) **Redox equilibria**. When a species gains electrons during a reaction, it undergoes **reduction** and, conversely, when a species loses electrons it undergoes **oxidation**. In the total reaction, these processes occur simultaneously, for example:

$$Ce^{4+} + Fe^{2+} = Ce^{3+} + Fe^{3+}$$

The cerium is reduced from oxidation state 4 to 3, while the iron is oxidized from 2 to 3. Any general **'redox process'** may be written:

$$Ox1 + Red2 = Red1 + Ox2$$

The equilibrium constant of redox reactions is generally expressed in terms of the appropriate electrode potentials (Topics C5, C8), but for the above reaction:

$$K = (a(Ce^{3+}) \,.\, a(Fe^{3+}))/(a(Ce^{4+}) \,.\, a(Fe^{2+})) = 2.2 \times 10^{12}$$

Summary

For ionic equilibria in solution, which are widely used in analytical chemistry, a large equilibrium constant for the reaction indicates that it will proceed practically to completion. If the equilibrium constant is of the order of 10^{10}, then the ratio of products to reactants will be much greater than 1000 to 1. For example:

$$H^+ + OH^- = H_2O \qquad K = 10^{14}$$

$$C(aq) + A(aq) = CA(solid) \qquad K = 10^{10}$$

$$M(aq) + L(aq) = ML(complex) \qquad K = 10^{10}$$

$$Ox1 + Red2 = Red1 + Ox2 \qquad K = 10^{12}$$

Therefore, these reactions may be used for quantitative measurements, for example by volumetric or gravimetric techniques (Topics C5, C7 and C8).

It should be noted that, in calculations involving solution equilibria, certain rules should always be considered.

- **Electroneutrality**. The concentrations of positive and negative charges must be equal. Sometimes, ions that do not react are omitted from the equations, although they must be present in the solution.
- **Stoichiometry**. The total amounts of all species containing an element must be constant, since no element can be created or destroyed.
- **Equilibria**. All possible equilibria, including those involving the solvent, must be taken into account.

C2 Electrochemical reactions

Key Notes

Electrochemical reactions	Reactions involving charged species in solution are important in many analyses. Their chemistry and the laws governing them must be known.
Electrochemical cells	The construction of an electrochemical cell using two electrodes allows the study of the cell reaction, changes in concentration, and electrolysis.
Electrode potentials	The potential difference between electrodes depends on the properties of the electrodes and the concentrations in the solution.
Electrolysis	If the cell reaction is driven by applying an external voltage supply, then useful chemical reactions may occur.

Related topics Solution equilibria (C1) Other topics in Section C (C3–C10)

Electrochemical reactions

Important electrochemical reactions have already been noted in Topic C1, for example:

- **acid–base** reactions, where the acid donates a proton to the base;
- **precipitation** reactions, where the reactants form an insoluble product;
- **complexation** reactions, where a ligand coordinates to an acceptor;
- **oxidation–reduction** reactions, where the oxidizing agent gains electrons from the reducing species.

All of these reactions involve charged species and all may be studied by electrochemical methods and used for analysis.

Electrochemical cells

In order to study an electrochemical reaction, the appropriate cell must be constructed. It is impossible to measure the electrical properties using a single contact. For example, connecting to one end of a resistor will not allow measurement of its resistance. Connections must be made to two electrodes, and a cell must be constructed. Electrical connection to the solution, whether to measure a cell emf or to conduct electrolysis reactions, must be made through two electrodes.

Cells with two similar, inert electrodes placed in the same solution may be used for measuring **conductance,** discussed in Topic C10. Cells where electrolytic reactions occur are used in **voltammetry**, which is discussed in Topic C9. For **potentiometric** methods, discussed here and in other parts of this section, two dissimilar electrodes are used to construct a cell whose emf depends on the solutions and electrodes used.

Many different types of electrodes are available and the most important are described in Topic C3. The simplest are made of a metal, such as zinc, copper, or platinum. Consider a cell set up with a zinc electrode dipping into a zinc sulfate solution, and connected through a porous disc to a copper sulfate solution in contact with a copper electrode, as shown in *Figure 1*.

The cell is conventionally written as:

$$Zn|Zn^{2+}\,SO_4^{2-}|Cu^{2+}\,SO_4^{2-}|Cu$$

The cell reaction will result in simultaneous reduction or addition of electrons to a species in solution at the one electrode and oxidation, or removal of electrons at the other.

$Cu^{2+} + 2e^- = Cu$ reduction (note: 'reduction on the right')

$Zn - 2e^- = Zn^{2+}$ oxidation (note: 'oxidation on the left')

Adding these gives the cell reaction (excluding the sulfate anions).

$$Cu^{2+} + Zn = Cu + Zn^{2+}$$

If the cell reaction is allowed to take place, by connecting the copper and zinc electrodes electrically, then it acts as a **galvanic cell** or **battery**, and does work. If the potential difference is measured with no current flow, it is defined as the **electromotive force**, or **emf,** E, and this is equivalent to reversible conditions. If an external voltage is applied to the cell it can cause **electrolysis**, driving the cell reaction.

In order to prevent the zinc sulfate and the copper sulfate mixing, a porous disc, or a conducting **salt bridge** of potassium sulfate in a tube, or a wick can be used to connect the solutions. This helps to reduce the **liquid junction potential**, which occurs when two solutions of unequal concentration, or containing dissimilar ions are placed in contact. Because of the different rates of **diffusion** of the ions, a liquid junction (or diffusion) potential, E_L, is set up and this affects the total cell emf. For example, between a solution of 0.1 M HCl and a solution of 0.01 M HCl, E_L is about 38 mV.

By using a third solution, which must not react, and in which the cation and anion have similar mobilities and diffusion characteristics, to connect the two electrode solutions the liquid junction potential may be minimized. Concentrated potassium chloride, about 4 M, is often used, since K^+ and Cl^- ions have almost the same mobility (see Topic C10). Potassium nitrate or ammonium nitrate is also suitable if chloride ions react. Use of a **salt bridge** reduces the effective E_L to about 1 mV.

The laws of thermodynamics, which are described in detail in textbooks of physical chemistry, provide the connection between the chemical reaction and the electrochemical quantities that are measured in analysis.

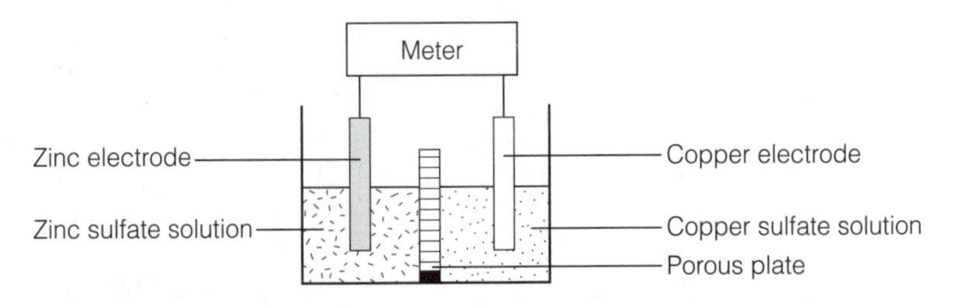

Fig. 1. Daniell cell.

The free energy G depends on the activity (or approximately the concentration) of the reactants and products. For a general reaction:

$$\text{Ox1} + \text{Red2} = \text{Red1} + \text{Ox2}$$

$$\Delta G = \Delta G^{\ominus} + RT \ln [(a(\text{Ox2}).a(\text{Red1}))/(a(\text{Ox1}).a(\text{Red2}))]$$

where the superscript, $\ominus$, means that it applies to the **standard state**. For gases, this is 1 atmosphere pressure; for liquids and solids it is less than 1 atmosphere pressure for the pure material; and for solutions, it is an activity of 1.

The electromotive force measures the free energy change under reversible conditions when n Faradays of electrical charge occur in the cell reaction and is related to it by the equation:

$$\Delta G = -nFE$$

For example, in the Daniell cell reaction, 2 moles of electrons are transferred, and the cell emf is about 1.1 V. Therefore:

$$\Delta G = -2 \times 96485 \times 1.1 = -212.3 \text{ kJ mol}^{-1}$$

Combining the above equations, for the Daniell cell:

$$E = E^{\ominus} + (RT/2F) \ln [(a(\text{Zn}^{2+}).a(\text{Cu}))/(a(\text{Zn}).a(\text{Cu}^{2+}))]$$

This is often referred to as the **Nernst equation** for the cell. (Note: the value of $E^{\ominus}$ measures the cell emf and relates to the free energy change under standard state conditions, while the value of E relates to the free energy change under other, specified conditions. It is important to recognize that these equations apply at *any* temperature, not just a standard temperature.)

Electrode potentials

In order to establish any scale it is necessary to have a reference point. Since measurement of emf is made using a cell, one electrode must be taken as a **reference electrode**. The standard hydrogen electrode (SHE) has been chosen as the primary reference.

A **hydrogen electrode** may be constructed using a platinum metal plate as contact, and bubbling hydrogen gas through so that it makes contact with both solution and platinum. When the gas is at 1 atmosphere pressure and the solution has an activity of hydrogen ions of 1, this is a **standard hydrogen electrode (SHE)**.

By convention, the standard hydrogen electrode is assigned an electrode potential of 0.000 V and all other electrode potentials are compared to it.

A cell can be constructed with a copper electrode in copper sulfate and a standard hydrogen electrode, as shown in *Figure 2*.

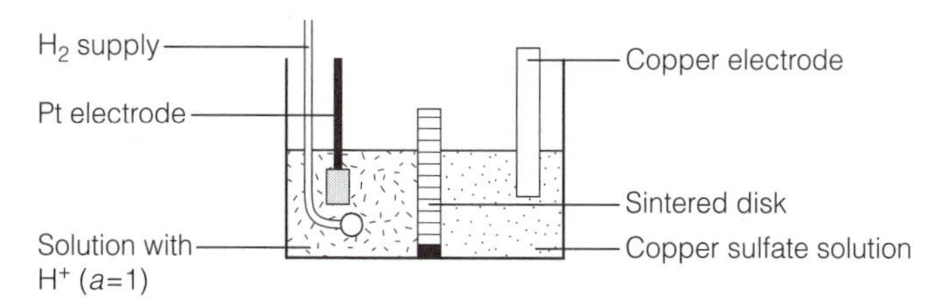

Fig. 2. Copper-hydrogen electrochemical cell.

The cell reaction is the reduction of copper ions by hydrogen:

$$Cu^{2+} + H_2 = Cu + 2\,H^+$$

Writing the Nernst equation for this cell:

$$E = E^{\ominus} - (RT/nF) \ln [(a\,(Cu).a\,(H^+)^2)/(a\,(Cu^{2+}).a\,(H_2))]$$

The superscript, $\ominus$, means the **standard** value, when the activities are all 1. Note that the quantity (RT/F) has the value 0.02569 V at 25°C. Here, $n = 2$, since two electrons are transferred at each electrode. For a standard hydrogen electrode, $a\,(H^+) = 1$ and $a\,(H_2) = p(H_2) = 1$, so, we may write:

$$E = E^{\ominus} - (RT/2F) \ln [a\,(Cu)/(a\,(Cu^{2+})]$$

which is the equation for the **copper electrode potential**. It may also be written:

$$E = E^{\ominus} + (RT/2F) \ln [a\,(Cu^{2+})/(a\,(Cu)]$$

and for a pure copper electrode, $a(Cu) = 1$, so at 25°C:

$$E = E^{\ominus} + 0.01284 \ln [a\,(Cu^{2+})]$$

This means that the electrode potential of a copper electrode depends logarithmically on the activity of the copper ions in solution. This is known as the **Nernst equation**. Standard electrode potentials are listed in Table 1 of Topic C3.

Although the SHE is essential to define electrode potentials, it is not suitable for many routine analytical uses, and the **saturated calomel electrode (SCE)** is often used in practical cells. This is described in Topic C3.

The Nernst equation is very useful in analytical measurements, since it allows the analyst to measure variations in concentration during reactions or experiments. It is important to note that, strictly, electrochemical cells measure the **activity**, although in many cases this may be related to the concentration.

Electrolysis

If an electric current is passed through a cell consisting of two electrodes in contact with an electrolyte solution, the current causes the ions to migrate towards the electrode of opposite charge, where a reaction takes place. For example, when acidified water is electrolyzed:

(i) the hydrogen ions are attracted to the negative electrode (**cathode**) and gain electrons to form hydrogen gas:

$$2H^+ + 2e^- = H_2$$

(ii) the hydroxyl ions are attracted to the positive electrode (**anode**) and lose electrons to become water and oxygen gas:

$$2OH^- - 2e^- = H_2O + \tfrac{1}{2}\,O_2$$

The total reaction is therefore:

$$2H_2O = 2H^+ + 2OH^- = H_2O + H_2 + \tfrac{1}{2}\,O_2$$

In order to cause 1 mole of reaction involving the transfer of n electrons, then n moles of electronic charge must be transferred or nF Coulombs cause 1 mole of such a reaction, where F is the Faraday equal to 96458 C.

Electrolysis is the basis of the analytical techniques of **polarography**, **voltammetry**, **amperometry** and **coulometry** (see Topic C9). Electrolysis may also be used for the deposition, production and purification of materials. For example, the copper electrolysis cell has two electrodes placed in an acidified aqueous solution of a copper salt.

Example
If a current of 0.123 amps is passed through a copper electrolysis cell for one hour, how much copper will deposit?

$$0.123\ \text{A} = 0.123\ \text{C s}^{-1}$$

Since 1 hour = 3600 s, therefore, $It = 442.8$ C are passed.

1 mole of copper is 63.5 g, which requires $2F = 2 \times 96485$ C, so 442.8 C deposit $63.5 \times 442.8/(2 \times 96485)$ g = 0.1457 g.

It is possible to deposit metals onto an electrode quantitatively, using the technique of **electrogravimetry,** although this is not often used. Electrolysis may also be used to generate reagents to react with analytes, and this is referred to as **coulometry**.

C3 Potentiometry

Key Notes

Cells

In order to make measurements of electrode potentials, or to study the changes that take place in a solution reaction, an appropriate electrochemical cell must be set up.

Indicator electrodes

The indicator electrode makes electrical contact with the solution and acts as a sensor, which responds to the activity of particular ions in solution and acquires a potential dependent on the concentration of those ions.

Selectivity

The ideal electrode should respond to a single ion, but this is not often the case. The effectiveness of any indicator electrode is determined by its selectivity.

Direct potentiometry

The direct measurement of concentrations is possible using electrodes of high selectivity and reproducibility. The measurement may also be used to follow titrations.

Related topics

pH and its control (C4)
Titrimetry I: acid–base titrations (C5)
Titrimetry II: complexation, precipitation and redox titrations (C7)

Cells

When a cell is set up, but not connected to any outside circuit, no reaction should take place. If the cell is now 'shorted out' by connecting the electrodes on right and left, electrons will flow and the cell reaction will occur. This reaction changes the concentrations of the original solutions. Therefore, to measure the original sample concentrations within the cell, a device must be used that does not allow current, and hence electrons, to flow. While older systems used a potentiometer, where the potential difference was balanced by adjusting an electrical circuit so that an external source gave exactly the same potential difference detected by the null-point of a galvonometer, modern potentiometry uses **digital voltmeters** (DVM), where the current used to take the measurement is extremely small. A suitable experimental arrangement is shown in *Figure 1.*

If a check is needed on the correctness of the measured value for an experimental cell, a **standard cell**, such as the Weston cadmium cell, may be used as a calibration, since the value of its emf is accurately known over a range of temperatures. The electrode potential is defined using the standard hydrogen electrode as reference, as described in Topic C2.

Indicator electrodes

There are many types of indicator electrode used in analyses to construct electrochemical cells. They may be classified as shown in *Table 1.*

When two electrodes are combined in a cell, the measured emf may be separated into 'half-cell emfs' that relate to the individual electrodes. For a Daniell cell discussed in Topic C2:

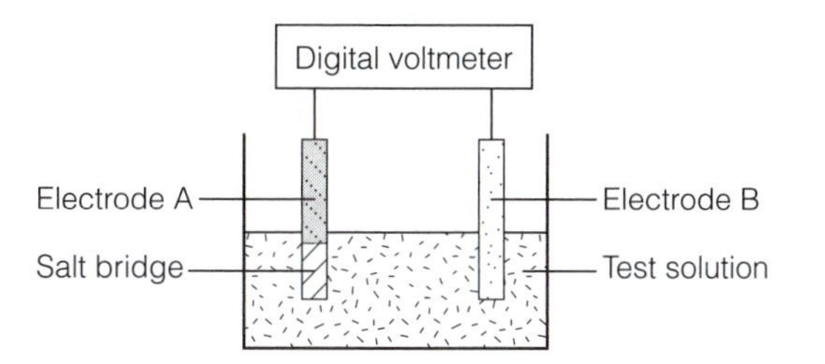

Fig. 1. Experimental arrangement with salt bridge junction and DVM.

Table 1. Potentiometric indicator electrodes

Class	Description	Example
Class 1	Metal/metal ion	Ag/Ag^+ (cation reversible)
Class 2	Metal/saturated metal salt/anion	$Ag/AgCl/Cl^-$ (anion reversible
Redox	Inert metal/redox couple	Pt/Ce^{4+}, Ce^{3+} Pt/H^+, H_2
Membrane	Inner electrode/solution/ ion selective membrane	Glass electrode Fluoride electrode
ISFET	Coated field-effect transistor	pH-sensitive
Gas-sensing electrodes	pH-electrode + membrane	For CO_2, NH_3

$$E(\text{cell}) = E(\text{right}) - E(\text{left}) = E(\text{Cu}^{2+}, \text{Cu}) - E(\text{Zn}^{2+}, \text{Zn})$$

Dividing the emf equation into separate electrode equations:

$$E(\text{Cu}^{2+}, \text{Cu}) = E^{\ominus}(\text{Cu}^{2+}, \text{Cu}) + (RT/2F) \ln (a\,(\text{Cu}^{2+})/(a\,(\text{Cu}))$$

$$E(\text{Zn}^{2+}, \text{Zn}) = E^{\ominus}(\text{Zn}^{2+}, \text{Zn}) + (RT/2F) \ln (a\,(\text{Zn}^{2+})/(a\,(\text{Zn}))$$

If these equations are obeyed, then the electrodes show **Nernstian response**.

Class 1 electrodes

These are the simplest electrodes, but not necessarily the easiest to use. A metal rod is immersed in a solution of ions of the same metal, for example silver with a solution containing silver ions. With some ions it is important to prevent hydrolysis or complexation taking place.

$$\text{Ag}^+ \text{(solution)} + e^- = \text{Ag(solid)}$$

$$E(\text{Ag}^+/\text{Ag}) = E^{\ominus}(\text{Ag}^+, \text{Ag}) + RT/F \ln (a\,(\text{Ag}^+)/(a\,(\text{Ag}))$$

if a pure silver rod is used, a(Ag) = 1, so we may write:

$$E(\text{Ag}^+/\text{Ag}) = E^{\ominus}(\text{Ag}^+, \text{Ag}) + RT/F \ln (a\,(\text{Ag}^+))$$

This is therefore an electrode reversible to silver ions. An example of the use of this electrode is given later. $E^{\ominus}(Ag^+, Ag) = 0.800$ V at 25°C is the **standard electrode potential** of the silver electrode.

Class 2 electrodes

When an insoluble salt of a metal (see also Topic C8) is present, the concentration of the metal ion depends on the concentration of the anion and on the solubility product K_{sp}. Therefore, if we put insoluble silver chloride, AgCl, in contact with a silver rod,

$$a\,(Ag^+) = K_{sp}/(a\,(Cl^-))$$

so that the concentration of silver ions is controlled by the silver chloride solubility.

$$E(AgCl/Ag) = E^{\ominus}(AgCl/Ag) - RT/F \ln (a(Cl^-))$$

Since K_{sp} (AgCl) ~ 1.6×10^{-10} at 25°C,

$$E^{\ominus}\,(AgCl/Ag) = 0.22\text{ V}$$

This electrode is reversible to chloride ions.

Another important electrode of this class is the **calomel electrode,** which is discussed below.

In order to make accurate potentiometric measurements, a cell must be constructed that is reproducible and reliable. The emf should depend chiefly on the particular species in the sample that it is intended to measure, and if necessary the system must be calibrated.

As noted in Topic C2, electrode potentials are conventionally referred to the standard hydrogen electrode (SHE), for which the standard electrode potential, $E^{\ominus}$ = 0.000V. Another electrode, for example the silver electrode, may be combined with the SHE to form a cell. *Table 2* gives a representative list of standard electrode potentials at 25°C.

Table 2. Standard reduction electrode potentials at 25°C

Electrode reaction		$E^{\ominus}$/V
Li^+	$+ e^- = Li$	−3.04
K^+	$+ e^- = K$	−2.92
Ca^{2+}	$+ 2e^- = Ca$	−2.87
Na^+	$+ e^- = Na$	−2.71
Mg^{2+}	$+ 2e^- = Mg$	−2.37
Al^{3+}	$+ 3e^- = Al$	−1.66
Mn^{2+}	$+ 2e^- = Mn$	−1.18
Zn^{2+}	$+ 2e^- = Zn$	−0.76
Fe^{2+}	$+ 2e^- = Fe$	−0.44
Sn^{2+}	$+ 2e^- = Sn$	−0.14
Pb^{2+}	$+ 2e^- = Pb$	−0.13
H^+	**$+ e^- = ½ H$**	**0.0000 exactly**
AgBr	$+ e^- = Ag + Br^-$	+0.10
Sn^{4+}	$+ 2e^- = Sn^{2+}$	+0.15
AgCl	$+ e^- = Ag + Cl^-$	+0.22
Cu^{2+}	$+ 2e^- = Cu$	+0.34
Hg_2Cl_2	$+ 2e^- = 2Hg + 2Cl^-$	+0.27
½ I_2	$+ e^- = I^-$	+0.54
Fe^{3+}	$+ e^- = Fe^{2+}$	+0.76
Ag^+	$+ e^- = Ag$	+0.80
IO_3^-	$+6H^+ + 5e^- = ½ I_2 + 3H_2O$	+1.19
½ O_2	$+2H^+ + 2e^- = H_2O$	+1.23
½ $Cr_2O_7^{2-}$	$+7H^+ + 3e^- = Cr^{3+} + 7/2\ H_2O$	+1.33
½ Cl_2	$+ e^- = Cl^-$	+1.36
Ce^{4+}	$+ e^- = Ce^{3+}$	+1.44
MnO_4^-	$+8H^+ + 5e^- = Mn^{2+} + 4H_2O$	+1.52

The SHE is rather inconvenient to use, since it requires a supply of inflammable hydrogen and has a tendency to change emf slightly as the bubbles of hydrogen cover the metal.

Reference electrodes should have a constant potential, should not react with the sample, and should have a very small liquid junction potential with the sample solution. Two reference electrodes are commonly used.

The calomel reference electrode is shown in *Figure 2(a)*. This is an electrode of Class 2, with liquid mercury in contact with mercury(I) chloride, or calomel, in a solution of potassium chloride of high, fixed concentration.

The electrode reaction is

$$Hg_2Cl_2 \text{ (solid)} + 2e^- = 2Hg \text{ (liquid)} + 2Cl^- \text{ (aq)}$$

The electrode potential is given by:

$$E(\text{cal}) = E^{\ominus}(\text{cal}) - RT/2F \ln (a(Cl^-)^2)$$

If a concentrated solution of KCl is used, either saturated or 3.5 M, then this electrode has a constant potential at 25°C. Changes in the solution outside the electrode have a very small effect on the potential of this electrode, since the chloride concentration is high and is not altered significantly by the external solution. Additionally, the concentrated KCl acts as a salt bridge.

$$E(\text{cal, sat}) = 0.244 \text{ V (saturated calomel electrode, SCE)}$$

$$E(\text{cal, 3.5M}) = 0.250 \text{ V (calomel electrode, 3.5M KCl)}$$

The silver/silver chloride reference is similar, having a silver wire coated with silver chloride and in contact with concentrated KCl solution.

$$E \text{ (AgCl, 3.5M)} = 0.204 \text{ V}$$

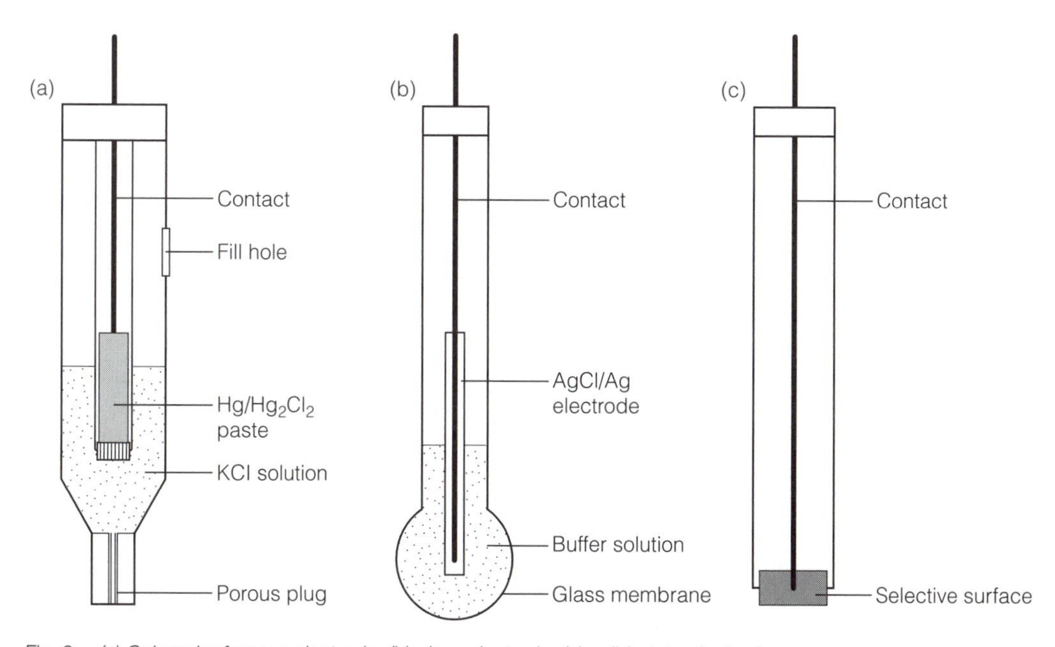

Fig. 2. (a) Calomel reference electrode; (b) glass electrode; (c) solid-state electrode.

If the sample solution might react with chloride ions, for example silver or lead salts, then a **double junction reference electrode** may be used, with an additional liquid junction of KNO_3.

Redox electrodes

If an inert wire, usually platinum, is placed into a solution containing both oxidized and reduced species, it will take up an equilibrium potential dependent on the ratio of their concentrations:

$$\text{Ox} + ne^- = \text{Red}$$

$$E(\text{Ox, Red}) = E^{\ominus}(\text{Ox, Red}) + (RT/nF) \ln (a(\text{Ox})/a\,(\text{Red}))$$

Ion-selective electrodes (ISE)

All of the above electrodes are sensitive to particular ions, but have certain disadvantages, such as loss of response due to poisoning, or a tendency to mechanical or other failure.

A large variety of special electrodes have been developed to test for a very wide range of ions. Many involve an insoluble membrane, capable of electrical conduction, whose surface potential depends chiefly on the concentration of a particular ion.

The **glass electrode** for measuring pH, shown in *Figure 2(b)* is an early example. A thin membrane of a conducting glass separates the inner solution of a silver–silver chloride electrode in a chloride-containing solution from the sample solution outside. Both sides of the glass membrane become hydrated, and the outer surface exchanges its cations with the hydrogen ions of the sample solution. The potential of the glass electrode depends on the ratio of the activities of H^+ ions inside and out, on the potential of the inner silver halide electrode and on the **asymmetry potential,** characteristic of each particular electrode.

This gives a total electrode potential:

$$E(\text{glass}) = E^*(\text{glass}) + RT/F \ln (a(\text{H+}))$$

$$= E^*(\text{glass}) - (2.303\ RT/F)\ \text{pH}$$

The unknown constant E^* must be determined by calibration with known buffers, as described in Topic C4. Modern glass electrodes for pH measurement often incorporate a reference electrode.

Although these electrodes work very well in the pH range 1–9, at high pH they suffer from an **alkaline error** due to the effect of other cations. This is discussed in the next section. Glass electrodes are also fragile and may suffer from slow response times.

Crystalline membrane electrodes, constructed either as the glass electrode, or with a direct contact as shown in *Figure 2(c),* have an outer crystal surface which responds to particular ions. The **fluoride electrode** has a crystal of LaF_3, treated with Eu(II) to increase conductivity, which responds selectively to the adsorption of free F^- ion on its surface. The selectivity is very good, due to the ability of the small F^- on to fit to the LaF_3 crystal lattice to the exclusion of larger ions. However, OH^- ions are also small and can interfere, and F^- may also form undissociated hydrofluoric acid. Therefore, it is necessary to use this electrode in a buffer solution at about pH 6.

Interferences occur with metals such as Al^{3+}, Fe^{3+} and Ca^{2+}, which combine with the fluoride. With a SCE reference electrode, the cell emf is:

$$E(\text{cell}) = E^*(\text{F}^-) - (RT/F)\ \ln\ (a\ (\text{F}^-)) - E(\text{SCE})$$

or, at 25°C:

$$E(\text{cell}) = \text{k} - 0.0257\ \ln\ (a(\text{F}^-))$$

If a cell is constructed with a fluoride ISE combined with a SCE, calibration with solutions of fluoride ions of known concentration enable the value of k to be found. When the same cell is filled with an unknown solution, the fluoride ion concentration may be determined. This is an example of **direct potentiometry**.

Other crystalline ion-selective electrodes use AgCl for Cl^-, Ag_2S for Ag^+ and S^{2-}, $Ag_2S + CuS$ for Cu^{2+} and many more.

Liquid membrane electrodes are shown in *Figure 3(a)*. Typically an organophilic porous plastic membrane separates the internal reference from the sample solution. Another reservoir contains a liquid ion-exchanger, which saturates the pores in the membrane. The calcium electrode uses calcium dodecyl phosphate $[(RO)_2\ PO_2]_2$ Ca where R is the dodecyl radical, dissolved in dioctyl phenylphosphonate.

The electrode response is given by:

$$E(\text{Ca}^{2+}) = \text{k} + (RT/2F)\ \ln\ (a(\text{Ca}^{2+}))$$

Ion selective field effect transistors (ISFETs)

One disadvantage of the glass electrode for measuring pH is its fragility. A modern development uses a field effect transistor where the gate surface is coated with a material, often aluminum oxide, whose surface is sensitive to pH. As the potential at the surface changes with pH, the current induced through the transistor varies. A temperature diode, incorporated in the electrode

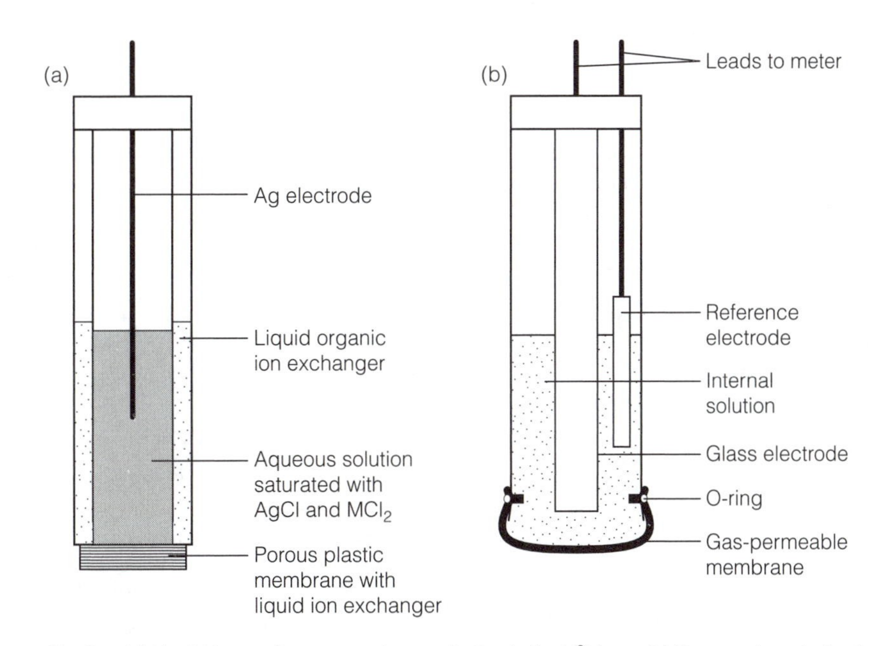

Fig. 3. (a) Liquid ion exchange membrane electrode for M^{2+} ions. (b) Gas sensing electrode using glass ISE.

simultaneously measures the temperature so that a rapid reading of pH, compensated for temperature is obtained. The electrodes are robust and the response time is a few seconds. An example is given in Topic H1.

Gas-sensing electrodes

If gases such as CO_2 or NH_3 are allowed to diffuse into the solution surrounding a pH electrode, they will alter the pH. The construction is shown in *Figure 3(b)*. The pH electrode, often incorporating a reference electrode as well, is separated from the sample solution by a microporous hydrophobic membrane, which will allow gases but not water to diffuse through rapidly. For CO_2 the overall equilibrium occurs in 3 stages:

(i) carbon dioxide gas diffuses from the outer solution through the membrane until inner and outer solutions are at the same concentration;

(ii) the solution of CO_2 forms the acid H_2CO_3, which dissociates to form hydrogen ions:

$$CO_2 + 2H_2O = H_2CO_3 + H_2O = H_3O^+ + HCO_3^-$$

$$K = a(H_3O^+).\ a(HCO_3^-)/(p(CO_2\ (\text{sample})))$$

(iii) if the internal solution within the membrane has a constant activity of HCO_3^-, for example sodium hydrogen carbonate, then the pH may be calculated:

$$\text{pH} = -\log\ (H_3O^+) = \log\ (K.\ p\ (CO_2\ (\text{sample})))$$

so that by measuring the pH we can find the concentration of CO_2.

Similar arguments apply with an ammonia-sensing electrode.

Selectivity

The ideal electrode should respond only to changes in concentration of a single ion *i*. This is rarely true and the response of a real electrode system can be described by the **Nikolsky–Eisenmann equation**:

$$E_i = E^{\ominus}{}_i + S\ \log[a(1) + \Sigma K_{\text{pot}1,2}\ (a(2))^{(z1/z2)}]$$

where S is the slope of the emf $-\log(a)$ plot, which for Nernstian behavior should be $0.0592/z(1)$ at 25°C, $a(1)$ is the activity of the ion selected of charge z(1), $a(2)$ is the activity of the interfering ion of charge z(2) and $K_{\text{pot}1,2}$ is the **selectivity coefficient** for ion 1 with respect to ion 2. The smaller the value of this selectivity coefficient, the smaller the interference by the second ion.

Direct potentiometry

By calibration, and the use of appropriate electrode response equations, it is possible to measure concentrations directly, as indicated in the example above for fluoride ion in tap water.

A calibration plot should indicate three things:

(i) whether the electrode responds to the correct ion;
(ii) whether that response is Nernstian;
(iii) what range of concentrations may be studied.

Figure 4 shows the calibration plot for a copper ion selective electrode, where a **total ionic strength adjustment buffer (TISAB),** such as 1M $NaNO_3$, has been added to each solution so that the response is effectively at constant ionic strength and constant activity coefficient.

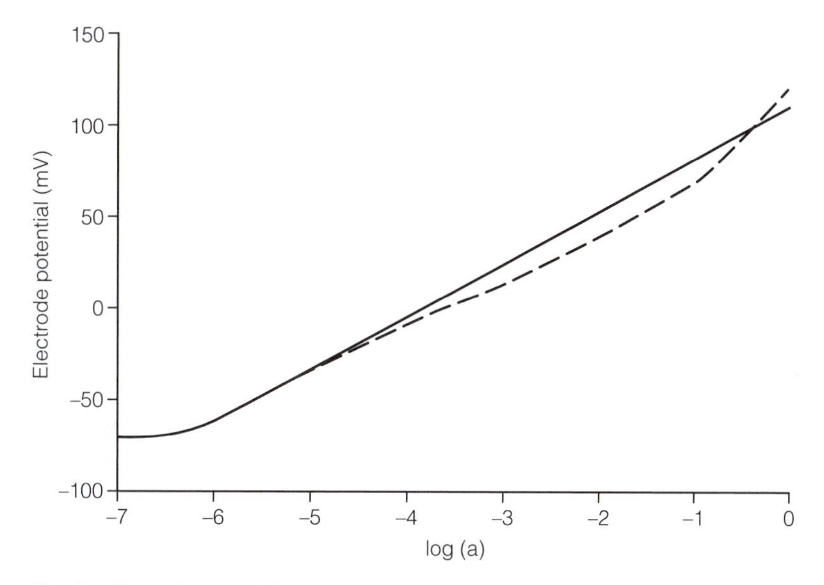

Fig. 4. Electrode potential response for a copper ISE. (a) Against activity (solid line); (b) against concentration (dashed line).

This graph shows that over the concentration range 10^{-1} to 10^{-5} M the calibration is linear with a slope of 0.029 V/log ($a(Cu^{2+})$). Below about 10^{-6} M the line curves, since the solubility of the crystal becomes significant. Interferences from Cu(I), Ag(I) and Hg(II) are troublesome.

If a calibration curve is constructed, direct measurement of solution concentrations within that range may be made.

Example

Calibration performed with a ISE selective to Mg^{2+} ions gave a value of S = 0.0296, and $E = 0.411$ V for a solution of $a(Mg^{2+}) = 1.77 \times 10^{-3}$ M.

What is the activity of Mg^{2+} ions when $E = 0.439$ V?

Substitution gives $a(Mg^{2+}) = 1.998 \times 10^{-4}$ M.

The method of standard additions (Topic B4) has the advantage that comparison is made in the same matrix solution, so that interferences and other effects are the same for the unknown and for the standard.

Potentiometric titrations are discussed in Topic C5.

C4 pH AND ITS CONTROL

Key Notes

Definition of pH

Since the concentrations of ions in aqueous solution vary over an enormous range from greater than 1 molar down to 10^{-14} molar or less, it is convenient to use a logarithmic function to describe them. For hydrogen ions in aqueous solution, denoted by H_3O^+, often called the hydronium ion, this is defined as:

$$pH = -\log(a(H_3O^+)) \approx -\log(c(H_3O^+)/\text{mol dm}^{-3})$$

The pH scale

Pure water contains equal amounts hydrogen ions and hydroxyl ions OH^-, at a concentration of 10^{-7} molar. Therefore, the pH of neutral water is $-\log(10^{-7}) = 7$. Acids have pH values less than 7, down to less than 1, while for alkalis, the pH is greater than 7, up to 13 or more.

Buffers

For many analytical measurements it is necessary to control the pH so that it varies as little as possible. Solutions that resist changes in pH are known as buffer solutions. They usually consist of a mixture of a weak acid and its salt for lower pH values, or a weak base and its salt for higher pH values.

pH measurement

pH may be estimated approximately using visual indicators, but more accurate determination requires the use of pH meters. These must be calibrated using standard buffers.

pH control

This is most usually achieved with buffers. Enzyme reactions, electrophoretic separations and also spectrometric and electrochemical determinations may all require pH control.

Related topics

Titrimetry I: acid–base titrations (C5)

Definition of pH

The acidity or alkalinity of a reaction mixture is most important. It can control the rate of reaction, the nature of the species present and even subjective properties such as taste. The original definition of pH (Sorensen, 1909) related it to the **concentration** of hydrogen ions. Two facts should be recognized. First, like many ions in solution, the hydrogen ion does not exist in aqueous solutions as a 'bare' H^+ species, but as a variety of **hydrated ions**, such as H_3O^+. Second, the determination of pH is often carried out by methods that measure the **activity** of the hydrogen ions, $a(H_3O^+)$

$$a(H_3O^+) = c(H_3O^+)\gamma_\pm \quad \text{or} \quad pH = -\log[a(H_3O^+)]$$

where $c(H_3O^+)$ is the molar concentration of hydrogen ions, and $\gamma_\pm$ is the mean ionic activity coefficient of the ions in solution (see Topic C1).

At low concentrations ($<10^{-2}$ molar), $\gamma_\pm$ is close to 1, and the difference between concentration and activity is small for uni-univalent electrolytes.

The **practical** (or **operational**) definition of pH recognizes that it is determined using electrochemical cells having an electrode selective to hydrogen ions. This has been discussed in Topics C2 and C3, but a typical cell is:

Reference electrode || Solution X | glass electrode.

This gives an output e.m.f , E_X,

$$E_X = E^* + (RT/F) \ln [a(H_3O^+)_X]$$

The constant E^* depends on the exact nature of the reference and glass electrodes, and is best eliminated by calibration with a standard solution S which has a pH that is accurately known.

$$E_S = E^* + (RT/F) \ln [a(H_3O^+)_S]$$

Subtracting these and converting the logarithms gives a practical definition of pH:

$$\mathrm{pH(X)} = \mathrm{pH(S)} + (E_S - E_X)/(RT \ln(10)/F)$$

Typical calibration buffers are discussed below.

The pH scale

In all aqueous solutions, pH values may range between about 0 and 14 or more as shown in *Figure* 1. Molar solutions of strong mineral acids, such as HCl, HNO_3 or H_2SO_4 have pH values less than 1. Weak acids, such as ethanoic or citric acid in decimolar solution have a pH of around 3.

A useful standard is 0.05 M potassium hydrogen phthalate which, at 15°C has a pH of 4.00. Although pure water is neutral and has a pH of 7.00, freshly distilled water rapidly absorbs carbon dioxide from the air to form a very dilute solution of carbonic acid, and therefore has a pH of around 6.

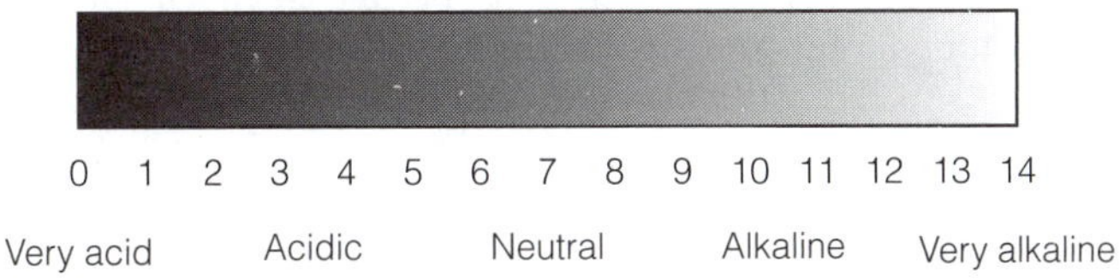

0 1 2 3 4 5 6 7 8 9 10 11 12 13 14

Very acid Acidic Neutral Alkaline Very alkaline

Fig. 1. The pH scale.

Another standard occasionally used is 0.05 M borax (sodium tetraborate, $Na_2B_4O_7$), which has a pH of 9.18 at 25°C.

Dilute alkalis such as ammonia or calcium hydroxide (lime water) have pH values near to 12, and for molar caustic alkalis, such as NaOH, the pH is over 13.

Buffers

As many reactions depend greatly upon the concentration of hydrogen ions in the solutions being used, it is important to control the pH. This is usually achieved by using a solution which has a pH that is accurately known and that resists any change in pH as solvent for the experiment. Such solutions are called **buffers**.

The equilibria that govern the reactions of weak acids or bases in aqueous solution will resist attempts to change them. This is known as **Le Chatelier's principle**. For example, the dissociation of ethanoic acid obeys the equation:

$$CH_3COOH \rightleftharpoons CH_3COO^- + H_3O^+$$

and an equilibrium constant is written (in terms of concentrations)

$$K_a = c(Ac^-) \times c(H_3O^+)/c(HAc) = 1.75 \times 10^{-5}$$

using the abbreviation Ac for the CH_3COO^- group. Converting to logarithmic form, and recalling that $pK = -\log(K)$:

$$pH = pK_a + \log [c(\text{salt})/c(\text{acid})]$$

This is the **Henderson-Hasselbalch equation**.

If we make a mixture containing both the free acid, HAc, and its salt sodium ethanoate, NaAc, then the equilbrium and the concentrations of acid and salt will determine the concentration of hydrogen ions and the pH.

Example 1

For a mixture of 50 cm^3 of 0.1 M HAc with 40 cm^3 of 0.1M NaAc, giving a total volume of 90 cm^3,

$$c(H_3O^+) = 1.75 \times 10^{-5} \times [(50 \times 0.1/90)/(40 \times 0.1/90)]$$
$$= 2.19 \times 10^{-5} \text{ M, so that}$$
$$pH = 4.66$$

Addition of acid to this buffer shifts the above equilibrium to the left and most of the added hydrogen ion combines with the anion. Adding 10 cm^3 of 0.1 M HCl lowers the pH only to about 4.45. If this amount of acid were added to 90 cm^3 of water, the pH would be 2.0. Similarly, when alkali is added, the hydroxyl ions react with the acid to produce more salt. 10 cm^3 of 0.1 M NaOH will raise the pH only to around 4.85. If this amount of alkali were added to 90 cm^3 of water, the pH would rise to 12.

Weak bases and their salts behave in much the same way. For example, ammonia and ammonium chloride:

$$NH_3 + H_2O \rightleftharpoons NH_4^+ + OH^-$$

$$K_b = c(OH^-) \times c(NH_4^+)/c(NH_3) = 1.75 \times 10^{-5}$$

or, rewriting the Henderson-Hasselbalch equation:

$$pOH = pK_b + \log[c\,(\text{salt})/c(\text{base})]$$

or, since $pH + pOH = 14.0$

$$pH = 14.0 - pK_b - \log[c(\text{salt})/c(\text{base})]$$

For a mixture of equal amounts of 0.1 M ammonia and 0.1 M ammonium chloride

$$pH = 14.0 - 4.75 = 9.25$$

A most useful range of buffers is obtained by using salts of a dibasic (or tribasic) acid such as phosphoric acid, H_3PO_4 – for example, potassium dihydrogen phosphate, KH_2PO_4, and disodium hydrogen phosphate, Na_2HPO_4. The equilibrium involved here is:

$$H_2PO_4^- \rightleftharpoons H_3O^+ + HPO_4^{2-}$$

For this equilibrium, the second dissociation constant of phosphoric acid, K_{a2}, is close to 1×10^{-7}, or $pK_{a2} = 7$. *Figure* 2 shows the effect of adding acid or alkali on the pH of a mixture containing 50 cm^3 each of 0.1 M KH_2PO_4 and 0.1 M Na_2HPO_4, which originally has a pH of 7.0. Adding 0.001 moles of acid to 100 cm^3 of water would lower the pH to 2.

The more concentrated the buffer, the greater will be its **buffer capacity**. This is the amount of acid (or alkali) that, when added to 1 liter of buffer, will change its pH by 1 unit. In the above example, the buffer capacity is about 0.04 moles. If we had used a more concentrated buffer, the capacity would be greater. Very dilute buffers have little buffer capacity, and hence have limited use.

Table 1 gives a selection of buffers and standard solutions that are useful for pH control. These solutions and others are often used to calibrate pH meters.

pH measurement

Two important methods exist for pH measurement: **visual**, using **indicators**, and **potentiometric**, by means of electrochemical cells.

Indicators for pH measurement are weak acids (or bases) where the color of the acid form is different from that of the salt.

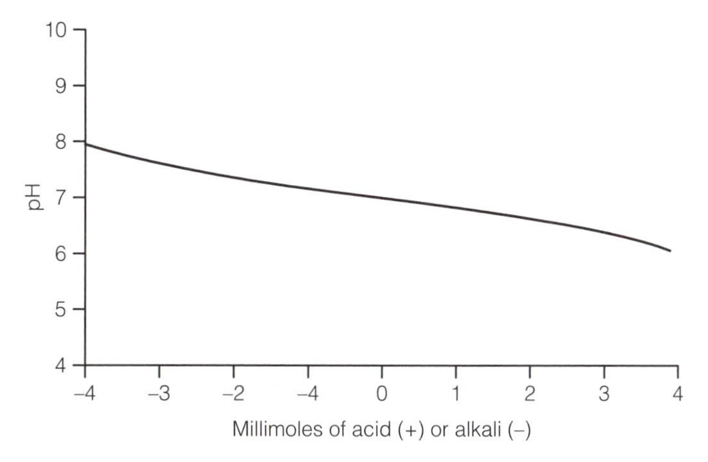

Fig. 2. The effect of adding acid (+) or alkali (–) to a phosphate buffer mixture, originally at pH 7.

Table 1. Buffer solutions

Solution	pH at 25°C
0.05 M potassium hydrogen phthalate	4.008
0.1 M ethanoic acid, 0.1 M sodium ethanoate	4.640
0.025 M potassium dihydrogen phosphate, 0.025 M disodium hydrogen phosphate	6.865
0.01 M borax	9.180
0.1 M ammonia, 0.1 M ammonium chloride	9.250
0.025 M sodium bicarbonate, 0.025 M sodium carbonate	10.012
Saturated calcium hydroxide	12.454

$$\underset{\text{Colour 1}}{\text{HIn}} \rightleftharpoons H_3O^+ + \underset{\text{Colour 2}}{\text{In}^-}$$

The dissociation constant of the indicator, K_{In}, is given by

$$K_{In} = c(H_3O^+) \times c(In^-)/c(HIn)$$

For example, for methyl orange (*Fig. 3*):

$$\text{Acid form, red} \rightleftharpoons \text{Alkaline form, yellow} + H^+$$

As with buffers, the equilibrium and the concentration of hydrogen ions will govern the ratio of color 2 to color 1.

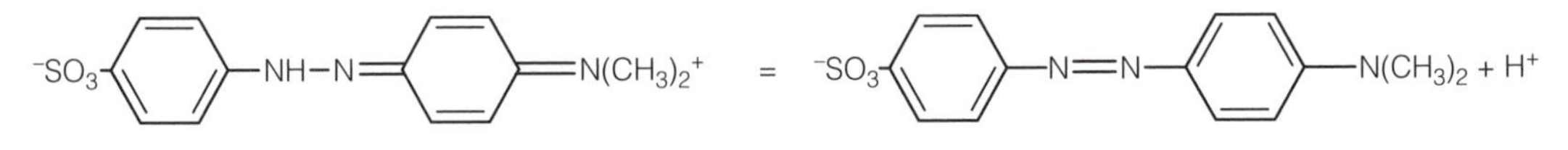

Fig. 3. Structures of methyl orange at low and high pH (only one tautomeric form is shown).

Example 2
With the indicator bromocresol green, where $K_{In} = 1.6 \times 10^{-5}$, or $pK_{In} = 4.8$ and color 1 (acid) is yellow, while color 2 (salt) is blue, a solution of pH 4.0 will give:

$$\log [c(In^-)/c(HIn)] = \log [c(\text{color 2})/c(\text{color 1})] = pH - pK_{In} = -0.8$$

Therefore, $c(\text{color 2})/c(\text{color 1}) = 0.16$, which means about 14% blue, 86% yellow, or visually a very yellowish green. When the pH is equal to the pK_{In}, there are equal amounts of each form making a green color.

A wide range of indicators is available for titrations and other purposes and these are discussed further in Topics C5 and C7.

This provides a useful and rapid method of **estimating** pH by eye; for example, using **litmus paper** which is red below about pH 6 and blue above pH 8. Both wide range and narrow range indicator papers are available to enable a rapid estimation of pH. However, to determine the pH accurately using indicators, careful spectrometric comparison would be needed and this is a time-consuming method that is rarely used.

The **pH meter** uses a reference electrode and a glass electrode with a high-resistance voltmeter and affords a rapid and accurate method of measuring pH (*Fig. 4*). The calomel reference electrode is decribed fully in Topic C3.

The glass electrode is an example of a **membrane ion-selective** electrode and is described in Topic C3. It responds to hydrogen ions:

$$E(\text{glass}) = E^* + (RT/F) \ln [a(H_3O^+)_X]$$

Therefore the complete cell

$$Pt \mid Hg \mid Hg_2Cl_2(s) \mid KCl\ (sat, aq) \parallel \text{Solution X} \mid \text{glass membrane} \mid AgCl \mid Ag$$

has an e.m.f. at 25°C equal to:

$$E\ (\text{cell}) = (E^* - 0.241) + (RT/F) \ln [a(H_3O^+)_X]$$

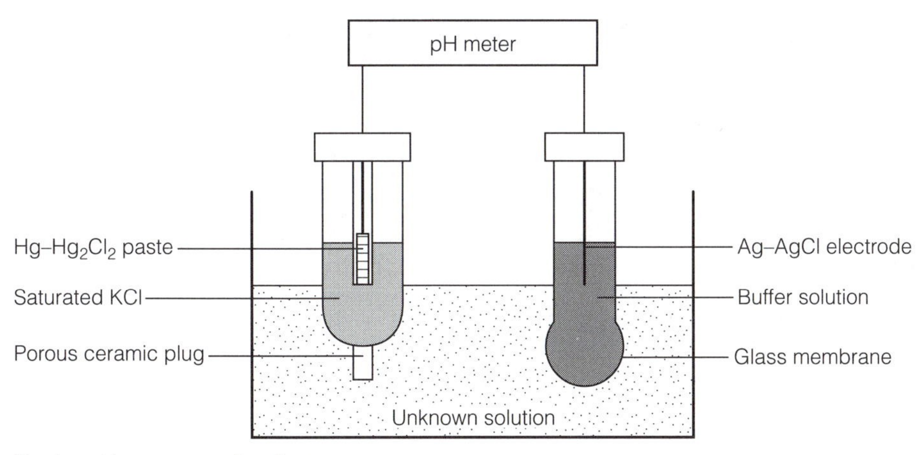

Fig. 4. pH measurement system.

By using one of the standards described above, for example, 0.05 M potassium hydrogen phthalate, which has a pH at 25°C of 4.008, we may eliminate E^* and measure the activity of hydrogen ions, and hence the pH, in the unknown solution X.

$$pH(X) = pH(S) + (E_S - E_X)/(RT \ln(10)/F)$$

pH control

It is often necessary to control the pH of a solution, especially if hydrogen ions are being generated or consumed, or if the nature of the species being analyzed changes with the pH. A few examples will illustrate this problem.

(i) In recording the UV spectrum of a solution of a weak acid, such as a phenol, the peak maxima occur at different wavelengths in an acid medium compared with those in a basic medium. Comparisons may best be made using a constant pH buffer.

(ii) In complexometric titrations (see Topic C7), such as the determination of magnesium by EDTA, the complex is formed readily and completely at high pH, so the titration is carried out using an ammonia-ammonium chloride buffer to keep the solution at pH 10.

(iii) As noted in Topic C3, the **fluoride electrode** detects free F^- ions very well, but OH^- ions interfere, and H^+ ions form undissociated H_2F_2. It is therefore essential to make measurements in a buffer of about pH 5–6.

pH is most often controlled by performing the analysis in a suitable buffer solution. Occasionally, where a reaction produces an acid (or alkali) the technique of **pH-stat titration** may be used. Here, the acid produced is detected, and sufficient alkali added to return the pH to the optimum value.

C5 TITRIMETRY I: ACID–BASE TITRATIONS

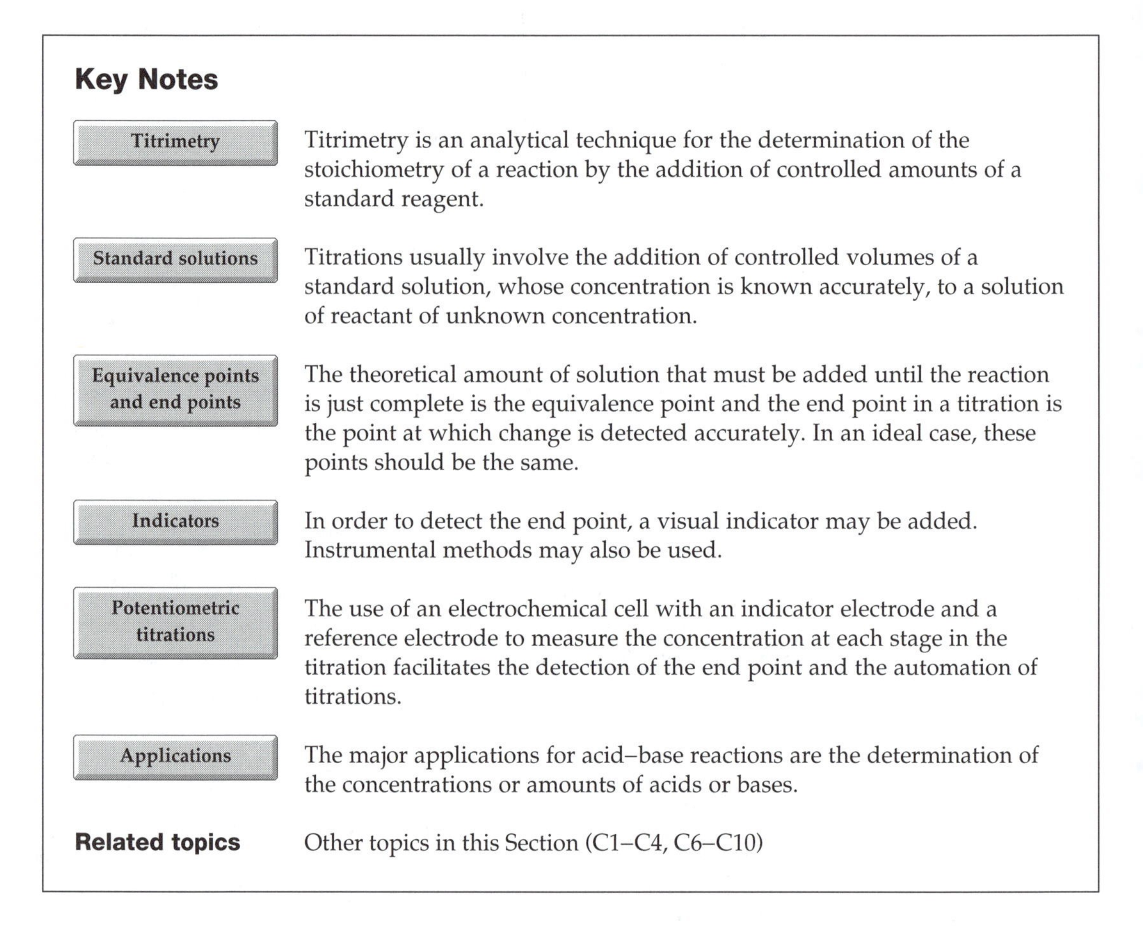

Key Notes

Titrimetry
Titrimetry is an analytical technique for the determination of the stoichiometry of a reaction by the addition of controlled amounts of a standard reagent.

Standard solutions
Titrations usually involve the addition of controlled volumes of a standard solution, whose concentration is known accurately, to a solution of reactant of unknown concentration.

Equivalence points and end points
The theoretical amount of solution that must be added until the reaction is just complete is the equivalence point and the end point in a titration is the point at which change is detected accurately. In an ideal case, these points should be the same.

Indicators
In order to detect the end point, a visual indicator may be added. Instrumental methods may also be used.

Potentiometric titrations
The use of an electrochemical cell with an indicator electrode and a reference electrode to measure the concentration at each stage in the titration facilitates the detection of the end point and the automation of titrations.

Applications
The major applications for acid–base reactions are the determination of the concentrations or amounts of acids or bases.

Related topics
Other topics in this Section (C1–C4, C6–C10)

Titrimetry

In order to obtain accurate quantitative data for a reaction in solution, it is necessary that the reaction be fast, complete and occur in fixed, reproducible amounts. The requirement for **fast reaction** is achieved readily when ionic species are involved, although in some other cases, it is necessary to warm the solutions or add a catalyst. The reaction will be **complete** provided the equilibrium constant is large (see Topic C1).

The technique of **volumetric analysis** is the simplest type of titrimetry, and involves the addition of controlled volumes of a reagent solution, the **titrant,** to a known volume of another solution, the **titrand** in a volumetric titration. This procedure may be automated, and the changes detected instrumentally (Topics C2, C3, C9 and C10). In some cases, excess of a reagent is added and the excess measured by **back titration**.

The volumes and concentrations can be measured with high accuracy. Calibration of volumetric glassware by discharge into weighed containers allows the determination of volumes to 0.01 cm^3. This corresponds to about 0.05% for a typical titration volume of 20 cm^3.

Weighing solutes prior to dissolution gives comparable accuracy. Pre-determined volumes, for example 25 cm^3 are measured with manual or mechanical **pipets,** while cumulative volumes of titrant during addition are measured by a manual or automated **buret**.

Any reaction used in titrimetry will cause the concentrations of the species in solution to change. For acid–base reactions, the concentration of hydrogen ions, and hence the pH, will alter and similar changes in the ionic concentrations occur with other reactions.

Figure 1 shows the pH values in the titration of a strong base by a strong acid as a function of the volume of acid added. pH is discussed in Topic C4. The figure shows that there is only a small change in pH before about 24 cm^3 where there is still an excess of base, and also after 26 cm^3 where there is an excess of acid. However, in the region around 25 cm^3, where the acid exactly neutralizes the base, the change in pH is very large. This signifies the **end point** of the titration.

Provided that the end point coincides with the **equivalence point** or the stoichiometry of the reaction studied, the amounts of titrant and titrand measured should correspond to the actual amounts present.

For a general reaction:

$$aA + bB = cC + dD$$

a moles of A react with b moles of B to produce c moles of C and d moles of D. A **molar** solution contains 1 mole of solute per 1000 cm^3 (or 1 l) of solution. Therefore, 1 cm^3 of a 1 M solution contains 1 mmol of solute.

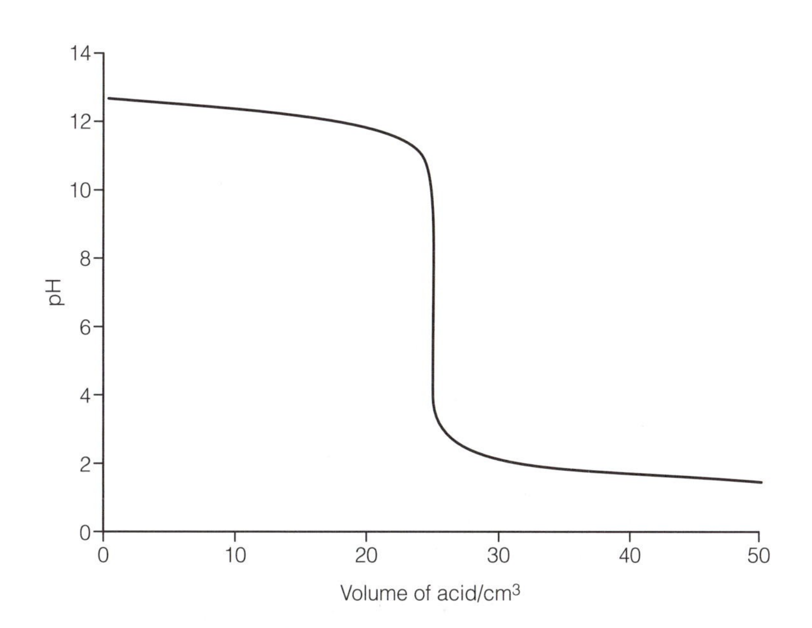

Fig. 1. Titration of 25 cm^3 of 0.1 M NaOH with 0.1 M HCl (pH against volume added).

Example 1

A 25.00 cm^3 aliquot of a solution of a base of known concentration 0.1057 M is titrated with an acid of unknown concentration. The reaction involved 1 mole of base and 1 mole of acid. The end point was determined as 24.88 cm^3 of acid added. What is the concentration of the acid?

25.00 cm^3 of the base solution contained 25.00 × 0.1057 = 2.6425 mmol base. From the known reaction, this should be equivalent to 2.6425 mmol of acid. Since the volume of acid at the end point was 24.88 cm^3, the concentration must be (2.6425/24.88) = 0.1062 M.

In some cases, the end point detected does not correspond exactly with the equivalence point. This may be due to problems with the reaction, or to the small amount of reagent needed to react with additional materials (for example, the added indicator) present in the titrand. In these cases, a **blank titration** must be performed, or allowance made for the **titration error**.

Standard solutions

In order to achieve the highest accuracy, it is necessary to use well-established standard materials as reagents in the primary calibration or **standardization** of the reacting solutions (see Topic A5).

The most important of these are called **primary standards,** and should be easy to obtain, purify and dry, should be stable and not hygroscopic, but should be readily soluble and react rapidly and stoichiometrically. They should ideally have a high relative molecular mass to minimize weighing errors.

The above criteria mean that reagents such as sodium hydroxide, which is hygroscopic and may react with carbon dioxide from the air, and potassium permanganate, which slowly decomposes in air, are unsuitable as primary standards.

Solutions used for quantitative analysis need to be checked and calibrated frequently. For example, hydrochloric acid solutions should be checked against sodium carbonate solution and sodium hydroxide against potassium hydrogen phthalate.

Equivalence points and end points

The end point of a titration is based upon experimental observation, whereas the equivalence point is the theoretical value dependent upon the reaction equation. In an ideal case, these should be the same, but a check may be needed to ensure that factors such as blank errors do not affect the results.

In any titration, the end point corresponds to rapid changes in the concentration of species. This may be detected in many ways. Instrumental methods are discussed later, and visual indicators are discussed below.

For reactions such as a strong acid neutralizing a strong base (*Fig. 1*), the change at the end point is large, and the rate of change with volume is very great, as shown by the derivative plot in *Figure 2(b)*. In other cases, such as the reaction of weak acids with weak bases, the titration curve shows a much less pronounced change, and the derivative plot may be needed to confirm the end point, as in *Figure 2(b)*. The second derivative plot is also useful, but relies greatly on very accurate measurements. Plots of the concentration of a species (for example H^+, or OH^-), against the volume added, give straight lines intersecting at the end point.

If a mixture of acids, or a polybasic acid such as maleic acid, is titrated, then **two** end points are obtained. Similarly, for mixtures of two titrands, two end points will be detected, provided the species have sufficiently different equilibrium constants for the reaction (e.g. $K_{a1} \gg K_{a2}$).

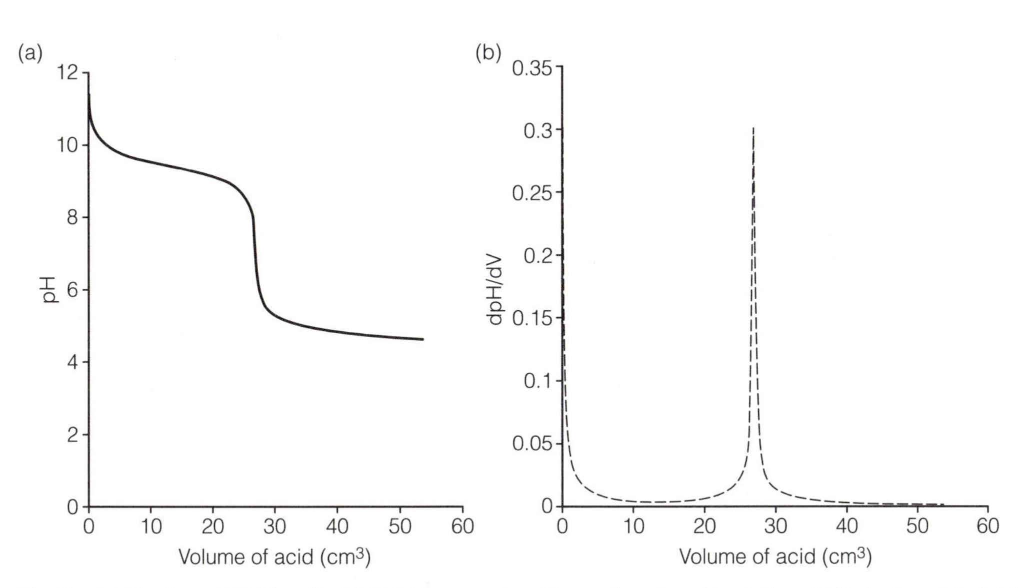

Fig. 2. (a) Titration of 25.00 cm^3 of 0.105 M ammonia with an ethanoic acid solution. (b) Derivative plot of d (pH)/dV showing the end point as 26.80 cm^3.

Indicators

In general, indicators have two forms, which possess different colors. Indicators for acid–base titrations are themselves weak acids or bases where the two forms differ in color as shown in *Figure 3* of Topic C4 for methyl orange.

$$HIn \text{ (color 1)} = H^+ + In^- \text{ (color 2)}$$

The choice of indicator depends upon the reaction to be studied. As noted in Topic C2, the equilibrium constant of the indicator must match the pH range, or electrode potential range of the species being titrated. *Table 1* shows a selection of indicators for acid–base reactions.

As a general rule it is noted that the color change takes place over the range:

$$pH = pK_{In} \pm 1$$

Similarly for other reactions, the concentration at which the indicator changes must match the concentrations at the end point.

For example, in the titration of the weak dibasic maleic acid with sodium hydroxide, the first end point, corresponding to sodium hydrogen maleate occurs at pH = 3.5, while the second end point for disodium maleate is at pH = 9. Since $pK_{In} = 3.7$ for methyl orange, this will change at the first end point, while phenolphthalein would change color around pH = 9.

Table 1. Typical visual indicators for acid-base titrimetry

Acid–base	Low pH color	High pH color	pK_{In}
Thymol blue	Red	Yellow	1.7
Methyl orange	Red	Yellow	3.7
Bromothymol blue	Yellow	Blue	7.0
Phenolphthalein	Colorless	Red	9.6
Alizarin yellow R	Yellow	Orange	11

Potentiometric titrations

The use of electrodes, particularly the glass electrode for pH measurements and the wide range of other ion selective electrodes (ISE) described in Topic C3, enables titrations to be studied throughout the addition of titrant, so that small changes may be detected. It also allows **automation** of the titration.

To compare potentiometric titrations with those where the end point is detected visually, it is useful to think of the ion selective electrode as an **indicator electrode** responding to the ion to be detected, and to remember that it must be combined with a **reference electrode**, the potential of which must not be affected by the titration reactions.

In general, the titration will produce a graph such as *Figure 2(a)* where pH, pIon or E(cell) is plotted against the volume of titrant added. The sharp change at the end point is readily observed. Derivative plots such as *Figure 2(b)* are an aid to finding the end point.

Three types of potentiometric titration may be recognized. Detailed discussion of the reactions is given in Topics C6 and C7, but the principles will be discussed here.

- **R,** or **reagent sensed.** A reagent is added, for example, some copper(II) EDTA complex, which is sensed by a copper ISE. Addition of EDTA to a solution containing Ca^{2+} (or Mg^{2+}, Ni^{2+} etc.) ions gives an abrupt change at the end point where the excess EDTA reduces the concentration of Cu^{2+} ions.
- **S,** or **sample sensed.** The ISE senses the ion in the solution being titrated. pH titrations using a glass electrode belong to this type, as do titrations of fluoride with La^{3+}.
- **T,** or **titrant sensed.** The titrant added to the sample is sensed by an ISE responsive to that ion; for example a silver ion ISE for the titration of halide ions by silver nitrate.

Potentiometric titrations for complex, precipitation and redox systems are discussed in Topic C7.

Applications

There are many applications for acid–base titrations, several of which are routinely used analytical methods described in the appropriate topics.

- The determination of the concentration of acid in foods and pharmaceuticals.
- The measurement of **acid number** (or base number) during the course of a reaction. For example, in the production of polyester resins by the reaction of a glycol with maleic and phthalic acids, the total acid remaining is determined by titration of a weighed sample with potassium hydroxide using phenolphthalein as indicator.
- The Kjeldahl method for nitrogen determination is a good example of a back titration. The sample (for example, a food product) is oxidized by concentrated sulfuric acid to remove carbonaceous matter. Excess sodium hydroxide solution is then added, and the ammonia released is carefully distilled off into a known volume of standard acid, such as 0.1 M boric acid. The excess acid is then titrated with standard alkali.

Automated titrations are important in producing rapid, reproducible results in commercial and research laboratories. Samples may be prepared and loaded using mechanical pipets or direct weighing and dissolution methods. Titrant is added to the sample titrand solution using peristaltic pumps, or burets driven by pressure or piston systems. The addition is very reproducible after accurate calibration. The progress of the titration is most often followed by potentiometric measurements, as outlined above (see also Topic H2).

C6 Complexation, solubility and redox equilibria

Key Notes

Complexation

A complex is formed by reactions between two or more species that are capable of independent existence. Often this is between a metal ion, M and a coordinating molecule L.

Solubility

Ionic reactions producing a compound that is insoluble in the chosen solvent used may be used for analysis.

Redox equilibria

Where one species is reduced while the other is simultaneously oxidized, the reactions are termed redox reactions and are useful in many analytical methods.

Related topics

Titrimetry II (C7)

Other topics in this Section (C1–C5, C8–C10)

Complexation

The formation of stable compounds and complexes is important in analytical chemistry, since many species may be formed in a real sample. The amounts and nature of the species present are analyzed to study **speciation**. For example, in a natural water sample, the metal ions may form complexes with water molecules, carbonate species, plant acids or pollutants. Complexes may be used for titrations, both directly and for masking unwanted reactions.

The formation of a complex compound between an acceptor species, most usually a metal ion, M, and a coordinating species, or donor **ligand**, L, involves the formation of coordinate bonds, for example, hexamino cobalt (III)

$$Co^{3+} + 6NH_3 = Co\ (NH_3)_6^{\ 3+}$$

The formation of such complexes involves interactions between the orbitals of the central atom and suitable orbitals or lone pair electrons of the ligands. The structure and stability of the complexes are discussed more fully in textbooks of inorganic chemistry.

The formation constant for the equilibrium may be represented in two ways. The **stepwise formation constants**, K_f, relates to each addition of a ligand molecule:

$$M + L = ML \qquad K_{f1} = c\,(ML)/(c\,(M)\,.\,c\,(L))$$

$$ML + L = ML_2 \qquad K_{f2} = c\,(ML_2)/(c\,(ML)\,.\,c\,(L))$$

or, generally:

$$ML_{n-1} + L = ML_n \qquad K_{fn} = c\,(ML_n)/(c\,(ML_{n-1})\,.\,c\,(L))$$

The **overall formation constant**, β, relates to the formation of the entire complex in one equation, so that for the complex with n ligands:

$$M + nL = ML_n$$

$$\beta n = c\,(ML_n)/(c\,(M)\,.\,c\,(L)^{\,n})$$

The overall formation constant is the product of the stepwise formation constants:

$$\beta n = K_{f1}\,.\,K_{f2.}\,\ldots\,K_{fn} \quad \text{or} \quad \log \beta n = \log K_{f1} + \log K_{f2} + \ldots + \log K_{fn}$$

When a ligand is used that can bond to several sites, it is referred to as a **multidentate ligand**. One of the most important examples is ethylenediamine tetracetic acid (EDTA):

$$(HOOCCH_2)_2\text{-}N\text{-}CH_2\text{-}CH_2\text{-}N\text{-}(CH_2COOH)_2$$

This tetrabasic acid, abbreviated to H_4Y, has four acetate group and two nitrogens, which may complex to the central metal ion, as shown in *Figure 1*. It is important to remember that for a satisfactory titration, the equilibrium constant K of the reaction must be greater than 1000.

Since the concentrations of the various species containing Y (H_4Y, H_3Y^-, H_2Y^{2-}, HY^{3-}, Y^{4-}) will vary with the pH, a formation constant K_{MY} may be written:

$$M^{n+} + Y^{4-} = MY^{(n-4)+}$$

$$K_{MY} = c(MY^{(n-4)+})/(c(M^{n+}).c(Y^{4-}))$$

$$\text{and} \quad c(Y^{4-}) = c_L \times \alpha_4$$

where α_4 depends on the pH and the acid dissociation constants (K_1, K_2, K_3, K_4) of EDTA, as shown in *Table 1*, and c_L is the total concentration of all the ligand species.

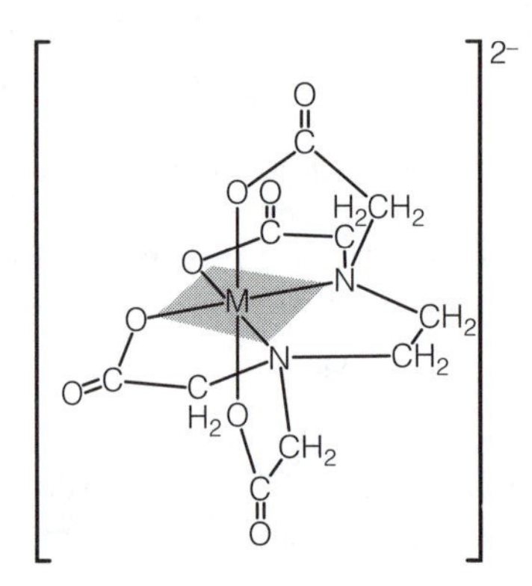

Fig. 1. The structure of a metal-EDTA chelate showing its octahedral geometry.

Table 1. Values of α_4 as a function of pH

pH	α_4	pH	α_4
2	3.7×10^{-14}	8	5.4×10^{-3}
3	2.5×10^{-11}	9	5.2×10^{-2}
4	3.6×10^{-9}	10	3.5×10^{-1}
5	3.5×10^{-7}	11	8.5×10^{-1}
6	2.2×10^{-5}	12	9.8×10^{-1}
7	4.8×10^{-4}		

To compare the formation constants at similar pH conditions the equilibrium may be written:

$$K'_{MY} = K_{MY}\,\alpha_4 = c(MY^{(n-4)+})/(c(M^{n+})\,.\,c_L)$$

Example

The formation constant K_{MY} for magnesium-EDTA is 5×10^8. Should pH 5, or pH 10 be used to titrate Mg?

At pH 5, for Mg:

$$K'_{MY} = 5 \times 10^8 \times 3.5 \times 10^{-7} = 1.75 \times 10^2$$

At pH 10, for Mg:

$$K'_{MY} = 5 \times 10^8 \times 3.5 \times 10^{-1} = 1.75 \times 10^8$$

Therefore, magnesium could be titrated only at pH 10, as the value of the equilibrium constant is too small below this.

The fact that EDTA forms a number of ring systems adds considerably to the stability of the complex. This is called the **chelate effect**.

Solubility

The formation of insoluble compounds by reaction between two soluble species is discussed in detail in Topic C8, where the amount of an insoluble product is measured **gravimetrically** (*Table 2*).

The solubility equilibrium is described by the **solubility product, K_{sp}**:

$$M^{n+}(\text{solvated}) + A^{n-}(\text{solvated}) = MA(\text{solid}) + \text{solvent}$$

$$K_{sp} = a\,(M^{n+})\,.\,a\,(A^{n-})$$

(Note: This equilibrium constant is written in the inverse way to most others, and thus a very small solubility product is desirable for complete reaction. The activities of the pure solid MA and of the solvent are taken as 1.)

There are some general rules governing which cations will form precipitates with which anions. Nitrates and perchlorates are generally soluble.

Various other reagents are also useful for gravimetric analysis, for example dimethylglyoxime (a) for nickel and oxine (b) for aluminum and magnesium (*Fig. 2(a) and (b)*).

If an excess of a precipitating ion is present, in order to keep the equilibrium constant unchanged, the concentration of the other ion must decrease. This is called the **common ion effect**. Occasionally, complex formation may occur, so that the precipitate formed redissolves when excess reagent is added.

Since equilibrium constants are thermodynamic quantities, they should really be written in terms of activities and the activity coefficient, depending on the ionic strength taken into account.

Table 2. Ions forming insoluble products (K_{sp} ~ < 10^{-5})

Anions	Cations
Cl^-, Br^-, I^-	Ag^+, Pb^{2+}, Cu^+, Hg_2^{2+}
CO_3^{2-}	Ca^{2+}, Sr^{2+}, Ba^{2+}, Mg^{2+}
SO_4^{2-}	Ca^{2+}, Ba^{2+}, Pb^{2+}
OH^-	Al^{3+}, Co^{3+}, Cr^{3+}, Fe^{3+}, Mg^{2+}, Mn^{2+}, Ni^{2+}, Zn^{2+}
S^{2-}	Ag^+, Cd^{2+}, Cu^{2+}, Fe^{2+}, Hg^{2+}, Mn^{2+}, Pb^{2+}, Zn^{2+}

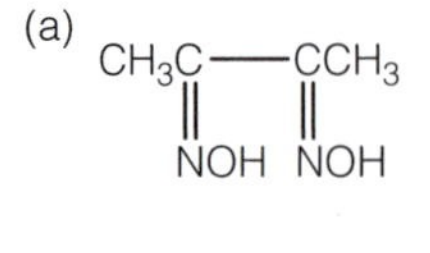

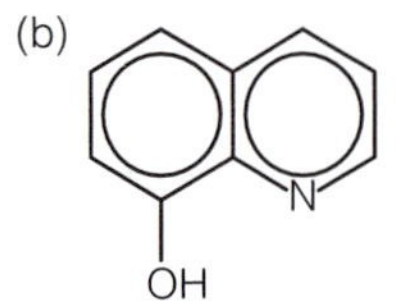

Fig. 2. Reagents for the precipitation of metal ions. (a) Dimethylglyoxime. (b) Oxine.

Redox equilibria

In order to establish a scale of oxidative power, it is necessary to have a standard, and since these reactions involve electrons, measurement of the **reduction electrode potential** is a convenient way to do this. The details are given more fully in Topic C3.

Some standard reduction electrode potentials, where the reagents are at unit activity, at 25°C are given in *Table 3*. These potentials allow the prediction of which ions will oxidize other ions, under standard conditions, that is when the concentrations are molar. A more poisitve electrode potential will oxidize a more negative potential.

It was shown in Topic C2 that the electrochemical cell e.m.f. is related to the free energy change, and hence to the equilibrium constant:

$$E^{\ominus} = (RT/nF) \ln K$$

Therefore, the larger the cell e.m.f, the larger the equilibrium constant, and the more complete the reaction.

Example

For the reaction of cerium(IV) ions with iron(II) ions, what is the likely reaction, and what is the equilibrium constant? Which reagent is the oxidizing agent, and which the reducing agent?

$$\text{Cell: Pt} \mid Fe^{2+}, Fe^{3+} \parallel Ce^{4+}, Ce^{3+} \mid \text{Pt}$$

$$E^{\ominus}(\text{cell}) = E^{\ominus}(\text{rhs}) - E^{\ominus}(\text{lhs}) = 1.44 - 0.77 = 0.67 \text{ V}$$

Table 3. Standard reduction electrode potentials of some common redox systems at 25°C

Reaction	$E^{\ominus}/V$
$H_2O_2 + 2H^+ + 2e^- = 2H_2O$	1.77
$MnO_4^- + 8H^+ + 5e^- = Mn^{2+} + 5H_2O$	1.51
$Ce^{4+} + e^- = Ce^{3+}$	1.44
$Cr_2O_7^{2-} + 14H^+ + 6e^- = 2Cr^{3+} + 7H_2O$	1.33
$I_2 + 2e^- = 2I^-$	0.54
$Fe^{3+} + e^- = Fe^{2+}$	0.77
$S_4O_6^{2-} + 2e^- = 2S_2O_3^{2-}$	0.08
$2CO_2 + 2e^- = C_2O_4^{2-}$	−0.49

$$\ln K = 0.67/(8.314 \times 298/(1 \times 96485)) = 26.09$$

$$\text{or } K = 2.1 \times 10^{11}$$

Cerium(IV) oxidizes iron(II) almost completely.

Examination of *Table 3* shows that permanganate will oxidize iron(II) and oxalate, iodine will oxidize thiosulfate to dithionite, but iodide will be oxidized by iron(III) to iodine.

C7 Titrimetry II: complexation, precipitation and redox titrations

Key Notes

Complexation titrations	The techniques of titrimetry, using both visual and potentiometric end point detection, are used to measure species, particularly metal ions, in a wide range of samples.
Precipitation titrations	Reactions producing an insoluble product are valuable analytical tools for the titrimetric determination of halide and other anions.
Redox titrations	Oxidation and reduction titrations may be used to measure many species, especially metals in high or low valency states, iodine and iodides, and easily oxidized organic compounds.
Related topics	Complexation, solubility and redox equilibria (C6)

Complexation titrations

During complexation reactions the concentration of the analyte ion (for example, a metal ion) changes most rapidly at the end point. As noted in Topic C6, the most widely used complexing agent is ethylenediaminetetracetic acid or EDTA, and *Table 1* gives a selection of metal EDTA formation constants.

Using the values of α_4 given in Topic C6, *Table 1*, we may calculate the practical, or **conditional**, formation constant at a particular pH

$$K'_{MY} = K_{MY}\,\alpha_4$$

From the data in the tables, it can be calculated that magnesium could be titrated at pH 10, but not at low pH. This has already been discussed in Topic C6.

Table 1. Metal-EDTA formation constants at 25°C

Cation	K_{MY}	log (K_{MY})
Ag^+	2.0×10^7	7.3
Mg^{2+}	4.9×10^8	8.7
Ca^{2+}	5.0×10^{10}	10.7
Fe^{2+}	2.1×10^{14}	14.3
Fe^{3+}	1.0×10^{25}	25.1
Zn^{2+}	3.2×10^{16}	16.5
Cd^{2+}	2.9×10^{16}	16.5
V^{3+}	8.0×10^{25}	25.9

It is possible to titrate two cationic species in a solution by performing the titration at different pH values. However, if a solution of high pH must be used, this might cause precipitation of metal hydroxides or other insoluble species. In order to prevent this, **secondary complexing agents** can be added to retain the metal ion in solution. Ammonium chloride and triethylamine are typical reagents for this purpose.

Zinc ions, which might otherwise form insoluble $Zn(OH)_2$ at pH of 10, may be converted to soluble zinc amine complexes. These are less stable than the EDTA complex and the zinc may then be reacted quantitatively.

Secondary complexing agents may also act as **masking agents**. Examples of this are the use of cyanide ions to form stronger complexes with heavy metal ions so that magnesium can be titrated, or masking Fe^{2+} and Mn^{2+} using hydroxylamine in water hardness determinations.

Standard solutions of EDTA may be prepared from the dry disodium salt (Na_2H_2Y, RMM 336 or the dihydrate, RMM 372), by dissolving a known amount in water free of heavy metals. Alternatively, the solution may be **standardized** by a standard magnesium salt solution.

In complex and precipitation titrations, as in others, the end point corresponds to a rapid change in the concentration of species. This may be detected by instrumental methods, particularly potentiometry (see Topic C3) and by visual indicators discussed below. Using suitable indicators, or potentiometric measurements, it is possible to detect two or more end points.

Complexometric indicators behave in a similar way to titrating complexing agents such as EDTA. They generally change color with pH, but one species, for example HIn^{2-}, will react with excess metal ions M^{n+}:

$$\underset{\text{blue}}{HIn^{2-}} + M^{n+} = \underset{\text{red}}{MIn^{(n-3)+}} + H^+$$

Some selected indicators for complexometric titrations are given in *Table 2.*

In order to use **potentiometric methods** to study complexometric titrations, an electrode specific to the metal ion may be used (see Topics C3 and C5), for example, a copper ISE to follow the reaction of copper with EDTA.

Alternatively, a 'J'-shaped electrode with a small mercury pool may be used together with a small amount of added Hg-EDTA complex. This acts in a similar way to the Class 2 electrodes, where the complexes determine the concentration of ions in contact with the mercury pool:

$$M^{n+} + HY^{3-} = MY^{(n-4)+} + H^+$$

$$Hg^{2+} + HY^{3-} = HgY^{2-} + H^+$$

$$Hg^{2+} + 2e^- = Hg$$

Table 2. Indicators for complexometric titrations

Indicator	Free color	Complex color*	Metal ions
Eriochrome black T	Blue	Red	Ba, Cd, Ca, Pb, Zn
Pyrocatechol violet	Yellow	Blue	Al, Bi, Cd, Co, Cu Fe, Mg, Mn, Ni, Zn
Xylenol orange	Yellow	Red	Bio, Cd, Pb, Th, Zn
Calcon carboxylic acid	Blue	Red	Ca, Cd, Mg, Mn, Zn

*varies with metal and pH

The major **application of complexation titrations** is for the determination of the concentrations or amounts of metallic elements in water, food and other industrial samples.

Example

The calcium and magnesium ions in hard water may be determined. The solution is adjusted to pH 12 with NaOH, when $Mg(OH)_2$ is precipitated. The calcium is then titrated with EDTA using calcon carboxylic acid as indicator. Both calcium and magnesium are then determined in by titrating a sample with EDTA at pH 10 using eriochrome black T, and finding the magnesium by difference.

Precipitation titrations

For **precipitation reactions,** the change in the concentration of either ion forming the precipitate may be considered. Since the changes often involve many orders of magnitude of concentration, it is again convenient to use the pX notation. For example, for the reaction of silver ions with chloride to form an insoluble silver chloride precipitate

$$Ag^+ + Cl^- = AgCl\ (s)$$

the concentration may be expressed as:

$$pAg = -\log (a\ (Ag^+)) \sim -\log (c\ (Ag^+))$$

Figure 1 shows the pAg values in the titration of sodium chloride by silver nitrate as a function of the volume of silver nitrate added. This figure shows that, before the end point pAg is very high (that is, the concentration of silver ions is small) and changes little, because there is still an excess of chloride and the silver is almost completely removed as precipitate. After the end point, there is an excess of silver ions, the concentration increases and pAg decreases. In the region around the end point, where the amounts are nearly equal, the change in pAg with volume added is very large. If a mixture of iodide and chloride ions is titrated (dashed line), the iodide, which is less soluble, precipitates first and pAg is even higher than for chloride. Then the chloride precipitates. Both end points can be found.

As noted in *Table 1* of Topic A5, silver nitrate, sodium chloride and potassium chloride are primary standards for silver halide precipitation reactions. Other

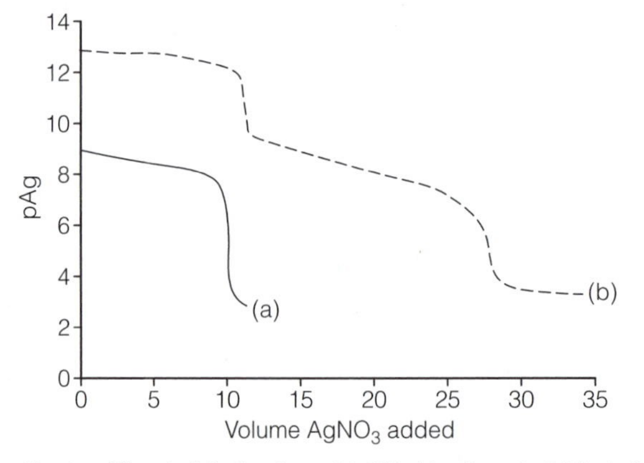

Fig. 1. Silver-halide titrations. (a) Chloride alone (solid line); (b) iodide plus chloride (dashed line).

precipitation titrations (e.g., barium with sulfate, zinc with ferrocyanide) are less commonly performed.

Indicators for silver-halide precipitation titrations are of two types. The first react specifically when an excess of titrant becomes present immediately after the end point – for example, if a small amount of potassium chromate is added, it will react with excess silver ions to produce deep red silver chromate in **neutral** solutions (**Mohr's method**). In acid solutions, the silver is titrated with potassium thiocyanate (KCNS) solution (**Volhard's method**). Iron (III) ammonium sulfate solution is added and reacts with an excesss of thiocyanate to produce a deep red iron thiocyanate species.

Adsorption indicators such as fluorescein adsorb onto the precipitate when excess silver ions are present and the precipitate takes on a pinkish color.

As with other indicators, the change of color is detectable by eye over a range:

$$\log (c(\text{ion})) = \pm 1$$

A selection of indicators for precipitation titrations is given in *Table 3*.

Table 3. Indicators for precipitation titrations

Precipitation	Color 1	Color 2	Ions detected
CrO_4^{2-}	Yellow	Deep red	Ag^+
Fe^{3+}	Light brown	Deep red	CNS^-
Fluorescein	Green-yellow	Pink	Ag^+

It is considerably easier to titrate mixed chloride and bromide in solution by potentiometry. The use of a silver ion selective electrode, or even a silver wire, together with a double junction reference electrode, since the chloride ions from a calomel electrode would react, allows the determination of the silver ion concentration. Other precipitation titrations may be followed using suitable ion selective electrodes.

The major applications of precipitation titrations involve the determination of halides with silver, or the reverse, or the determination of silver in acid solutions with thiocyanate.

Example

An insecticide containing chlorine was digested in nitric acid to convert the chlorine to soluble chloride. Silver nitrate was added in excess, and the excess titrated with potassium thiocyanate by Volhard's method. It is important to know whether **all** the chlorine is converted to chloride.

Redox titrations

Oxidation–reduction or **redox** titrations are used for determining metals with two well-defined oxidation states, and indirect methods for the determination of organic compounds.

For **redox reactions** the concentrations both of the oxidized species, Ox, and of the reduced species, Red, will change simultaneously. Considering a cell with a redox electrode and a reference electrode:

$$\text{SCE} \,||\, a\,(\text{Ox}), a\,(\text{Red}) \,|\, \text{Pt}$$

the cell emf is given by:

$$E = E^{\ominus} + (RT/F) \ln (a\,(\text{Ox})/a\,(\text{Red})) - E_{\text{SCE}}$$

Therefore, as the reaction proceeds during titration, the ratio of the concentrations will change and the emf will alter. The potentiometric titration curve will resemble those described in Topic C5. A summary list of redox reagents is given in *Table 3* of Topic C6.

For standard solutions, sodium oxalate and iron(II) ammonium sulfate and potassium iodate are suitable, but potassium permanganate and iodine solutions decompose on standing and must be standardized before use.

The indicators for redox reactions are reagents whose oxidized and reduced forms differ in color:

$$\mathrm{In(Ox)\ (color\ 3) + ne^- = In(Red)\ (color\ 4)}$$

An example of this is 1,10-phenanthroline iron (II)

$$[\mathrm{Fe}\ (\mathrm{C_{12}H_8N_2})_3]^{3+} + e^- = [\mathrm{Fe}\ (\mathrm{C_{12}H_8\ N_2})_3]^{2+}$$

oxidized form, pale blue reduced form, deep red

In several cases, the indicator reaction additionally involves hydrogen ions, so the change is pH dependent. *Table 4* lists commonly used redox indicators.

Table 4. Indicators for redox titrations

Redox	Oxidized color	Reduced color	E_{In}/V	Solution
1,10-phenanthroline iron(II) complex	Pale blue	Red	1.11	1 M H_2SO_4
Diphenylamine	Violet	Colorless	0.76	Dilute acid
Methylene blue	Blue	Colorless	0.53	1M acid
Phenosafranine	Red	Colorless	0.28	1M acid

For a redox indicator where one electron is involved, at 25°C, the color change takes place at electrode potentials in the range

$$E = E_{\mathrm{In}} \pm 0.059$$

One further useful indicator employed in redox titrations involving iodine is **starch,** or more synthetic equivalent materials. The starch forms a blue-black complex with iodine, which is rendered colorless when all the iodine has been removed.

The applications of redox titrations include the determination of metals, with two well-defined oxidation states, which are present in metallurgical samples and ores. In order to dissolve the material, it may be necessary to use oxidative conditions, for example, concentrated nitric acid. This will convert the majority of the ions into their higher oxidation state, and in order to titrate them they must first be reduced quantitatively. This may be done by passing the acidified solution through a **Jones reductor,** which contains a zinc-mercury amalgam. The effluent may then be titrated using a suitable oxidant. Some organic compounds, such as phenols, may be determined by bromination with a bromate/bromide mixture, followed by back titration of the excess using thiosulfate.

C8 GRAVIMETRY

Key notes

Gravimetry Gravimetry is the analytical technique of obtaining a stable solid compound, of known stoichiometric composition so that the amount of an analyte in the sample may be found by weighing.

Precipitation This involves treatment of an analytical sample, usually in solution, to obtain a quantitative amount of an insoluble compound of known composition.

Purification The precipitate must be as pure as possible. Substances that are similar and might precipitate under the same conditions must be removed, and the analysis must be carried out so that no impurities are co-precipitated.

Drying and heating If precipitation is carried out from solution, the solid precipitate will have solvent associated with it. This must be removed. Heating near the boiling point of the solvent will do this, and further heating may be needed to obtain a more stable compound whose formula is known.

Weighing The procedures of weighing the container initially and with the final sample are most important.

Related topics Complexation, solubility and redox equilibria (C6); Thermogravimetry (G1)

Gravimetry

Gravimetry is one of the 'classical' techniques of analysis, and although less frequently used now, it is of value when an accurate reference method is required for comparison with an instrumental technique.

If an element is present in a mixture, for example, silver in a sample of nickel, one way of separating it is to dissolve the metal completely in a suitable solvent. In this example, the metal mixture could be dissolved in concentrated nitric acid and a reagent added that would react with the silver to produce a precipitate, which for silver might be a sodium chloride solution:

$$Ag(s) + Ni(s) + HNO_3(sol) = AgNO_3(sol) + Ni(NO_3)_2(sol)$$

$$AgNO_3(sol) + NaCl(sol) = AgCl(s) + NaNO_3(sol)$$

The silver chloride is precipitated completely, and may be filtered off since both nickel nitrate and nickel chloride are very soluble in water. The precipitate will be wet and may contain traces of nickel in solution, so must be thoroughly **washed** and **dried,** as discussed below.

Since weighing may be carried out readily and accurately in almost all laboratories, gravimetry is often used as a reference method. Analysis of major components of metal samples such as steel, and of minerals and soils may be carried out by gravimetric methods, but they often involve lengthy separations and are time consuming. Newer instrumental methods may determine several components simultaneously, rapidly and are generally applicable down to trace levels.

Precipitation

As noted in Topic C1, many elements will form compounds insoluble in water or other solvents. Provided the compound is stable, or may be converted into a stable compound easily, these insoluble precipitates may be used for analysis. The technique for obtaining a precipitate may be summarized as follows:

(i) The sample should be dissolved as completely as possible in a suitable solvent. Any residue that does not dissolve (for example, silica present in the metal sample of the above example), may be filtered off at this stage.
(ii) Unless there are undesirable changes when the sample is heated, the solutions should be warmed. This speeds up reactions and helps to form a more granular precipitate.
(iii) The precipitating reagent must be chosen to give as insoluble a precipitate as possible. Preferably, a reagent that will produce the largest mass of precipitate should be used. For example, **aluminum** may be precipitated and heated to give the oxide, Al_2O_3 when 10.0 mg of aluminum will produce 18.9 mg precipitate. If 'aluminon' (8-hydroxyquinoline, C_9H_6ON) is used, 170.0 mg of precipitate results.
(iv) The precipitating reagent should be added slowly, with stirring to the warm solution. To check whether precipitation is complete, the precipitate is allowed to settle, and more reagent added. If further precipitate does not form, the reaction is complete. If the solution appears cloudy, it is possible that a **colloidal form** of the solid is present. This may be coagulated by further warming or adding more reagent.
(v) The reaction mixture is filtered. Various means may be used for this. The simplest is a **quantitative filter paper** (ashless), which has been dried and weighed previously. These may be dried, or burnt ('ashed').

Another useful filter is a **sintered glass** or **porcelain crucible**, dried and weighed as before. Glass will withstand heating to about 300°C and porcelain to over 800°C.

Purification

The precipitate should be washed to remove traces of solution. This may cause difficulties, as the washing may redissolve the solid. Using a wash liquid with a **common ion** reduces the solubility.

For example, if 0.18 g of a precipitate of lead sulfate is washed with 1 dm^3 of distilled water, it will dissolve 0.046 g of precipitate or 25%, as the solubility product is $K_{sp} = 2.3 \times 10^{-8}$ (mol $dm^{-3})^2$. If the precipitate is washed with 1 dm^3 of 0.01 M sulfuric acid, then the amount dissolved is much less, 0.7 mg or 0.4%.

Drying and heating

Drying can be done in stages. To remove water, the filtered solid in its container is placed in a **desiccator** and left for a few hours. A **vacuum desiccator** is even more efficient for removing solvents at low temperature. Heating in ovens, furnaces or directly with burners will raise the temperature to remove materials or to decompose the precipitate to a more stable form. For example, 'basic aluminum succinate' is a good precipitate for aluminum, but must be ignited to constant weight at about 1200°C to convert to aluminum oxide.

The stages involved in drying and decomposition can be studied using **thermogravimetry** (see Topic G1).

Weighing

Modern balances can readily weigh samples directly, and masses from several grams down to a few micrograms can be weighed accurately and quickly. It is important that the conditions are the same for the initial weighing (crucible, filter paper) as for the final weighings. Temperature is especially important and hot samples should **never** be placed directly onto a balance pan.

C9 VOLTAMMETRY AND AMPEROMETRY

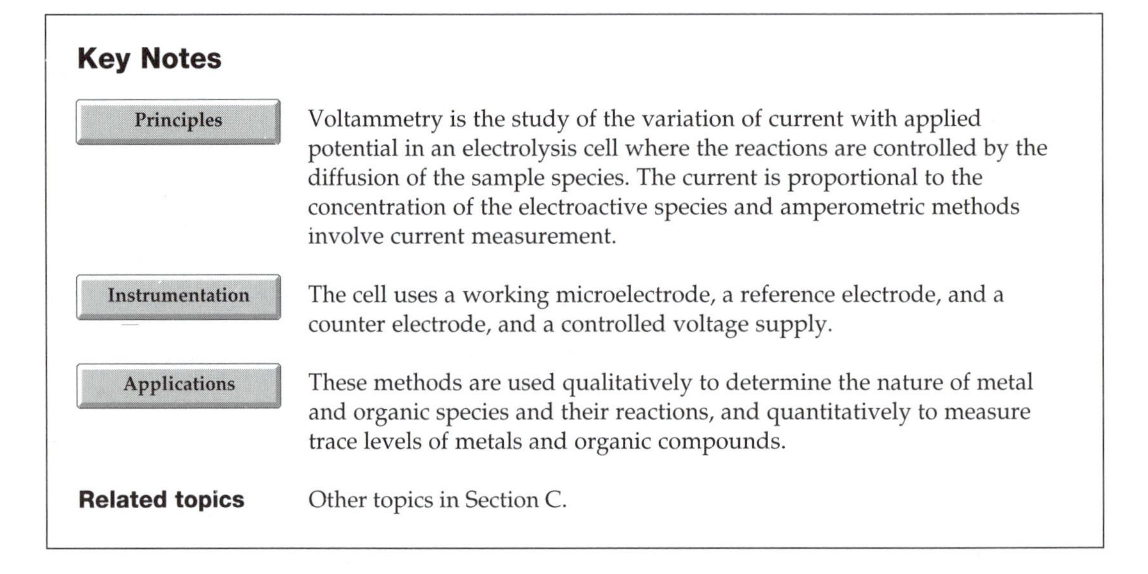

Key Notes

Principles	Voltammetry is the study of the variation of current with applied potential in an electrolysis cell where the reactions are controlled by the diffusion of the sample species. The current is proportional to the concentration of the electroactive species and amperometric methods involve current measurement.
Instrumentation	The cell uses a working microelectrode, a reference electrode, and a counter electrode, and a controlled voltage supply.
Applications	These methods are used qualitatively to determine the nature of metal and organic species and their reactions, and quantitatively to measure trace levels of metals and organic compounds.
Related topics	Other topics in Section C.

Principles

Voltammetric techniques involve the electrolysis of the solution to be analyzed using a controlled external power source and measuring the resultant current-potential or current time curves to obtain information about the solution.

The species to be determined undergoes oxidation or reduction at a **working electrode**. The voltage between the working electrode and an **auxillary** or **counter electrode** is controlled by the external circuitry in order to maintain a preselected potential difference at the working electrode, with respect to the reference electrode, as a function of time. A typical voltammetric cell, shown in *Figure 1*, has a working electrode, reference electrode and an auxillary electrode and contains the solution to be analyzed. Often the solution is deaerated with nitrogen to prevent interference due to the reduction reactions of oxygen.

If there is no reaction at the working electrode, the potential changes greatly for a very small increase in current. A mercury drop electrode, for example, has a polarization range between +0.3 V and –2.7 V against the SCE and in the absence of oxygen so that many reactions that occur in that range may be studied.

By controlling the potential of the working electrode, a particular reaction may be selected. Suppose a cell has two inert, solid electrodes and a reference electrode, which dip into an aqueous solution containing copper ions.

In order to cause any reaction, the applied potential must exceed the **decomposition potential**. This may be calculated by considering the reactions at each electrode, and adding the extra potentials or **overpotentials** due to polarization effects at the electrodes and to the voltage needed to drive the current against the resistance of the solution. In this example, in order to drive the cell reaction, a voltage greater than about –2.54 V must be applied.

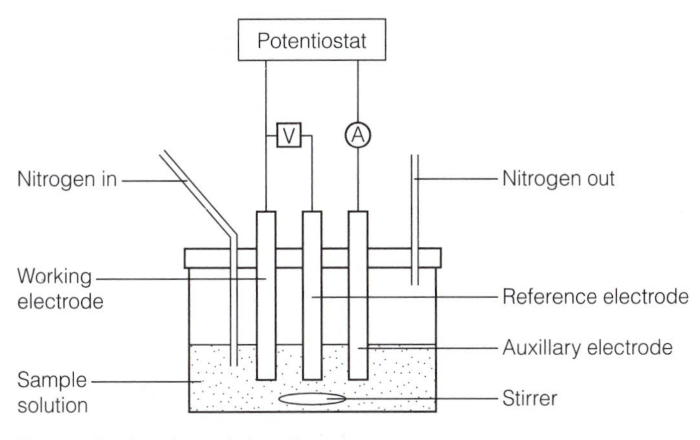

Fig. 1. Basic voltammetric cell.

The reaction that occurs is:

$$Cu^{2+} + H_2O = Cu(s) + \tfrac{1}{2} O_2 + 2H^+$$

The copper is then deposited on the **cathode** and oxygen is evolved at the **anode**. This is the basis for **electrogravimetry,** where the copper is completely deposited from solution and the increase in weight of the cathode determined. The analysis may be conducted using either a controlled potential or a controlled current.

Coulometric methods of analysis involve measuring the quantity of electricity in coulombs needed to convert the analyte to a different oxidation state. If the electrolysis occurs at 100% efficiency, Faraday's laws may be applied and each 96485 C will bring about the reaction corresponding to 1 mole of electron transfer.

For example, using a silver anode, the passage of a current produces silver ions, which react with any chloride in the solution. Bromine and acids may also be generated coulometrically.

Polarographic methods employ a microelectrode, often a dropping mercury electrode (DME), as the working electrode, plus a reference electrode (SCE) and a mercury pool as auxillary electrode. The simplest potential–time regime, where the potential increases regularly (**linear potential sweep dc voltammetry**) is applied to the cell containing the analyte and a **supporting electrolyte** to carry the majority of the current.

In these methods, the transport of ions to the electrodes depends on three factors: diffusion, convection or stirring, and conduction. The effects of conduction of the ion that reacts at the electrode is minimized by using a concentration of supporting electrolyte such as KCl about 50-times higher than that of the analyte. Stirring and convection are minimized. The resulting polarographic curve shows three regions.

(i) If a potential difference is applied across a cell and no reaction occurs, only the residual current I_r will flow.

(ii) If a reducible ion, say Cd^{2+}, is present, it will migrate to the dropping mercury cathode. If the applied potential exceeds its decomposition potential, E_D, it will be reduced to the metal which dissolves in the mercury:

$$Cd^{2+} + 2e^- \overset{Hg}{=} Cd(Hg)$$

As the cadmium plates out, the layer around the electrode is depleted and more cadmium ions must **diffuse** in from outside through the **diffusion layer** of thickness δ. This will cause a **current**, I, to flow, which depends on the concentration gradient between the bulk solution and the surface. Eventually, the surface concentration becomes zero, and the **limiting diffusion current** is reached:

$$I_d = \text{constant}\,(c(\text{bulk}))/\delta = k_S\,(c(\text{bulk}))$$

The constant, k_S, depends on the number of electrons transferred, the diffusion coefficient of the ion in the solution, and the characteristics of the cathode.

(iii) If the potential is increased further, the current does not increase unless other reducible ions are present. These three regions are shown in *Figure* 2.

The potential difference, E, across the cell at any stage is:

$$E = E_{SCE} - E_{DME} \qquad \text{or}$$

$$E = E_{SCE} - (E^{\ominus}{}_{Cd} + (RT/2F)\ln[(a(Cd^{2+}, \text{surface})/a(Cd(Hg))]$$

From the equations above, the concentrations may be substituted by the currents, since the concentration of reduced species in the mercury depends on the current I and the diffusion constant in the amalgam, k_A

$$I = k_A(c(Cd(Hg))$$

$$E = E_{SCE} - (E^{\ominus}{}_{Cd} + (RT/2F)\ln(k_A/k_S)) + (RT/2F)\ln[(I_d - I)/I]$$

When $I = \frac{1}{2} I_d$, that is at the half-wave position, the DME has the **half-wave potential, $E_{1/2}$**

$$E_{DME,1/2} = E_{SCE} - (E_{1/2} + (RT/2F)\ln[\tfrac{1}{2} I_d/\tfrac{1}{2} I_d] = E_{1/2}$$

The half-wave potential is usually quoted relative to the SCE, and, like the standard electrode potential, is characteristic of the electrode reaction. Typical values are shown in *Table 1*.

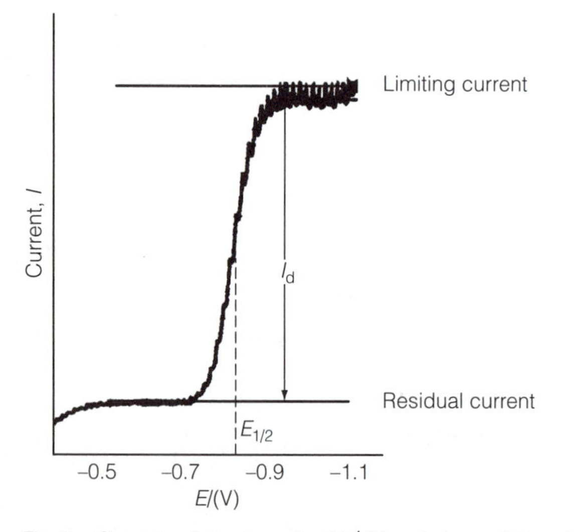

Fig. 2. Current-voltage curve for 10^{-4} M cadmium sulfate solution.

Table 1. Value of the half-wave potential

Ion	$E_{½}$/V with respect to the SCE in 0.1 M KCl	in 1 M NH_3, NH_4Cl
Cd^{2+}	−0.60	−0.81
Cu^{2+}	+0.04	−0.24
	−0.22	−0.51
Zn^{2+}	−1.00	−1.35
K^+	−2.14	
$C_6H_5NO_2$	−0.22	

Note that copper(II) is reduced in two stages.

Instrumentation

For most voltammetric and amperometric methods, the instrumentation includes a working microelectrode, a reference electrode and an auxillary or counter electrode, together with electronic equipment to control the voltage and voltage sweep, plus a computer or recorder to collect data.

The earliest microelectrode used was the **dropping mercury electrode (DME)**, where pure mercury flows through a fine capillary, either due to gravitational force, or by applied pressure. Drop times of a few seconds are usual. This electrode has the advantages that:

- the surface area is small and is constantly refreshed so that products of electrolysis do not accumulate;
- mercury has a high overpotential for hydrogen formation, which allows the reduction of other species.

Other electrodes used are the **static**, or **hanging mercury drop electrode**, where the drop is dislodged at a particular time and size, and **solid microelectrodes**, such as platinum and glassy carbon, which may be incorporated into a **rotating disc electrode**.

Dissolved **oxygen** in the sample solution must be removed, since oxygen may be reduced in two steps, giving waves that overlap with those of the sample.

$$O_2 + 2H^+ + 2e^- = H_2O_2 \qquad E_{½} = -0.05 \text{ V}$$
$$H_2O_2 + 2H^+ + 2e^- = H_2O \qquad E_{½} = -0.9 \text{ V}$$

This is usually done by passing oxygen-free nitrogen through the sample solution during the experiment.

Maxima on the waves are due to surface effects, and may be suppressed by adding a small amount of surface-active agents, such as gelatin or Triton-X100.

Anodic stripping voltammetry is designed to measure trace amounts by preconcentrating them onto a suitable electrode. The experiment has two stages:

(i) The sample is electrolyzed onto a hanging mercury drop, or a mercury film deposited on a carbon electrode. By Faraday's laws, (Topic C2), passing a current of I amps for t seconds will produce a concentration c_R in a mercury film of thickness l, area A:

$$c_R = It/nF\,lA$$

Because the current is limited by diffusion:

$$I = mnFD\,c_B\,A$$

where m is a mass transfer coefficient, D the diffusion coefficient and c_B the bulk concentration.

(ii) The reduced species (that is, the metal) is then oxidized out of the film by making the electrode increasingly anodic. A peak appears on the current-potential plot, and the peak current can be shown to be:

$$I_p = k(c_B \, \nu t)$$

where the constant k includes the diffusion and other constants, and ν is the rate of increase of the anodic potential. The peak potential at which an active species is oxidized is characteristic of that species, and is close to its half-wave potential.

Applications

Polarographic techniques may be used in both **qualitative** and **quantitative modes**.

Since the half-wave potential is characteristic of the particular reaction that is occurring at that potential, it is possible to identify the species involved. A simple case is shown in *Figure 3* where a mixture of metal ions was analyzed. The two reduction waves for copper occur at –0.1 and –0.35 V, cadmium at –0.69, nickel at –1.10 and zinc at –1.35 V. This illustrates an analysis that may identify the species qualitatively and, by using a standard addition method, can also determine the ions quantitatively.

Organic substances may be determined either in an aqueous or a nonaqueous medium. For example, the concentration of nitrobenzene in commercial aniline may be found by studying the reaction:

$$C_6H_5NO_2 + 4H^+ + 4e^- = C_6H_5NHOH + H_2O$$

The **oxygen electrode** is based on voltammetric principles and depends on the diffusion and reduction of oxygen. It is also called the **Clark sensor**.

The cell has a lead anode and a silver cathode set close together in an alkaline solution, often 1M KOH. At the anode, the reaction is

$$Pb(s) + 4OH^- (aq) = PbO_2^{2-} (aq) + 2H_2O + 2e^-$$

The silver cathode is inert, unless oxygen or another reducible species can diffuse to it. A semipermeable membrane through which only oxygen can diffuse surrounds the electrodes, and then the reduction reaction takes place.

$$O_2(aq) + 2H_2O + 4e^- = 4OH^-$$

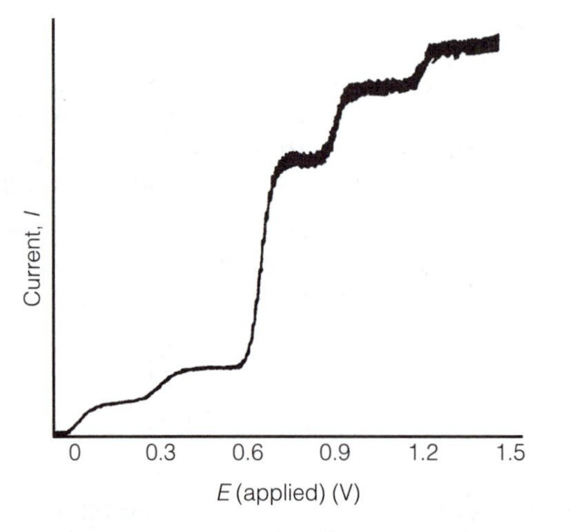

Fig. 3. Polarogram of Cu^{2+}, Cd^{2+}, Ni^{2+} and Zn^{2+} ions at ~10^{-4} M.

Since the current depends on the diffusion of the oxygen to the electrode from the external solution, and this diffusion is proportional to the concentration of oxygen in the external solution, this electrode may be used to measure dissolved oxygen.

Amperometric titrations are used to determine substances by measuring the limiting diffusion current of a species as a function of the volume of a reagent added to react with that species. Since I_d is proportional to the concentration, it will decrease as a species is used up, or increase as the excess of a species becomes greater. For example, for the determination of Pb^{2+} with $Cr_2O_7^{2-}$:

$$2Pb^{2+} + Cr_2O_7^{2-} + H_2O = 2PbCrO_4 + 2H^+$$

At an applied potential of 0.0 V, and at pH4, dichromate is reduced, but Pb^{2+} is not, giving the graph shown in *Figure 4*.

Applications of **anodic stripping voltammetry,** are chiefly for the determination of trace amounts of amalgam-forming metals (*Fig. 5*), while **cathodic stripping voltammetry** is used for determining species that form insoluble salts with mercury. The preconcentration stage allows determination in the concentration range 10^{-6} to10^{-8} M.

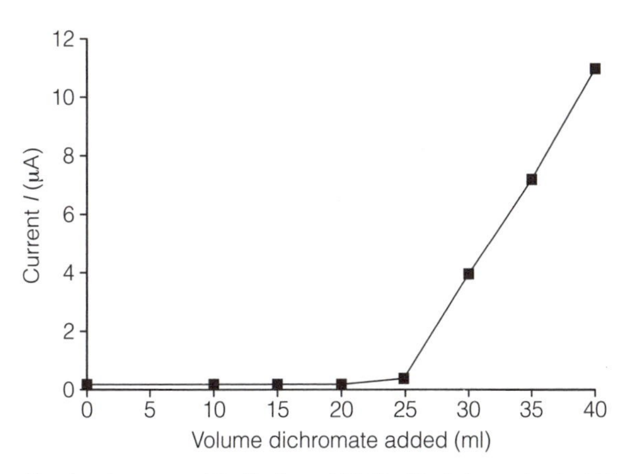

Fig. 4. Amperometric titration of Pb(II) with dichromate at pH 4 and 0.0 V.

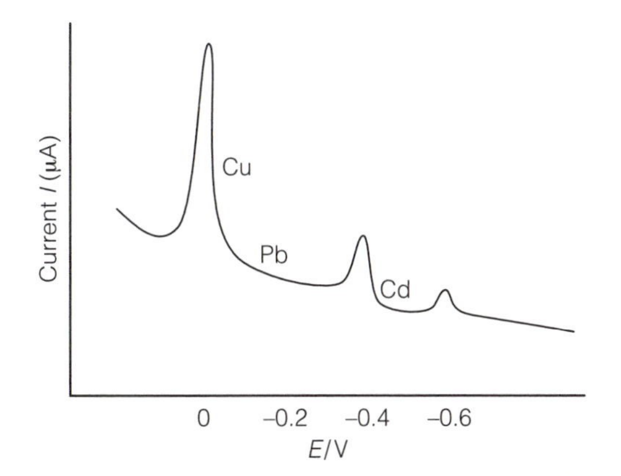

Fig. 5. Anodic stripping voltammogram for 7.5 ppb Cu(II), 2.6 ppb Pb(II) and trace amounts of Cd(II).

C10 CONDUCTIMETRY

Key Notes

Movement of ions	Ions in solution or in molten ionic solids will move when an electric field is applied. The speed of movement will depend on the size and charge of the ion.
Conductance	In ionic solutions, the ions carry the current, but Ohm's law still applies. Increasing the number of ions increases the conductance.
Conductivity	Since analysts are concerned with concentrations, it is preferable to compare molar conductivities, which depend on the chacteristics of the ions.
Conductimetric titrations	If the number or nature of ions present in a cell change, then the conductance will change. This is useful for many types of titration, including some in nonaqueous media.
Related topics	Other topics in Section C.

Movement of ions

When an electric field is applied across a solution, a force is exerted on the ions that will cause them to move. Positive ions will move toward the more negative electrode, negative ions toward the more positive, but each will carry current. The speed with which the ions move usually depends on:

(i) the electric field strength (V m^{-1});
(ii) the charge z on the ion;
(iii) the size of the ion in solution; and
(iv) the viscosity of the solvent.

In water, the hydrogen ion, H_3O^+, and the hydroxyl ion, OH^-, are small, and move most rapidly by a very fast exchange with the molecules of water. In dilute solutions, the ions move independently of each other.

For an electric field of 1 V m^{-1}, the ionic speed is called the **mobility**, u_i. *Table 1* gives the ionic size and mobility of selected ions in aqueous solutions at 25°C.

It is worth noting that despite their larger size in crystals, **hydrated** potassium ions are smaller than hydrated sodium ions when in solution and move faster.

Table 1. Selected vales of mobility and ionic radius

Ion	$10^8\ u_i/m^2\ s^{-1}\ V^{-1}$	r_i/nm
Na^+	5.2	0.18
K^+	7.6	0.13
F^-	5.5	0.17
Cl^-	7.9	0.12
Br^-	8.1	0.12

Similar rules apply to anions. Also, potassium ions and chloride ions have very similar mobilities. This is useful when 'salt bridges' are needed (see Topic C2).

If ions are present in other solvents, such as liquid ammonia, they will conduct in a similar way, but their mobilities will be different.

Conductance

Ohm's law still applies to relate the current, I, to the applied voltage, E, and the resistance, R, or to the **conductance**, $G = 1/R$.

$$I = E/R = EG$$

G is related to the concentration of ions and has units of ohms^{-1} or siemens (Ω^{-1} or S).

In order to measure the conductance of a solution, a conductance cell is used. It has two electrodes, often platinum, coated with platinum black, separated by a fixed distance. Connecting these to a **conductance meter**, or more simply to a Wheatstone bridge, the conductance may be measured directly. In order to prevent electrolysis (see Topics C2, C9) taking place, which would change the concentrations in the solution, the bridge uses **alternating current**. Plastic and flow-through conductance cells are also available

In any conductor, the conductance depends on the cell dimensions and on the characteristics of the conducting medium. For a cell with electrodes of area A, set a distance l apart:

$$G = \kappa/(l/A)$$

where κ is called the **conductivity** of the solution with units of Ω^{-1} m^{-1} or S m^{-1}. So that the results from different cells and solutions may be compared, the **cell constant**, $K = (l/A)$ is measured using a solution of known conductivity, often 0.1 M aqueous potassium chloride, which has a conductivity κ =1.288 Ω^{-1} m^{-1} at 25°C.

Conductivity

Although it is often possible to work directly with conductance, it is useful to know the conductivity and molar conductivity when results are to be compared. From the equations above:

$$\kappa = G(l/A)$$

Since the analyst is concerned with molar concentrations, and the conductivity depends mostly on the amounts of ions present, the **molar conductivity** Λ_m is defined by:

$$\Lambda_m = \kappa/c$$

where Λ_m has units of Ω^{-1} m^2 mol^{-1} provided that the concentration c is expressed in mol m^{-3} (= 1000 M).

Example

When immersed in a 0.1 M aqueous solution of KCl at 25°C, a conductance cell had $G = 8.59 \times 10^{-3}\ \Omega^{-1}$. For 0.1 M HCl, using the same cell, the conductance was measured as $2.57 \times 10^{-2}\ \Omega^{-1}$. Calculate the conductivity and molar conductivity of the HCl.

For the KCl calibration:

$$\kappa = G(l/A) = 8.59 \times 10^{-3} \times (l/A) = 1.288\ \Omega^{-1}\text{m}^{-1}$$

Thus $(l/A) = 149.9$ m^{-1} and for HCl, $\kappa = 3.85\ \Omega^{-1}$ m^{-1}

$$\Lambda_m = \kappa/c = 3.85/100 = 0.0385\ \Omega^{-1}\ m^2\ mol^{-1}$$

For **strong electrolytes**, it must be noted that, because of ionic interaction and the effects of the **ionic atmosphere,** the molar conductivity decreases as the concentration increases, according to the **Onsager equation**:

$$\Lambda_m = \Lambda^o{}_m - \mathrm{K}\, c$$

where $\Lambda^o{}_m$ is the molar conductivity at infinite dilution, and K is a constant, dependent on temperature and the solute.

For **weak electrolytes,** the ionic concentration depends on the dissociation constant, and since the relationship of the dissociation constant, K to the degree of ionization α is, for a weak acid, approximately:

$$K = c\alpha^2/(1 - \alpha)$$

The molar conductivity of a weak electrolyte may be written:

$$\Lambda_m = \alpha\Lambda^o{}_m$$

These equations show that, while Λ_m for a strong electrolyte decreases almost linearly with the $(c)^{1/2}$, for weak electrolyes such as ethanoic acid, it decreases approximately as $(1/c)^{1/2}$. More details will be found in textbooks of physical chemistry.

Direct measurement of concentration is possible. The presence of sodium chloride, or other ionic species will increase conductance, and by knowing the molar conductivity, the concentration may be found by measuring the conductivity of the solution.

When very pure water is required for the preparation of pure organics, or the manufacture of semiconductor components, conductance measurements are most useful. Provided dissolved CO_2 and other intrinsic factors are eliminated, extremely pure water should have a conductivity of around 150 μS m^{-1} (or 1.5 μS cm^{-1}) at 25°C. In order to comply with regulations for some work, it must be less than 200 μS m^{-1}.

Conductimetric titrations

Whenever titrations of ionic solutions are carried out, the number and nature of the ions change throughout the entire titration. If a strong base, say NaOH, is titrated with a strong acid, HCl, the reaction

$$Na^+ + OH^- + H^+ + Cl^- = Na^+ + Cl^- + H_2O$$

will first of all remove the OH^- as nonconducting water, so the conductance will decrease until the end point. Then excess HCl will increase the conductance. This gives a 'V' shaped graph of conductance versus titer. If a mixture of acids is used, they will be titrated in the order of their strength. *Figure 1(a)* shows the conductimetric titration of a mixture of HCl, boric acid and ammonium chloride using NaOH.

For precipitation reactions:

$$Ag^+ + NO_3^- + Na^+ + Cl^- = AgCl(s) + Na^+ + NO_3^-$$

The sodium and silver ions have similar conductivities, so adding NaCl solution does not change the conductance much until the end point is reached. Then the conductance rises with excess NaCl. This is shown in *Figure 1(b)*. The titrant is often about ten-times more concentrated than the sample solution, to avoid dilution errors.

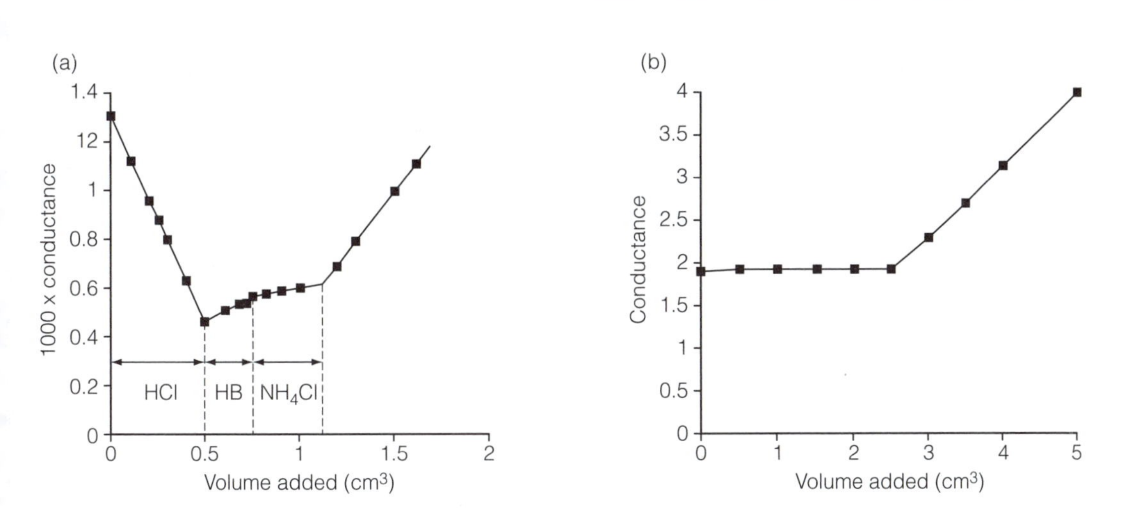

Fig. 1. (a) Conductimetric titration of mixed acids with 1.0 M NaOH. (b) Conductimetric titration of 0.1 M $AgNO_3$ with 1.0 M KCl.

Conductimetric titrations may be carried out in nonaqueous media. For example, phenols, which are very weak acids in water, have enhanced acidity when liquid ammonia is used as the solvent, and may be titrated conductimetrically with an ammoniacal solution of KOH.

D1 Solvent and solid-phase extraction

Key Notes

Extraction techniques

Solvent and solid-phase extraction are two techniques for separating mixtures of substances, either by selective transfer between two immiscible liquid phases or between a liquid and a solid phase.

Extraction efficiency and selectivity

Extraction efficiency is defined as the fraction or percentage of a substance that can be extracted in one or more steps. Selectivity is the degree to which a substance can be separated from others in a mixture.

Solvent extraction

Procedures are based on the extraction of nonpolar, uncharged species from an aqueous solution into an immiscible organic solvent, or the extraction of polar or ionized species into an aqueous solution from an organic solvent.

Solid-phase sorbents

Sorbents are particulate materials such as silica, chemically-modified silica, alumina and organic resins that can interact with and retain substances from solutions. Retained substances can be subsequently released back into a suitable solvent under controlled conditions.

Solid-phase extraction

Sample solutions are passed through the sorbent under conditions where either the analyte(s) are retained and matrix components washed through or the reverse. Retained analytes are removed with an alternative solvent before completing the analysis.

Related topic Solution equilibria (C1)

Extraction techniques

The purpose of an extraction technique is physically to separate components of a mixture (**solutes**) by exploiting differences in their relative solubilities in two immiscible liquids or between their affinities for a solid **sorbent**. Substances reach an equilibrium distribution through intimate contact between the two phases, which are then physically separated to enable the species in either phase to be recovered for completion of the analysis. An **equilibrium distribution** of the solutes between the two phases is established by dissolving the sample in a suitable solvent, then shaking the solution with a second, immiscible, solvent or by passing it through a sorbent bed or disk. Where the equilibrium distributions of two solutes differ, a separation is possible. The principal factors that determine how a solute will distribute between two phases are its polarity and the polarities of each phase. Degree of ionization, hydrogen bonding and other electrostatic interactions also play a part.

Most solvent-extraction procedures involve the extraction of solutes from an aqueous solution into a nonpolar or slightly polar organic solvent, such as hexane, methylbenzene or trichloromethane, although the reverse is also possible. Readily extractable solutes, therefore, include covalent neutral molecules with few polar and no ionized substituents; polar, ionized or ionic species will remain in the aqueous phase. A wide range of solid sorbents with different polarities are used in solid phase extraction making this a very versatile technique, but the factors affecting the distribution of solutes are essentially the same as in solvent extraction. Two examples illustrate these points.

Example 1
If an aqueous solution containing iodine and sodium chloride is shaken with a hydrocarbon or chlorinated hydrocarbon solvent, the iodine, being a covalent molecule, will be extracted largely into the organic phase, whilst the completely ionized and hydrated sodium chloride will remain in the aqueous phase.

Example 2
If an aqueous solution containing a mixture of weakly polar vitamins or drugs and highly polar sugars is passed through a hydrocarbon-modified silica sorbent which has a nonpolar surface, the vitamins or drugs will be retained on the surface of the sorbent whilst the sugars pass through.

Solvent extraction is governed by the **Nernst Distribution** or **Partition Law**. This states that at constant temperature and pressure, a solute, S, will always be distributed in the same proportions between two particular immiscible solvents. The ratio of the equilibrium concentrations (strictly their activities (Topic C1)) in the two phases defines a **distribution** or **partition coefficient**, K_D, given by the expression

$$K_D = \frac{[S]_{org}}{[S]_{aq}} \quad (1)$$

where [] denotes concentrations (activities) of the distributing solute species, S, and the value of K_D is independent of the total solute concentration. In practice, the solute often exists in different chemical forms due to dissociation (ionization), protonation, complexation or polymerization, so a more practically useful expression which defines a **distribution** or **partition ratio**, D, is

$$D = \frac{(C_S)_{org}}{(C_S)_{aq}} \quad (2)$$

where (C_S) represents the total concentration of all forms of the distributing solute S in each phase. If no interactions involving S occur in either phase, then D and K_D would be identical. However, the value of D is determined by the experimental conditions and it can be adjusted over a wide range to suit the requirements of the analytical procedure.

A distribution ratio can also be defined for solid phase extraction, i.e.

$$D = \frac{(C_S)_{sorb}}{(C_S)_{liq}} \quad (3)$$

where the numerator represents the solute concentration in the solid sorbent and the denominator the solute concentration in the liquid phase.

Solutes with large values of D (e.g. 10^4 or more) will be essentially quantitatively extracted into the organic phase or retained by the sorbent, although the nature of an equilibrium process means that 100% extraction or retention can never be achieved.

Extraction efficiency and selectivity

The **efficiency** of an extraction depends on the value of the distribution ratio, D. For solvent extraction, it also depends on the relative volumes of the two liquid phases and for solid-phase extraction on the surface area of the sorbent. With solvent extraction, the percentage of a solute extracted, E, is given by the expression

$$E = \frac{100D}{[D + (V_{aq}/V_{org})]} \tag{4}$$

where V_{aq} and V_{org} are the volumes of the aqueous and organic phases, respectively.

For solutes with small values of D, multiple extractions will improve the overall efficiency, and an alternative expression enables this to calculated

$$(C_{aq})_n = C_{aq}[V_{aq}/(DV_{org} + V_{aq})]^n \tag{5}$$

where C_{aq} and $(C_{aq})_n$ are the amounts of solute in the aqueous phase initially and remaining after n extractions, respectively.

The following example demonstrates the use of these formulae.

Example 1

A water sample contains 10 mg each of a halogenated pesticide and an ionic herbicide which are to be separated by extraction of the pesticide into methylbenzene. Given that the pesticide distribution ratio, D, for methylbenzene/water is 50, calculate the extraction efficiency for

(i) one extraction of 20 cm^3 of water with 10 cm^3 of methylbenzene
(ii) one extraction of 20 cm^3 of water with 30 cm^3 of methylbenzene
(iii) three extractions of the same 20 cm^3 of water with 10 cm^3 portions of methylbenzene (30 cm^3 in total)

(i) Substitution of the values for D, V_{aq} and V_{org} in equation (4) gives

$$E = \frac{100 \times 50}{[50 + (20/10)]} = \mathbf{96.15\%}$$

(ii) Substitution of the values for D, V_{aq} and V_{org} in equation (4) gives

$$E = \frac{100 \times 50}{[50 + (20/30)]} = \mathbf{98.68\%}$$

(iii) Substitution of the values for D, V_{aq}, V_{org}, C_{aq} and $(C_{aq})_n$ in equation (5) gives

$$\begin{aligned}(C_{aq})_3 &= 10 \times [20/((50 \times 10) + 20)]^3 \\ &= 10 \times [0.03846]^3 = 5.6896 \times 10^{-4}\ \text{mg} \\ E &= \mathbf{99.99\%}\end{aligned}$$

An extraction of 99.99%, as achieved in (iii), would be considered quantitative, although the lower efficiencies obtained in (i) and (ii) might be acceptable in the context of a defined analytical problem. It is clear that increasing the volume of organic solvent, or extracting with the same volume divided into several smaller portions, increases the overall efficiency of an extraction.

NB the ionic herbicide has a negligibly small distribution ratio, being very polar and highly water soluble.

Selectivity in extraction procedures is the degree to which solutes in a mixture can be separated by virtue of having different distribution ratios. For two solutes with distribution ratios D_1 and D_2, a **separation** or **selectivity factor**, β, is defined as

$$\beta = D_1/D_2 \quad \text{where } D_1 > D_2 \tag{6}$$

Selectivity factors exceeding 10^4 or 10^5 ($\log\beta$ values exceeding four or five) are necessary to achieve a quantitative separation of two solutes, as for most practical purposes, a separation would be considered complete if one solute could be extracted with greater than 99% efficiency whilst extracting less than 1% of another. The extraction of many solutes can be enhanced or supressed by adjusting solution conditions, e.g. pH or the addition of complexing agents (*vide infra*).

Solvent extraction

Solvent extraction (**SE**) is used as a means of sample pre-treatment or clean-up to separate analytes from matrix components that would interfere with their detection or quantitation (Topic A4). It is also used to pre-concentrate analytes present in samples at very low levels and which might otherwise be difficult or impossible to detect or quantify. Most extractions are carried out batchwise in a few minutes using separating funnels. However, the efficient extraction of solutes with very small distribution ratios (<1) can be achieved only by continuously exposing the sample solution to fresh solvent that is recycled by refluxing in a specially designed apparatus.

Broad classes of organic compounds, such as acids and bases, can be separated by pH control, and trace metal ions complexed with organic reagents can be separated or concentrated prior to spectrometric analysis (Topic E10).

Extraction of organic acids and bases

Organic compounds with acidic or basic functionalities dissociate or protonate in aqueous solutions according to the pH of the solution. Their extraction can, therefore, be optimized by pH adjustments. The relation between pH and the distribution ratio, D, of a weak acid can be derived in the following way:

A weak acid, HA, dissociates in water according to the equation

$$HA = H^+ + A^- \tag{7}$$

The acid dissociation constant, K_a, is defined as

$$K_a = \frac{[H^+]_{aq}[A^-]_{aq}}{[HA]_{aq}} \tag{8}$$

Only the undissociated form, $[HA]$, can be extracted into a nonpolar or slightly polar solvent such as diethyl ether, the distribution or partition coefficient, K_D, being given by

$$K_D = \frac{[HA]_{ether}}{[HA]_{aq}} \tag{9}$$

However, the distribution ratio takes account of both the dissociated and undissociated forms of the acid in the aqueous phase and is given by

$$D = \frac{[HA]_{ether}}{([HA]_{aq} + [A^-]_{aq})} \tag{10}$$

Re-arrangement of equation (8) and substitution for $[A^-]_{aq}$ in equation (10) gives

$$D = \frac{[HA]_{ether}}{[HA]_{aq}\,(1 + K_a/[H^+]_{aq})} \qquad (11)$$

and substituting K_D for $[HA]_{ether}/[HA]_{aq}$ from equation (9) gives

$$D = \frac{K_D}{(1 + K_a[H^+])} \qquad (12)$$

Equation (12) shows that at low pH, when the acid is undissociated, it is extracted with the greatest efficiency as $D \cong K_D$, whereas as the pH is increased the value of D decreases until at high pH the acid is completely dissociated into the anion A^-, and none will be extracted. This is shown graphically in *Figure 1* as pH versus E plots for two weak acids (curves 1 and 2) with different acid dissociation constants (K_a or pK_a values), curve 1 being for the stronger of the two acids. For a weak organic base, such as an amine, protonation occurs at low pH according to the equation

$$RNH_2 + H^+ = RNH_3^+ \qquad (13)$$

The relation between pH and the distribution ratio for a weak base (curve 3) is therefore the opposite of a weak acid so that it is possible to separate acids from bases in a mixture either by extracting the acids at low pH or the bases at high pH. Separating mixtures of acids or mixtures of bases is possible only if their dissociation or protonation constants differ by several pK units.

Extraction of metals

Metal ions in aqueous solutions are not themselves extractable into organic solvents, but many can be complexed with a variety of organic reagents to form extractable species. Some inorganic complex ions can be extracted as neutral ionic aggregates by association with suitable ions of opposite charge (counter ions). There are two principal metal extraction systems:

- Uncharged metal **chelate complexes** (ring structures that satisfy the coordination requirements of the metal) with organic reagents
  e.g. 8-hydroxyquinoline (oxine)
  acetylacetone (AcAc)

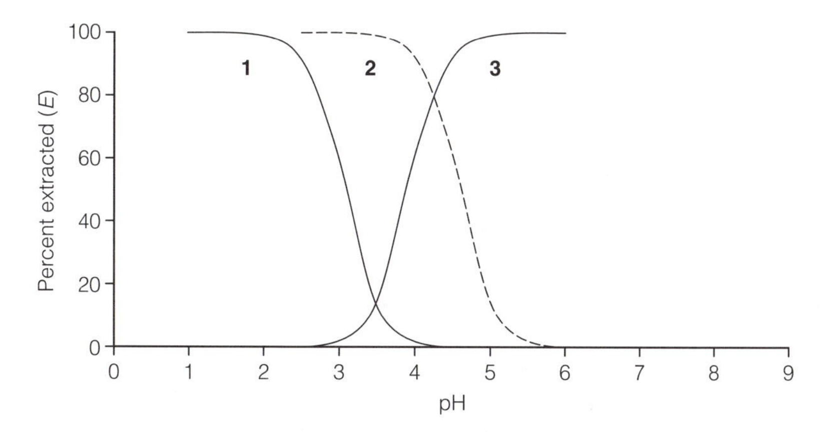

Fig. 1. Solvent extraction curves for two organic acids having different pK_a values (curves 1 and 2) and a base (curve 3). Curve 1 represents the acid with the larger K_a (smaller pK_a).

1-(2-pyridylazo)-2-naphthol (PAN)
sodium diethyldithiocarbamate (NaDDTC)
(see also *Table 1*);

- Electrically neutral **ion-association complexes**
  e.g. $Cu(neocuproine)_2^+$, NO_3^-
  $(C_6H_5CH_2)_3NH^+$, $GaCl_4^-$
  $[(C_2H_5)_2O]_3H^+$, $FeCl_4^-$
  (see also *Table 1*).

The Nernst distribution law applies to metal complexes, but their distribution ratios are determined by several interrelated equilibria. As in the case of organic acids and bases, the efficiency of extraction of metal chelates is pH dependent, and for some ion-association complexes, notably **oxonium systems** (hydrogen ions solvated with ethers, esters or ketones), inorganic complex ions can be extracted from concentrated solutions of mineral acids.

Reagents that form neutral metal chelate complexes (5- or 6-membered ring structures) are weakly acidic and contain one or more additional co-ordinating sites (O, N or S atoms). Protons are displaced according to the general equation

$$\underset{\text{not extractable}}{M^{n+}(H_2O)_x} + nHR = \underset{\text{extractable}}{MR_n} + nH^+ + xH_2O \qquad (14)$$

where HR is a weakly acidic and co-ordinating reagent (ligand), and the metal ion M^{n+} has a formal valency n. The removal of hydrogen ions is necessary to drive the reaction to completion, and pH control, which is essential, is achieved by buffering the aqueous solution. It is sometimes possible to improve the selectivity of the procedure by adding an additional reagent, known as a **masking agent**, that reacts preferentially with one of the metals to form a nonextractable water-soluble complex. Typical masking agents include EDTA (ethylenediaminetetraacetic acid), citrate, tartrate, fluoride, cyanide and thiourea. Solvents commonly used to extract metal chelate complexes include trichloromethane, 4-methyl-pentan-2-one and methylbenzene.

Electrically neutral ion-association complexes consist of cationic (positively charged) and anionic (negatively charged) species that form an overall neutral aggregate extractable by an organic solvent. Either the cations or the anions

Table 1. A selection of reagents and extraction systems for the solvent extraction of metals

Reagent	Type of metal complex
8-Hydroxyquinoline (oxine) Di-alkyldithiocarbamates e.g. sodium diethyldithiocarbamate (NaDDTC)	Neutral metal chelate complexes, extractable into organic solvents. Intense color of many facilitates colorimetric determinations.
1,10-Phenanthroline (*o*-phen) 2,9-Dimethyl-1,10-phenanthroline (neocuproine) Ethylenediaminetetraacetic acid (EDTA)	Ion-association complexes. Metals as cationic or anionic chelated complexes extracted with suitable counter ion.
Oxonium systems: i.e. protons solvated with alkyl ethers, ketones, esters or alcohols	Ion-pairs with anionic metal halide or thiocyanate complexes. Chloride complexes extractable from strong HCl solutions.

should be bulky organic hydrophobic groups to provide high solubility in nonpolar solvents. The metal ion can be a cationic or an anionic complex, and can be an inorganic species such as $FeCl_4^-$, MnO_4^- or a chelated organic complex such as Fe(1,10-phenanthroline)$_3^{2+}$ or UO_2(oxine)$_3^-$. Suitable counter-ions of opposite charge to the metal-complex ions include $(C_4H_9)_4N^+$, ClO_4^- $(C_6H_5CH_2)_3NH^+$, $[(C_4H_9O)_3P{=}O]H^+$ and an oxonium ion such as $[(C_2H_5)_2O]_3H^+$.

A list of metal chelate complexing agents and ion-association complexes is given in *Table 1*. Most complexing agents react with a large number of metals (up to 50 or more), but pH control, the use of masking agents and a variety of ion-association systems can enable the selective extraction and separation of just one or two metals to be accomplished.

Solid-phase sorbents

These are generally either silica or chemically-modified silica similar to the bonded phases used in high-performance liquid chromatography (Topics D6 and D7) but of larger particle size, typically 40–60 μm diameter. Solutes interact with the surface of the sorbent through van der Waals forces, dipolar interactions, H-bonding, ion-exchange and exclusion. The four chromatographic sorption mechanisms described in Topic D2 can be exploited depending on the sorbent selected and the nature of the sample. Sorbents can be classified according to the polarity of the surface. Hydrocarbon-modified silicas are nonpolar, and therefore hydrophobic, but are capable of extracting a very wide range of organic compounds from aqueous solutions. However, they do not extract very polar compounds well, if at all, and these are best extracted by unmodified silica, alumina or Florisil, all of which have a polar surface. Ionic and ionizable solutes are readily retained by an ion-exchange mechanism using cationic or anionic sorbents. Weak acids can be extracted from aqueous solutions of high pH when they are ionized, and weak bases from aqueous solutions of low pH when they are protonated. It should be noted that this is the opposite way around compared to solvent extraction into non-polar solvents. However, by suppressing ionization through pH control, extraction by hydrocarbon-modified silica sorbents is possible. Sorbents of intermediate polarity, such as cyanopropyl and aminopropyl modified silicas may have different selectivities to nonpolar and polar sorbents. Some SPE sorbents are listed in *Table 2* along with the predominant interaction mechanism for each one.

Solid-phase extraction

Compared to solvent extraction, **solid-phase extraction (SPE),** is a relatively new technique, but it has rapidly become established as the prime means of sample pre-treatment or the clean-up of **dirty** samples, i.e. those containing high levels of matrix components such as salts, proteins, polymers, resins, tars etc. In addition to being potential sources of interference with the detection and quantitation of analytes, their presence can be detrimental to the stability and performance of columns and detectors when a chromatographic analysis is required. The removal of interfering matrix components in general and the pre-concentration of trace and ultra-trace level analytes are other important uses of SPE which is versatile, rapid and, unlike solvent extraction, requires only small volumes of solvents, or none at all in the case of **solid phase microextraction** (*vide infra*). Furthermore, SPE sorbents are cheap enough to be discarded after use thus obviating the need for regeneration. The analysis of environmental, clinical, biological and pharmaceutical samples have all benefited from the rapid growth in the use of SPE where it has largely replaced solvent extraction. Specific examples include the determination of pesticides and herbicides in polluted surface

Table 2. Typical SPE sorbents and interaction mechanisms

Sorbent		Polarity	Interaction mechanisms
Silica	SiO_2	Polar	Adsorption; H-bonding
Florisil,	$MgSiO_3$	Polar	H-bonding
alumina	Al_2O_3	Polar	H-bonding
Bonded phases (modified silica)			
$-C_{18}H_{37}$	(C18 or ODS)	Nonpolar	Van der Waals interactions
$-C_8H_{17}$	(C8 or octyl)	Nonpolar	Van der Waals interactions
$-C_6H_5$	(phenyl)	Nonpolar	Van der Waals interactions and π–π interactions
$-(CH_2)_3CN$	(cyanopropyl)	Polar	Polar interactions; H-bonding
$-(CH_2)_3NH_2$	(aminopropyl)	Polar	H-bonding
$-(CH_2)_3C_6H_4SO_3H$		Ionic	Cation exchange
$-(CH_2)_3N(CH_3)_3Cl$		Ionic	Anion exchange
Chiral	(cyclodextrin)	Polar	Adsorption; H-bonding dipolar interactions steric effects
Styrene/divinyl benzene co-polymer		Nonpolar	Size exclusion

waters and soils, polycyclic aromatic hydrocarbons (PAHs) in drinking water, polluted industrial and urban atmospheres, and drugs in biological fluids.

Sorbents are either packed into disposable **cartridges** the size of a syringe barrel, fabricated into **disks** or incorporated into plastic **pipette tips or well plates**. Most SPE is carried out using a small packed bed of sorbent (25–500 mg) contained in a cartridge made from a polypropylene syringe barrel, the sorbent being retained in position by polyethylene fritted disks. The sorbent generally occupies only the lower half of the cartridge, leaving space above to accommodate several millilitres of the sample solution or washing and eluting solvents. A typical cartridge procedure is illustrated in *Figure 2* and consists of four distinct steps:

- **Sorbent conditioning.** The cartridge is flushed through with the sample solvent to wet the surface of the sorbent and to create the same pH and solvent composition as those of the sample, thus avoiding undesirable chemical changes when the sample is applied.
- **Sample loading or retention.** The sample solution is passed through the cartridge with the object of *either* retaining the analytes of interest whilst the matrix components pass through **or** retaining the matrix components whilst the analytes pass through. In some procedures, the analyte(s) and one or more of the matrix components are retained whilst the remainder of the matrix components pass through.
- **Rinsing.** This is necessary to remove all those components not retained by the sorbent during the retention step and which may remain trapped in the interstitial solvent.
- **Elution.** This final step is to recover retained analytes, otherwise the matrix-free solution and rinsings from the second and third steps are combined for quantitative recovery of the analytes before completion of the analysis.

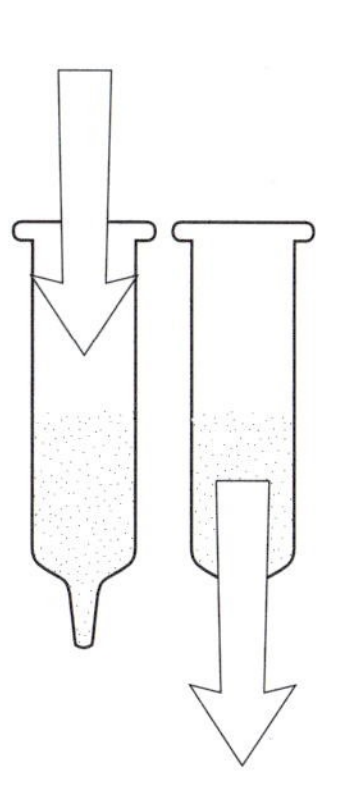

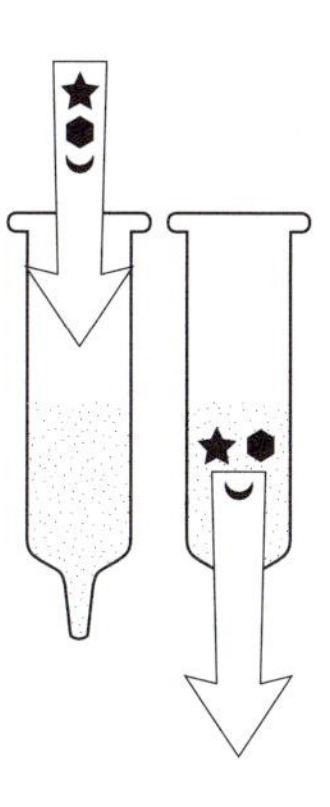

Conditioning
Conditioning the sorbent prior to sample application ensures reproducible retention of the compound of interest (the isolate)

Retention
★ Adsorbed isolate
⬢ Undesired matrix constituents
◡ Other undesired matrix components

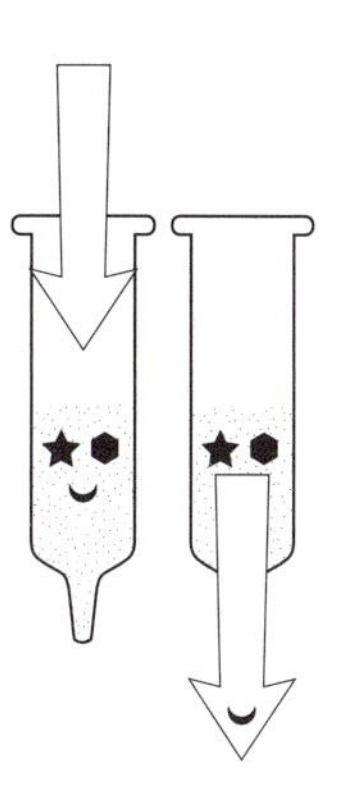

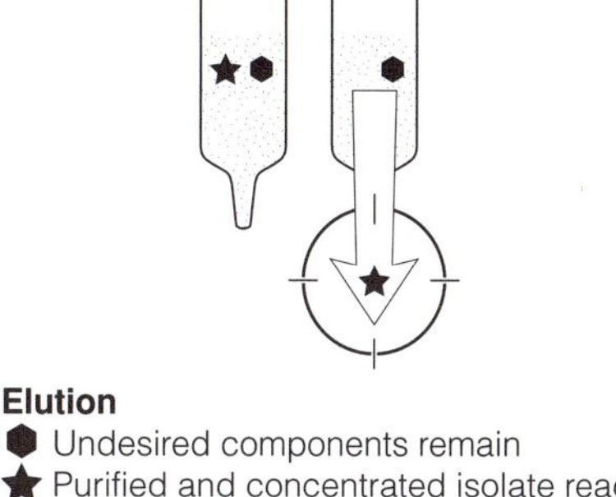

Rinse
◡ Rinse the columns to remove undesired matrix components

Elution
⬢ Undesired components remain
★ Purified and concentrated isolate ready for analysis

Fig. 2. Diagrammatic representation of a cartridge-based solid phase extraction procedure.

SPE can be semi- or fully-automated to increase sample throughput and to improve both precision and accuracy. The degree of automation ranges from the parallel off-line processing of batches of up to about 10 samples using a vacuum manifold to provide suction, to on-line autosamplers, xyz liquid handlers and robotic workstations.

Several alternative formats for SPE are available. These include

- **Disks**, which have relatively large cross-sectional areas compared to packed-bed cartridges and thin sorbent layers (0.5–1 mm thick) containing about 15 mg of material. This reduced bed-mass results in low void volumes (as little as 15 μl) thus minimizing solvent consumption in the rinsing and elution steps, improving selectivity and facilitating high solvent flow rates when large volumes of sample are to be processed.
- **Plastic pipette-tips** incorporating small sorbent beds designed for processing

very small volumes of sample and solvents rapidly, and having the advantage of allowing flow in both directions if required.

- **Well plates** containing upwards of 96 individual miniature sample-containers in a rectangular array and fitted with miniature SPE packed beds or disks. Well plates are used in xyz liquid handlers for processing large numbers of samples prior to the transfer of aliquots to analytical instruments, particularly gas and liquid chromatographs and mass spectrometers.
- **Solid-phase microextraction (SPME)** is an important variation of SPE that allows trace and ultra-trace levels of analytes in liquid or gaseous samples to be concentrated. The sorbent is a thin layer of a polymeric substance such as polydimethylsiloxane (PDMS) coated onto a fused-silica optical fibre about 1 cm long and attached to a modified microsyringe (*Fig. 3*). The fibre is exposed to the sample and then inserted directly into the injection port of a gas or liquid chromatograph to complete the analysis. An advantage of SPME over SPE is the avoidance of solvents, but good precision for quantitative determinations is more difficult to achieve, and automated systems are only just being developed. SPME is finding particular use in water analysis, the analysis of fragrances and volatiles in foodstuffs by headspace sampling (Topic D5), the detection of drugs and their metabolites in urine, blood and breath samples, and the monitoring of air quality in working environments.

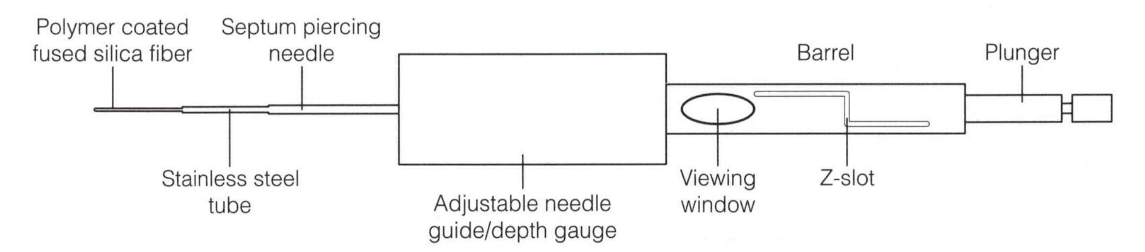

Fig. 3. Diagram of a solid phase microextraction device.

D2 PRINCIPLES OF CHROMATOGRAPHY

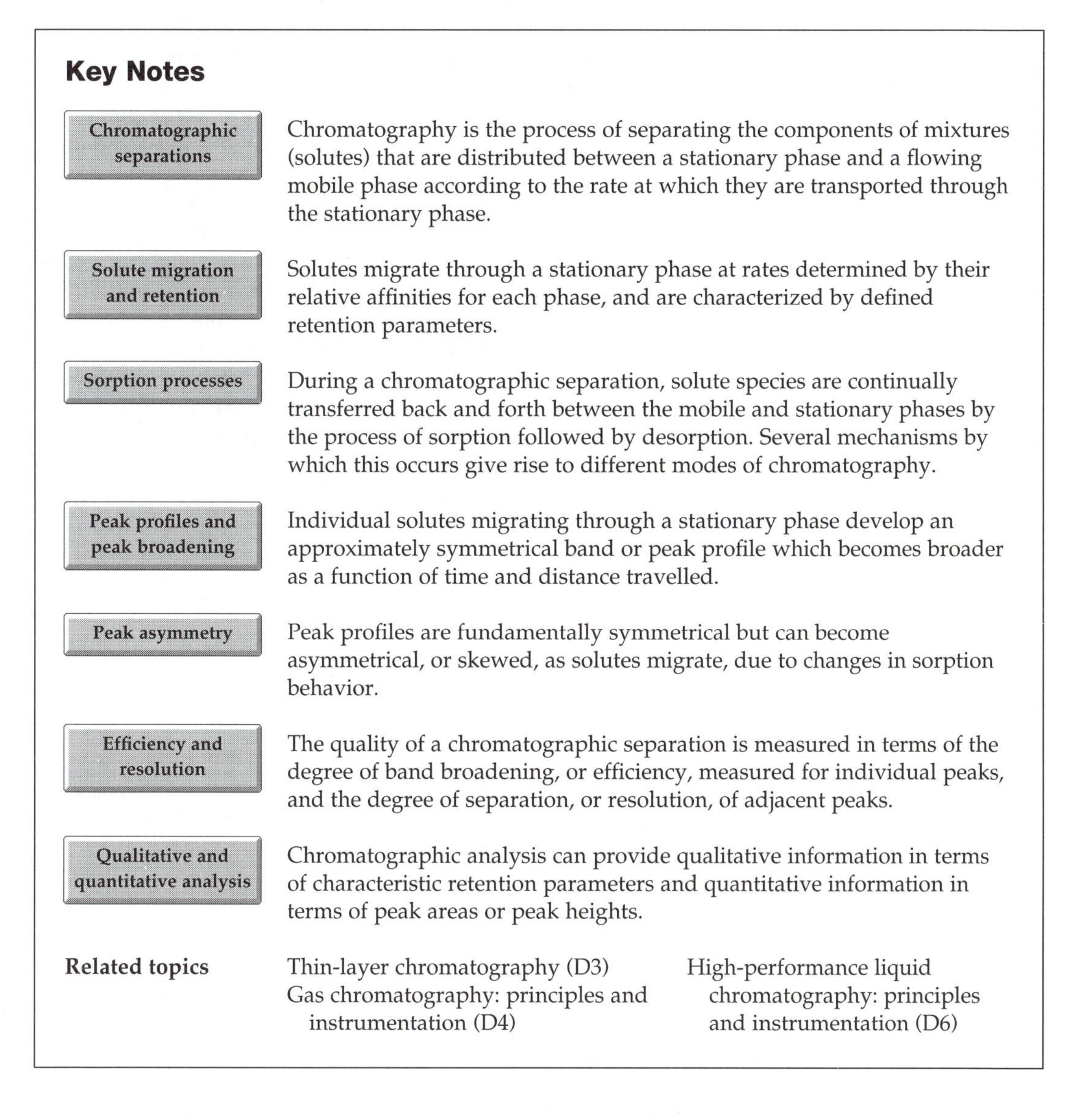

Key Notes

Chromatographic separations	Chromatography is the process of separating the components of mixtures (solutes) that are distributed between a stationary phase and a flowing mobile phase according to the rate at which they are transported through the stationary phase.
Solute migration and retention	Solutes migrate through a stationary phase at rates determined by their relative affinities for each phase, and are characterized by defined retention parameters.
Sorption processes	During a chromatographic separation, solute species are continually transferred back and forth between the mobile and stationary phases by the process of sorption followed by desorption. Several mechanisms by which this occurs give rise to different modes of chromatography.
Peak profiles and peak broadening	Individual solutes migrating through a stationary phase develop an approximately symmetrical band or peak profile which becomes broader as a function of time and distance travelled.
Peak asymmetry	Peak profiles are fundamentally symmetrical but can become asymmetrical, or skewed, as solutes migrate, due to changes in sorption behavior.
Efficiency and resolution	The quality of a chromatographic separation is measured in terms of the degree of band broadening, or efficiency, measured for individual peaks, and the degree of separation, or resolution, of adjacent peaks.
Qualitative and quantitative analysis	Chromatographic analysis can provide qualitative information in terms of characteristic retention parameters and quantitative information in terms of peak areas or peak heights.

Related topics

Thin-layer chromatography (D3)
Gas chromatography: principles and instrumentation (D4)
High-performance liquid chromatography: principles and instrumentation (D6)

Chromatographic separations

Chromatography was originally developed by the Russian botanist Michael Tswett in 1903 for the separation of colored plant pigments by percolating a petroleum ether extract through a glass column packed with powdered calcium carbonate. It is now, in general, the most widely used separation technique in analytical chemistry having developed into a number of related but quite different forms that enable the components of complex mixtures of organic or inorganic components to be separated and quantified. A chromatographic separation involves the placing of a sample onto a liquid or solid **stationary phase** and passing a liquid or gaseous **mobile phase** through or over it, a process known as **elution**. Sample components, or **solutes**, whose **distribution ratios** (*vide infra*) between the two phases differ will migrate (be eluted) at different rates, and this **differential rate of migration** will lead to their separation over a period of time and distance.

Chromatographic techniques can be classified according to whether the separation takes place on a **planar surface** or in a **column**. They can be further subdivided into **gas** and **liquid chromatography**, and by the physical form, solid or liquid, of the stationary phase and the nature of the interactions of solutes with it, known as **sorption mechanisms** (*vide infra*). *Table 1* lists the most important forms of chromatography, each based on different combinations of stationary and mobile phases and instrumental or other requirements.

Paper chromatography (PC) is simple and cheap but lacks the separating power and versatility of **thin-layer chromatography (TLC)** which has largely replaced it. Both require only inexpensive equipment and reagents, and, unlike the various forms of column chromatography, comparisons can be made between a number of samples and standards chromatographed simultaneously. **Gas (GC)** and **high performance liquid chromatography (HPLC)** are complementary techniques best suited to the separation of volatile and nonvolatile mixtures, respectively. Both these techniques are instrumentally-based and computer-controlled, with sophisticated software packages and the ability to separate very complex mixtures of up to 100 or more components. HPLC is particularly versatile, having several alternative **modes** suited to different types of solute. For example, **ion-exchange (IEC)** and **ion chromatography (IC)** are

Table 1. A classification of the principal chromatographic techniques

Technique	Stationary phase	Mobile phase	Format	Principal sorption mechanism
Paper chromatography (PC)	Paper (cellulose)	Liquid	Planar	Partition (adsorption, ion-exchange, exclusion)
Thin-layer chromatography (TLC)	Silica, cellulose, ion-exchange resin, controlled porosity solid	Liquid	Planar	Adsorption (partition, ion-exchange, exclusion)
Gas chromatography (GC)				
Gas-liquid chromatography (GLC)	Liquid	Gas	Column	Partition
Gas-solid chromatography (GSC)	Solid	Gas	Column	Adsorption
Liquid chromatography (LC)				
High-performance liquid chromatography (HPLC)	Solid or bonded-phase	Liquid	Column	Modified partition (adsorption)
Size-exclusion chromatography (SEC)	Controlled porosity solid	Liquid	Column	Exclusion
Ion-exchange chromatography (IEC) Ion chromatography (IC)	Ion-exchange resin or bonded-phase	Liquid	Column	Ion-exchange
Chiral chromatography (CC)	Solid chiral selector	Liquid	Column	Selective adsorption

modes that enable mixtures of either anionic or cationic solutes to be separated. **Size-exclusion (SEC)** and **chiral chromatography (CC)** are two additional modes used for separating mixtures of high relative molecular mass solutes and enantiomers, respectively.

Earlier forms of liquid chromatography, used for relatively large scale separations and known as **classical column chromatography**, are based on large glass columns through which the mobile phase flows by gravity compared to the pressurized systems used in **HPLC**.

Solute migration and retention

The rate of migration of a solute through a stationary phase is determined by its **distribution ratio**, D, which in turn is determined by its relative affinity for the two phases. In the context of chromatography, D is defined as the ratio of the total solute concentration, C_S, in the stationary phase to that in the mobile phase, C_M, i.e.

$$D = \frac{C_S}{C_M} \tag{1}$$

Thus, large values of D lead to slow solute migration, and small values of D lead to rapid solute migration. Solutes are **eluted** in order of increasing distribution ratio. The larger the differences between the distribution ratios of the solutes in a mixture, the more easily and quickly they can be separated. Because the interaction of solutes with the stationary phase slows down their rate of migration relative to the velocity of the mobile phase, the process is described as **retardation** or **retention**.

- **Column separations** (**GC** and **LC**). For the separation of mixtures on columns, solutes are characterized by the length of time they take to pass through, i.e. their **retention time**, t_R, or by a **retention factor**, k, that is directly proportional to D. The retention time and the retention factor are related by the expression

$$t_R = t_M(1 + k) \tag{2}$$

where t_M (sometimes written as t_0 and known as the **dead time**) is the time taken by a **nonretained** solute to pass through the column. A nonretained solute migrates at the same velocity as the mobile phase, having a distribution ratio and retention factor of zero; hence $t_R = t_M$. Solutes whose D and k values are greater than zero are proportionately retarded, having retention times longer than t_M, e.g.

$$\text{if } k = 1,\ t_R = 2\,t_M$$
$$\text{if } k = 2,\ t_R = 3\,t_M \text{ etc.}$$

Chromatographic conditions are generally adjusted so that k values are less than about 20, otherwise retention times become unacceptably long. Practical values of k are easily calculated using a re-arranged equation (2)

$$k = \frac{(t_R - t_M)}{t_M} \tag{3}$$

For size-exclusion chromatography (Topic D7), the **retention volume**, V_R, is used to characterize solutes. This is the volume of mobile phase required to elute the solute from the column. Retention times are directly proportional to retention volumes at a constant flow rate, so equation (2) can be re-written as

$$V_R = V_M\left(1 + D\frac{V_S}{V_M}\right) \tag{4}$$

or

$$V_R = V_M + DV_s \tag{5}$$

where D has been substituted for k using the relation

$$k = D\left(\frac{V_S}{V_M}\right) \tag{6}$$

and the volumes of the stationary and mobile phases in the column are V_S and V_M, respectively.

Equation (5) is regarded as a fundamental equation of column chromatography as it relates the retention volume of a solute to its distribution ratio.

- **Planar separations (PC** and **TLC)**. Separations are normally halted before the mobile phase has travelled completely across the surface, and solutes are characterized by the distance they have migrated relative to the leading edge of the mobile phase (**solvent front**). A solute **retardation factor**, R_f, is defined as

$$R_f = \frac{1}{1 + k} \tag{7}$$

The maximum value of R_f is 1, which is observed for a solute having a distribution ratio and retention factor of zero, and therefore migrating at the same velocity as the mobile phase. Solutes whose D and k values are greater than zero are proportionately retarded, the minimum value for R_f being zero, which is observed when the solute spends all of the time in the stationary phase and remains in its original position on the surface. Practical values of R_f are evaluated from the ratio

R_f = *(distance moved by solute) / (distance moved by the solvent front)*

Sorption processes

Sorption is the process whereby solute species are transferred from the mobile to the stationary phase, **desorption** being the reverse process. These processes occur continually throughout a chromatographic separation, and the system is therefore described as being in a state of **dynamic equilibrium**. A solute is repeatedly re-distributed between the phases as the mobile phase advances, in an attempt to maintain an equilibrium corresponding to its distribution ratio, D.

There are four basic **sorption mechanisms**, and it is common for two or more to be involved simultaneously in a particular mode of chromatography, *viz*: adsorption; partition; ion-exchange; exclusion.

- **Adsorption** is a surface effect, not to be confused with a*b*sorption, which is a bulk effect. Surface adsorption involves electrostatic interactions such as hydrogen-bonding, dipole–dipole and dipole-induced dipole attractions. Solute species compete with the mobile phase for a limited number of polar sites on the surface of the adsorbent of which **silica gel** is the most widely used. Its surface comprises **Si-O-Si** and **Si-OH (silanol)** groups, the latter being slightly acidic as well as being polar, which readily form hydrogen bonds with slightly-polar to very-polar solutes. Water in the atmosphere can de-activate an adsorbent surface by itself being adsorbed, thereby blocking adsorption sites. This can be overcome by drying the adsorbent if a more active material is required, although reproducibility may be difficult to achieve unless ambient temperature and humidity are carefully controlled. Some common adsorbents are listed in *Table 2*. Suitable

Table 2. Adsorbents for chromatographic separations (listed in order of decreasing polarity)

Alumina	(*most polar*)
Charcoal	
Silica gel	
Molecular sieve	
Magnesium silicate	
Cellulose	
Polymeric resins (styrene/divinyl benzene)	(*least polar*)

mobile phases for **TLC** and **HPLC** are to be found listed in *Table 1*, Topic D6.

The more polar the solute, the more tenaciously it will be adsorbed onto the surface of an adsorbent. Nonpolar solutes (e.g. saturated hydrocarbons) have little or no affinity for polar adsorbents, whilst polarizable solutes (e.g. unsaturated hydrocarbons) have weak affinities arising from dipole/induced dipole interactions. Polar solutes, especially those capable of hydrogen-bonding, are adsorbed strongly and require fairly polar mobile phases to elute them. An approximate order of increasing polarity and therefore order of elution (increasing distribution ratio) among classes of organic solutes is

alkanes<alkenes<aromatics<ethers<esters<ketones and aldehydes<thiols<amines and amides<alcohols<phenols<acids

Adsorption-based chromatography is particularly useful for separating mixtures of positional isomers such as 1,2-, 1,3- and 1,4-di-substituted aromatic compounds with polar substituents, whereas members of a homologous series have similar polarities and cannot generally be separated at all. **Chiral chromatography**, a mode of **HPLC**, depends on differences in the adsorptive interactions of two or more enantiomers of a compound with a chiral stationary phase. **Gas solid chromatography** (**GSC**) is an adsorption-based mode of **GC**.

- **Partition** is a sorption process analogous to solvent extraction (Topic D1), the liquid stationary phase being thinly coated or chemically bonded onto an inert solid. Where the liquid is bonded to the supporting solid, it is debateable as to whether it behaves as a liquid and whether the sorption process should be described as **modified partition**, because adsorption may also be involved. In true partition, solutes are distributed according to their relative solubilities in the mobile and stationary phases, but the exact mechanism for **bonded phases** is not clear. The use of bonded phases has become widespread in all forms of chromatography, and a pure partition mechanism probably occurs only in **gas liquid chromatography** (**GLC**) where the stationary phase is not chemically bonded to the column wall (Topic D4).

  Bonded phases are described in more detail in Topics D4 and D6.
- **Ion-exchange** is a process whereby **solute ions** in the mobile phase can exchange with **counter-ions** carrying the same charge and associated with oppositely charged groups chemically bound to the stationary phase. The stationary phase is a permeable polymeric solid, such as an insoluble organic resin or a chemically modified silica, containing fixed charge groups and

mobile counter-ions. Both **cationic** and **anionic** ion-exchangers are available, the exchange processes being represented by the following equations

cation-exchange: $n\mathbf{R}^-H^+ + X^{n+} = (\mathbf{R}^-)_nX^{n+} + nH^+$
anion-exchange: $n\mathbf{R}^+Cl^- + Y^{n-} = (\mathbf{R}^+)_nY^{n-} + nCl^-$

where **R** represents the polymeric resin or silica, and X^{n+} and Y^{n-} are solute **cations** and **anions** respectively of valency *n*.

The factors affecting ion-exchange equilibria and selectivity are described in Topic D7.

- **Exclusion** differs from the other sorption mechanisms in that no specific interactions between solute species and the stationary phase are necessary or desirable. The stationary phase is a controlled-porosity silica or polymer gel with a range of pore sizes, and solutes remain in the mobile phase throughout the separation, merely diffusing through the porous structure to different extents depending on their size and shape. Solutes whose size exceeds the diameter of the largest pores are entirely excluded from the structure and migrate at the same rate as the mobile phase. Solutes smaller than the diameter of the smallest pores can diffuse throughout the structure and have the slowest rate of migration. Solutes of an intermediate size can diffuse through some pores but not others, migrating at rates between those of the largest and smallest species.

  Size-exclusion chromatography (**SEC**) is a mode of **HPLC** (Topic D7) and is also a classical technique employing large columns of silica or polymeric gel particles and gravity flow of the mobile phase. It is sometimes described as **gel permeation** or **gel filtration chromatography**.

Peak profiles and band broadening

During a chromatographic separation, individual solutes develop a symmetrical or **Gaussian concentration profile** (Topic B2) in the direction of flow of the mobile phase. The profiles, known as **bands** or **peaks,** gradually broaden and often become asymmetrical as the solutes continue to migrate through the stationary phase. The principal underlying reasons accounting for the peak shapes and the observed broadening can be summarized as follows:

- continual sorption and desorption of a solute between a mobile and a stationary phase inherently produces a **Gaussian concentration profile** which broadens as the solute migrates further. (This can be demonstrated by a mathematical treatment of a solvent extraction procedure to separate mixtures, developed in 1952, and known as Craig Countercurrent Distribution);
- solute species travel slightly different total distances through a particulate stationary phase, causing concentration profiles to broaden symmetrically, this being known as the **multiple-path effect**;
- solute species spread by **diffusion** in all directions when they are in the mobile phase. Diffusion in both the direction of flow of the mobile phase and directly counter to it (**longitudinal** or **axial diffusion**) contributes to the symmetrical broadening of the peak profile;
- sorption and desorption, or **mass transfer,** between the stationary and mobile phases, are not instantaneous processes, and are sometimes kinetically slow. Because the mobile phase moves continuously, a true equilibrium distribution of a solute is never established, and the concentration profile in the stationary phase lags slightly behind that in the mobile phase causing further peak broadening. Slow desorption can also result in the peak becoming **asymmetrical** or **skewed** (*vide infra*);

- variations in the distribution ratio of a solute with its total concentration also leads to **asymmetrical** or **skewed** peaks (*vide infra*).

Figure 1 illustrates the symmetrical nature of a chromatographic peak and symmetrical broadening. *Figure 2* illustrates the mutiple-path, longitudinal diffusion and mass transfer effects.

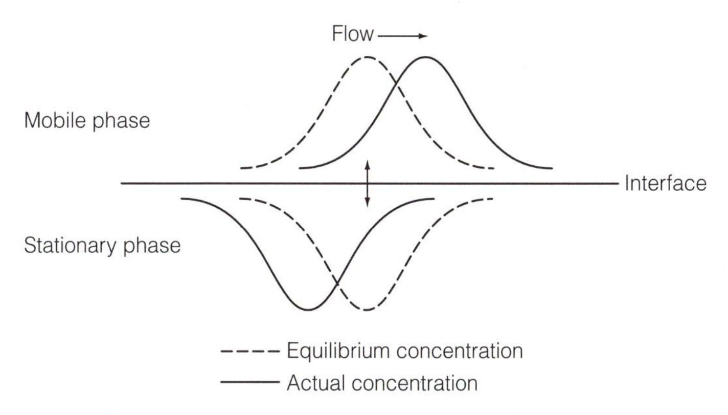

Fig. 1. The symmetrical nature and broadening of a chromatographic peak.

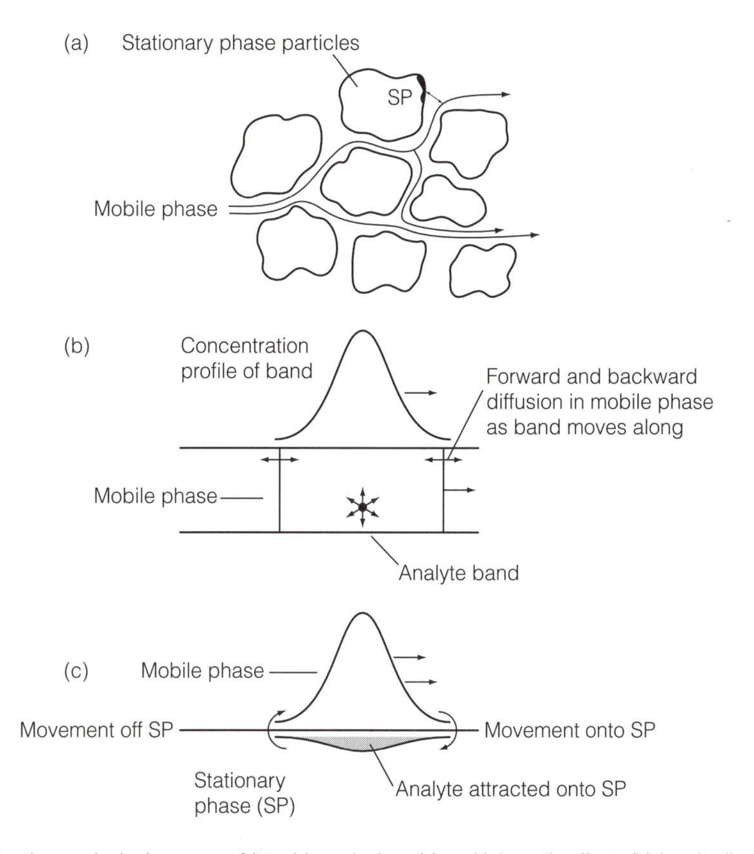

Fig. 2. Illustration of the three principal causes of band broadening: (a) multiple-path effect; (b) longitudinal diffusion effect; (c) mass-transfer (non-equilibrium) effect. Reproduced from A. Braithwaite & F.J. Smith, Chromatographic Methods, 5th edn, *1996, first published by Blackie Academic & Professional.*

Peak asymmetry

The concentration profile of a migrating solute is fundamentally symmetrical (Gaussian), only if the solute distribution ratio, D (defined by eqn. (1)), remains constant over the concentration range of the entire peak, as shown by a **linear sorption isotherm**, which is a plot of the solute concentration in the stationary phase, C_S, against that in the mobile phase, C_M (*Fig. 3(a)*). However, curved isotherms, resulting from changes in the solute distribution ratio at higher concentration levels, lead to two types of peak **asymmetry**, or **skew**, described as **tailing** and **fronting**.

Both tailing and fronting are undesirable, as closely eluting peaks will be less well separated and retention data less reproducible. Where either occur, reducing the amount of solute chromatographed will generally improve the peak shape, but slow de-sorption may still cause some tailing.

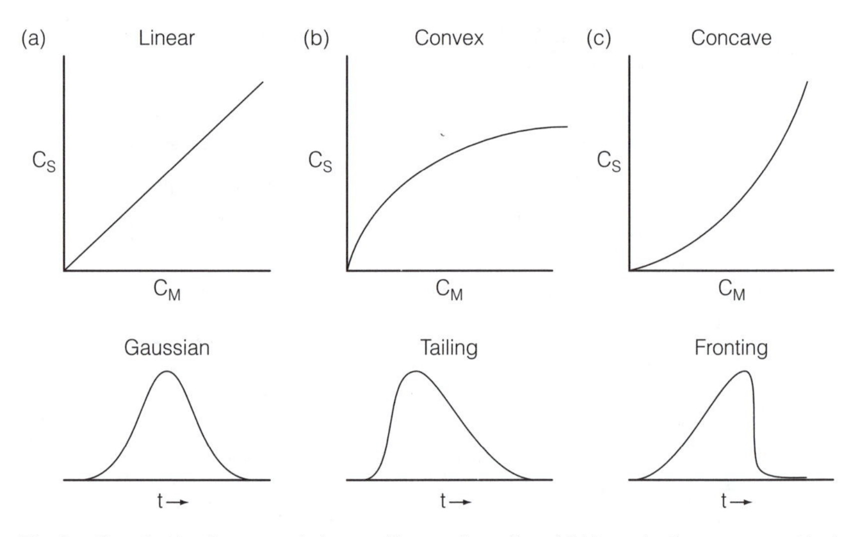

Fig. 3. Sorption isotherms and the resulting peak profiles. (a) Linear isotherm, symmetrical peak. (b) Curvature towards the C_M axis, tailing peak. (c) Curvature towards the C_S axis, fronting peak.

Efficiency and resolution

Two means of assessing the quality of a chromatographic separation are to measure the extent of band broadening of individual solute peaks (**efficiency**) and the degree of separation of adjacent peaks (**resolution**).

For column chromatography, a **plate number**, N (based on the theoretical plate concept of distillation columns), is used as a measure of **efficiency**. Assuming a Gaussian peak profile, N is defined in terms of the solute retention time, t_R, and the peak width as given by the standard deviation, σ_t (*Fig. 4*), i.e.

$$N = \left(\frac{t_R}{\sigma_t}\right)^2 \tag{8}$$

In practice, it is much easier to measure either the baseline width of a peak, W_b, or the width at half height, $W_{h/2}$, and two alternative expressions derived from equation (8) are

$$N = 16\left(\frac{t_R}{W_b}\right)^2 \tag{9}$$

and

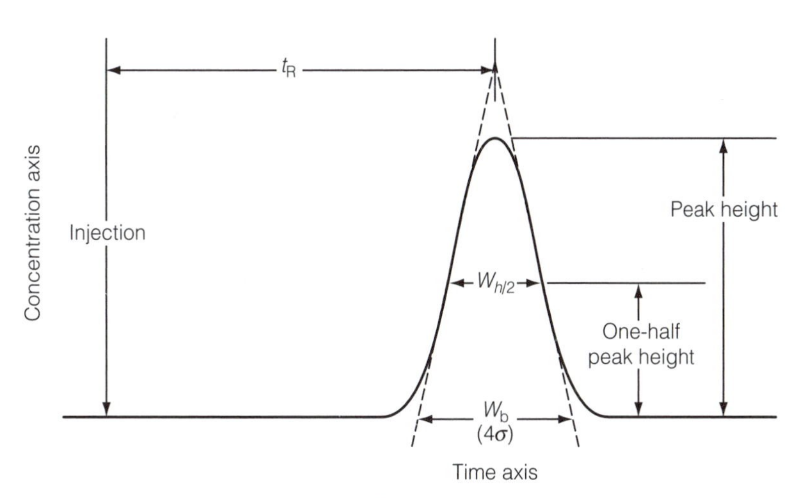

Fig. 4. Measurement of chromatographic efficiency from a Gaussian peak.

$$N = 5.54\left(\frac{t_R}{W_{h/2}}\right)^2 \tag{10}$$

Some laboratories favor the use of eqn. (9), but some favour eqn. (10) on the grounds that peak width at half height can be measured with greater accuracy than the base width. To make valid comparisons, the same formula should always be used.

An alternative measure of efficiency, which is independent of the length of a chromatographic column, is the plate height, H (or **H**eight **E**quivalent of a **T**heoretical **P**late, *HETP*), and given by

$$H = \frac{L}{N} \tag{11}$$

where L is the column length, usually expressed in millimetres or centimetres.

NB N is a **dimensionless number** and to ensure correct computation, the units of t_R and W_b or $W_{h/2}$ **must** be the same. Most chromatography data systems have software to perform the calculations. Plate numbers for solutes separated by **GC** and **HPLC** are of the order of several thousands to several hundreds of thousands.

Columns giving very high plate numbers are capable of separating multicomponent mixtures, but it is their **resolving power**, as measured by the **resolution**, R_S, that is of prime importance. This is defined as the difference between the retention times of two adjacent solute peaks, Δt_R, divided by their average base-widths, $(W_1 + W_2)/2$ (*Fig. 5*)

$$R_S = \frac{2\Delta t_R}{(W_1 + W_2)} \tag{12}$$

As in the case of the efficiency formulae, Gaussian peaks are assumed and all measurements **must** be in the same units as R_S is a dimensionless number. Values of R_S approaching or exceeding **1.5**, which is defined as **baseline resolution**, are deemed satisfactory for most purposes.

The following is an example of the calculation of efficiency and resolution using the above formulae:

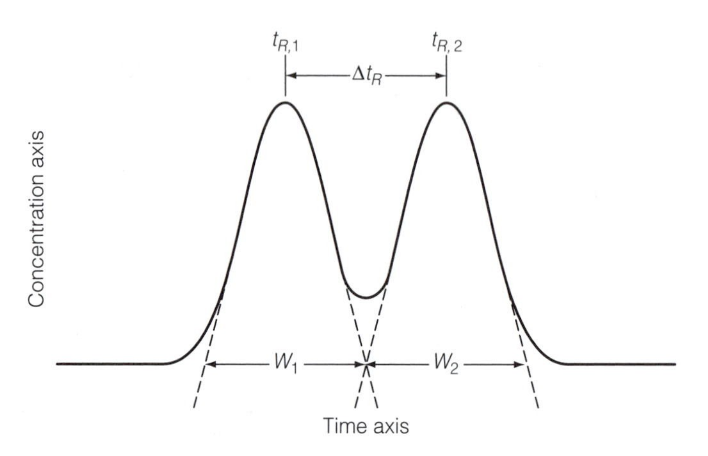

Fig. 5. Measurement of the resolution of two adjacent Gaussian peaks.

Example. A gas chromatographic method for the separation of a mixture of cyclohexane, *t*-butanol and benzene on a 10 m capillary column gave the following data:

Parameter	Cyclohexane	*t*-Butanol	Benzene
t_R	3 m 20 s	3 m 30 s	3 m 45 s
W_b	8 s	9 s	11 s
$W_{h/2}$	4.6 s	5.1 s	6.2 s

Calculate (a) the plate numbers, using both plate number formulae, for each solute, (b) the plate heights, and (c) the resolution between adjacent pairs of solutes.

(a) plate numbers $N = 16\left(\frac{t_R}{W_b}\right)^2$ $N = 5.54\left(\frac{t_R}{W_{h/2}}\right)^2$

cyclohexane $N = 16\left(\frac{200}{8}\right)^2 = \mathbf{10\,000}$ $N = 5.54\left(\frac{200}{4.6}\right)^2 = \mathbf{10\,473}$

t-butanol $N = 16\left(\frac{210}{9}\right)^2 = \mathbf{8711}$ $N = 5.54\left(\frac{210}{5.1}\right)^2 = \mathbf{9393}$

benzene $N = 16\left(\frac{225}{11}\right)^2 = \mathbf{6694}$ $N = 5.54\left(\frac{225}{6.2}\right)^2 = \mathbf{7296}$

(b) plate heights $H = \frac{10\,000}{N}$

cyclohexane $H = \mathbf{1.0\ mm}$ $H = \mathbf{0.95\ mm}$

t-butanol $H = \mathbf{1.15\ mm}$ $H = \mathbf{1.06\ mm}$

benzene $H = \mathbf{1.49\ mm}$ $H = \mathbf{1.37\ mm}$

(c) resolution $R_S = \frac{2\Delta t_R}{(W_1 + W_2)}$

cyclohexane/*t*-butanol $R_S = \frac{2\Delta t_R}{(W_1 + W_2)} = 20/17 = \mathbf{1.2}$

t-butanol/benzene $$R_S = \frac{2\Delta t_R}{(W_1 + W_2)} = 30/20 = \mathbf{1.5}$$

Note that (i) The plate numbers are slightly higher when half-height peak widths are used to calculate them due to peak tailing increasing W_b, hence comparisons of efficiencies are valid only if the same formula is used throughout.

(ii) The cyclohexane and t-butanol peaks are not fully resolved (R_S = **1.2**), but the t-butanol and benzene peaks have baseline resolution (R_S = **1.5**).

Qualitative and quantitative analysis

There are three approaches to **qualitative chromatographic analysis**, *viz*

- **Comparison of retention data** for unknown solutes with corresponding data for standards (known substances) obtained under identical conditions.
  For planar chromatography (**PC** and **TLC**), retardation factors, R_f values, for standards and unknowns are compared by chromatographing them simultaneously so as to eliminate variations in laboratory materials and conditions. For column separations, retention times, t_R, or volumes, V_R, are compared by chromatographing standards and unknowns sequentially under stable conditions with as little time between runs as possible.
- **Spiking** samples with known solutes.
  For column separations where samples are known to contain certain solutes, a comparison is made between two or more chromatograms run under identical conditions. The first is of the original sample, and subsequent ones are obtained after adding a **spike** of one of the known solutes. Any peak in the original chromatogram that is of increased size in a subsequent one can then be identified as the corresponding spiked solute. However, unambiguous identifications may not be possible.
- **Interfacing** the chromatograph with a spectrometer (Section F).
  For column separations, this provides spectral information for each separated solute in addition to retention data. Spectra of unknown solutes can be compared with those in computerized library databases or interpreted manually, even when pure standards are not available.

Comparisons of retention data alone are not always reliable as many substances can have identical chromatographic behavior. Two or more comparisons made under different chromatographic conditions, e.g. different stationary or mobile phases, reduce the chances of making an incorrect identification.

For **quantitative chromatographic analysis**, in addition to ensuring stable and reproducible conditions for sample preparation and chromatography, further specific requirements must also be met *viz*:

- the analyte (solute) must be identified and completely separated from other components in the chromatogram;
- standards of known purity must be available;
- a recognized calibration procedure must be used.

For planar chromatography, solute **spot areas** or **densities** can be measured *in situ*, or the solute spots can be removed, dissolved and measurements made by another analytical technique such as **UV spectrometry** (Topic E9).

For column separations, quantitation can be by **peak area**, **peak height or**

peak height × retention time. Integrated peak areas are directly proportional to the amount of analyte chromatographed when working within the linear range of the detector, and are the most reliable. They can be measured by computing-integrators, by triangulation (1/2-base × height of the triangle that approximates to the Gaussian peak profile) or by cutting out and weighing peaks drawn by a chart recorder. Detector responses for an analyte are established by the preparation of a **calibration graph** using chromatographed standards, with or without the addition of an **internal standard**, by **standard addition** or by **internal normalization**. These calibration procedures are described in Topics A5 and B4.

D3 THIN-LAYER CHROMATOGRAPHY

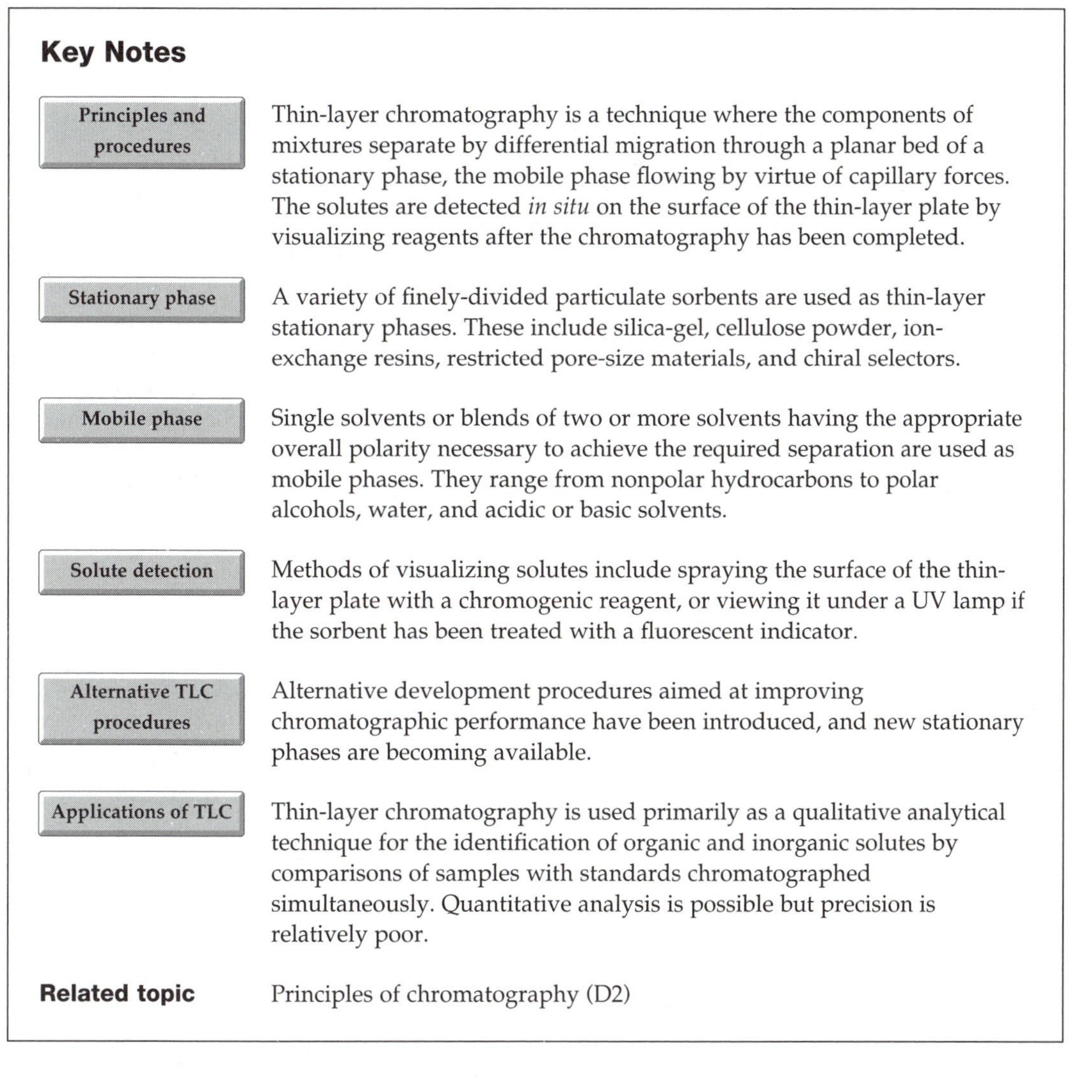

Key Notes

Principles and procedures	Thin-layer chromatography is a technique where the components of mixtures separate by differential migration through a planar bed of a stationary phase, the mobile phase flowing by virtue of capillary forces. The solutes are detected *in situ* on the surface of the thin-layer plate by visualizing reagents after the chromatography has been completed.
Stationary phase	A variety of finely-divided particulate sorbents are used as thin-layer stationary phases. These include silica-gel, cellulose powder, ion-exchange resins, restricted pore-size materials, and chiral selectors.
Mobile phase	Single solvents or blends of two or more solvents having the appropriate overall polarity necessary to achieve the required separation are used as mobile phases. They range from nonpolar hydrocarbons to polar alcohols, water, and acidic or basic solvents.
Solute detection	Methods of visualizing solutes include spraying the surface of the thin-layer plate with a chromogenic reagent, or viewing it under a UV lamp if the sorbent has been treated with a fluorescent indicator.
Alternative TLC procedures	Alternative development procedures aimed at improving chromatographic performance have been introduced, and new stationary phases are becoming available.
Applications of TLC	Thin-layer chromatography is used primarily as a qualitative analytical technique for the identification of organic and inorganic solutes by comparisons of samples with standards chromatographed simultaneously. Quantitative analysis is possible but precision is relatively poor.
Related topic	Principles of chromatography (D2)

Principles and procedures

Thin-layer chromatography is a form of planar chromatography similar to paper chromatography, but the stationary phase is a finely-divided sorbent spread as a thin layer on a supporting flat plastic, aluminum or glass plate. Solutes migrate through the stationary phase at rates determined by their **distribution ratios** (Topic D2), those with the largest values moving the least, if at all, whilst those with the smallest values moving with the advancing mobile phase, or **solvent front**. A typical **TLC** procedure consists of the following steps:

- sufficient mobile phase to provide about a 0.5 cm depth of liquid is poured into a **development tank,** or chamber, which is then covered and allowed to stand for several minutes to allow the atmosphere in the tank to become saturated with the solvent vapor;

- small volumes of liquid samples and standards, or solutions, are **spotted** onto the sorbent surface of a **TLC** plate along a line close to and parallel with one edge (the **origin**). The plate is then positioned in the tank with this edge in contact with the mobile phase and the cover replaced (*Fig. 1(a)*);
- the mobile phase is drawn through the bed of sorbent from the edge of the plate, principally by **capillary action**, and this **development** process is haited shortly before the **solvent front** reaches the opposite side of the plate. Sample components and standards migrate in parallel paths in the direction of flow of the mobile phase, separating into discrete zones or spots;
- the plate is removed from the development tank, dried in a current of warm air, and solute spots located by appropriate methods (*vide infra*);
- each solute is characterized by the distance migrated relative to the solvent front, i.e. its R_f value, which will lie between 0 and 1 (Topic D2), and unknowns are identified by comparisons with standards run simultaneously.

Figure 1(b) illustrates a developed and visualized **TLC** plate with R_f values shown alongside. Note that the shapes of some spots have become slightly elongated in the direction of flow of the mobile phase, which is an example of **tailing** (Topic D2). This is caused by slow desorption as the solute migrates, or saturation of adsorption sites by high concentrations of the solute leading to a **convex sorption isotherm**, and is most likely to occur where **adsorption** is the principal chromatographic **sorption mechanism** (Topic D2).

Stationary phase

Stationary phases used in TLC are **microparticulate sorbents** with particle diameters of between 10 and 30 μm. The smaller the mean particle size and the narrower the size range, the better the chromatographic performance in terms of **band spreading (efficiency)** and **resolution** (Topic D2). **Thin-layer chromatography plates** are prepared by coating sorbents onto rectangular plastic, aluminum or glass sheets in adherent uniform layers approximately 250 μm

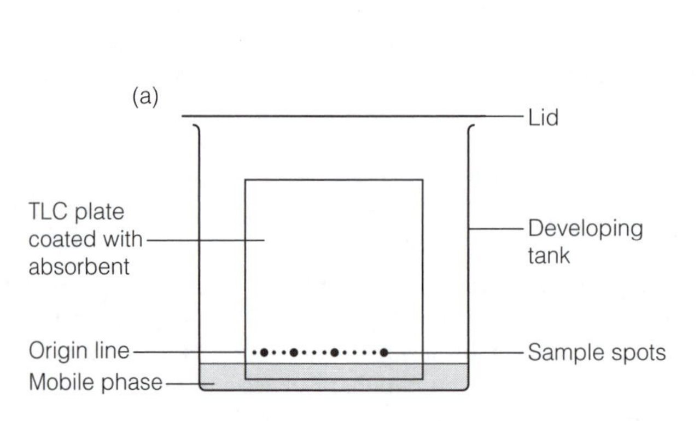

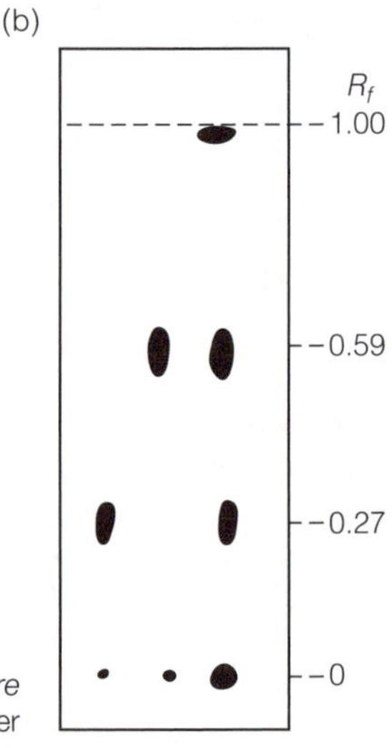

Fig. 1. TLC plates (a) during and (b) after development and visualization; R_f *values are shown alongside. a, Reproduced from R.J. Hamilton & S. Hamilton,* Thin-Layer Chromatography, *1987. b, Reproduced from R.M. Smith,* Gas and Liquid Chromatography in Analytical Chemistry, *1988. © John Wiley & Sons Ltd. Reproduced with permission.*

thick. Commercially produced plates are available in several sizes between 5 cm and 20 cm square and may incorporate an insoluble fluorescent reagent to facilitate the detection of solute spots (*vide infra*). The most commonly used sorbents are **silica** and **powdered cellulose**, and the corresponding sorption mechanisms are adsorption and partition, respectively. Thin layers can also be made of chemically-modified silicas, ion-exchange resins, exclusion gels and cyclodextrins that display chiral selectivity. Some of these sorbents are similar to the bonded phases used in **HPLC** which are discussed in Topics D6 and D7. Some TLC sorbents are listed in *Table 1*.

Table 1. Stationary phases (sorbents) for thin-layer chromatography

Sorbents	Chromatographic mechanism	Typical applications
Silica gels	Adsorption	Amino acids, hydrocarbons, alkaloids, vitamins
Hydrocarbon modified silicas	Modified partition	Nonpolar compounds
Cellulose powder	Partition	Amino acids, nucleotides, carbohydrates
Alumina	Adsorption	Hydrocarbons, alkaloids, food dyes, lipids, metal ions
Kieselguhrs (diatomaceous earths)	Partition	Sugars, fatty acids
Ion-exchange celluloses	Ion-exchange	Nucleic acids, nucleotides, halide and metal ions
Sephadex gels	Exclusion	Polymers, proteins, metal complexes
β-Cyclodextrins	Stereo-adsorptive interactions	Mixtures of enantiomers

Mobile phase

The range of mobile phases used in **TLC** is extremely wide and they are often selected empirically. Blends of two solvents are common because the **solvent strength**, or **eluting power**, can be easily adjusted to optimize a separation by altering solute distribution ratios. Some general guidelines in selecting and optimizing the composition of a mobile phase are

- Solvents should be of the highest purity as **TLC** is a very sensitive analytical technique;
- Mobile phase eluting power should be adjusted so that solute R_f values fall between 0.2 and 0.8 so as to maximize resolution;
- For separations on silica gel and other polar adsorbents, the overall polarity of the mobile phase determines solute migration rates and hence their R_f values; small additions of a slightly polar solvent, such as diethyl ether, to a nonpolar solvent, such as methylbenzene, will significantly increase R_f values;
- Polar and ionic solutes are best separated using a blend of a polar organic solvent, such as *n*-butanol, with water; the addition of small amounts of ethanoic acid or ammonia to the water increases the solubilities of basic and acidic solutes, respectively.

A helpful guide to solvent strength in adsorption and partition-based separations is an **eluotropic series**, an example of which is given for **HPLC** in *Table 1* of Topic D6. The solvents are listed in order of increasing solvent strength for adsorption-based separations. The order for partition-based separations is broadly similar.

Solute detection

Solutes separated by **TLC** remain on the surface of the plate after development. As the majority of solutes are colorless, their spots must be **located** using a chemical or physical means of **visualization**. Chemical methods include:

- Spraying the plate with a **locating**, or **chromogenic** reagent that reacts chemically with all solutes or those containing specific functional groups or structural features to give colored spots. The plates sometimes need to be warmed to accelerate the chromogenic reaction and intensify the spots. Examples are given in *Table 2*;
- Viewing the plate under a **UV lamp** set at an emission wavelength of 254 nm or 370 nm to reveal the solutes as dark spots or bright fluorescent spots on a uniform background fluorescence. Commercial plates can be purchased with an insoluble fluorescent substance incorporated into the stationary phase to provide the background fluorescence, or the plate can be sprayed with a fluorescent reagent after development;
- Spraying the plate with concentrated sulphuric or nitric acid and heating to oxidize and char organic solutes which are revealed as brown to black spots;
- Exposing the plate to iodine vapor in a sealed chamber when many organic solutes develop dark brown colorations;
- Scanning across the surface of the plate with a **densitometer**, an instrument that measures the intensity of radiation reflected from the surface when irradiated with a UV or visible lamp. Solutes that absorb radiation are recorded as peaks on a recorder trace or VDU screen;
- Radiolabelled solutes can be detected by autoradiography (blackening of a photographic film sensitive to X-rays), liquid scintillation counting on scraped-off areas of the stationary phase, or monitoring the surface of the plate with a Geiger-Müller tube.

Table 2. Some examples of TLC locating reagents

Method of detection	Color of solute spots	Application
General reagents		
Phosphomolybdic acid + heat	Dark blue	Many organics
Conc. sulphuric acid + heat	Charred brown-black	All organics
Iodine vapor	Brown	Many organics
Selective reagents		
Ninhydrin	Pink to purple	Amino acids and amines
2,4-Dinitrophenylhydrazone	Orange/red	Carbonyl compounds
Bromocresol green/ blue	Yellow	Organic acids
2,7-Fluorescein	Yellow-green	Most organics
Vanillin/ sulphuric acid	Blue, green, pink	Alcohols, ketones
Rhodamine-B	Red fluorescence	Lipids
Anisaldehyde/antimony trichloride	Various	Steroids
Diphenylamine/zinc	Various	Pesticides

Alternative TLC procedures

Variations of the basic development procedure and new stationary phases aimed at improving resolution, sensitivity, speed, reproducibility and selectivity have been developed from time to time. Two of the more significant and useful ones are:

- **Two-dimensional TLC.** This is a means of improving resolution in samples where the component solutes have similar chemical characteristics and hence

R_f values, e.g. amino acids. A single sample is spotted close to one corner of the plate and, after development, there is a partial separation of the solutes along one edge. The plate is dried, turned through 90° and developed a second time, but using a different mobile phase composition for which the solutes have different distribution ratios compared to those for the first mobile phase. This results in the partially-separated solutes separating further because of changes to their migration rates. Visualization gives a 2-D **map** or **fingerprint** of the sample components. The procedure for the separation of 14 amino acids is shown diagrammatically in *Figure 2*.

- **High-performance plates (HPTLC).** These are coated with a thinner layer (100 μm thick) of a 5 μm particle diameter stationary phase with a very narrow range of sizes. Sensitivity and resolution are improved because solute spots are compact, development is much faster, partly because smaller plates can be used, and solvent consumption is reduced. **HPTLC** plates with silica and bonded-phase silicas are commercially available.

Applications of TLC

TLC is applicable to a very wide range of mainly organic solutes. It is used primarily in the biochemical, pharmaceutical, clinical and forensic areas for **qualitative analysis** by the comparison of R_f values of sample solutes with those of standards run on the same plate. It is especially useful for checks on purity, to monitor the course of reactions and production processes, and to characterize complex materials such as natural products. The screening of samples for drugs

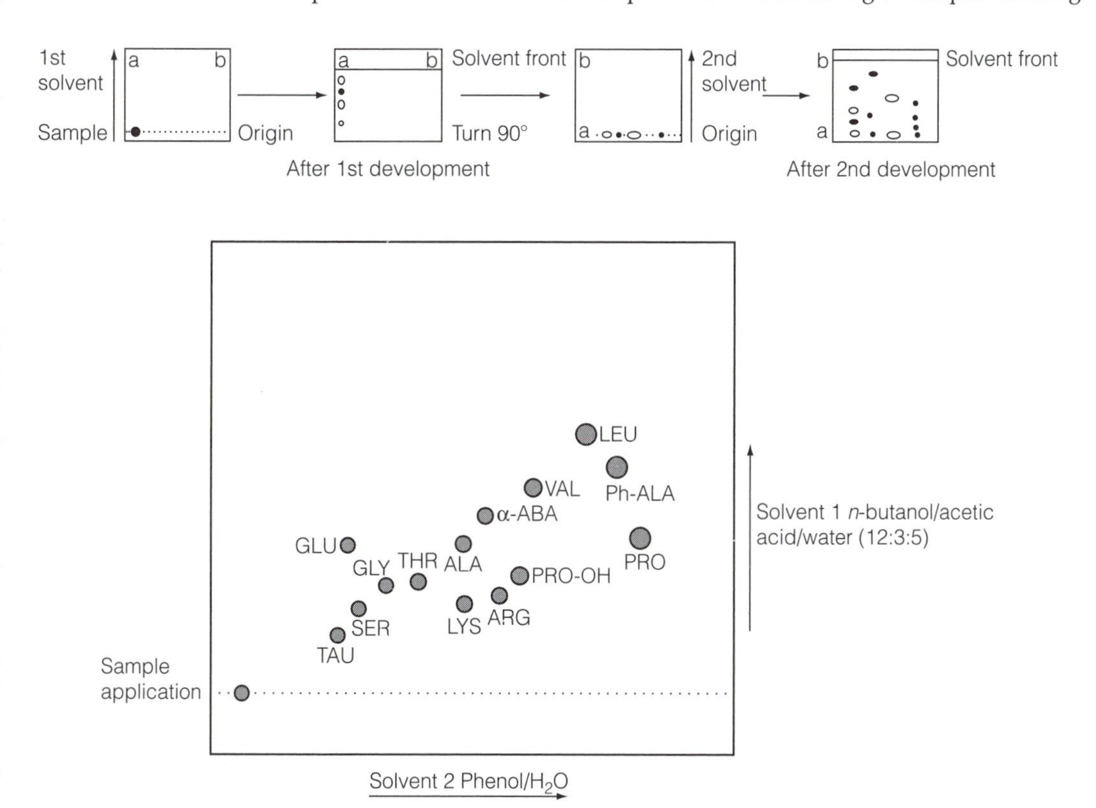

Fig. 2. Two-dimensional TLC of a mixture of 14 amino acids. Top panel, reproduced from R.J. Hamilton & S. Hamilton, Thin-Layer Chromatography, *1987. © John Wiley & Sons Ltd. Reproduced with permission. Bottom panel, reproduced from A. Braithwaite & F.J. Smith,* Chromatographic Methods, 5th edn, *1996, first published by Blackie Academic & Professional.*

in clinical and forensic studies, and testing for the presence or absence of specific substances (limit tests) are additional applications. *Tables 1 and 2* include further examples.

Quantitative TLC by the measurement of spot areas or by computerized scanning reflectance densitometry (Topic E8) is possible, but, unless **HPTLC** plates are used, the relative precision attainable is generally only 5–10%. The principal source of error is in applying sample spots to the plate, although automated systems can reduce this.

TLC has a number of advantages over **GC** and **HPLC**:

- the ability to run 10–20 or more samples simultaneously for immediate and direct comparison with standards, which represents a considerable saving in time;
- the basic technique is very cheap, versatile and quick;
- all solutes, including those that do not migrate from the origin, are detectable.

Disadvantages are the limited reproducibility of R_f values due primarily to changes in the composition of the mobile phase during development of the plate, and increasing spot diffusion (band broadening) because the flow rate of the mobile phase slows as it travels across the plate.

D4 Gas chromatography: principles and instrumentation

Key Notes

Principles — Gas chromatography is a technique for the separation of volatile components of mixtures by differential migration through a column containing a liquid or solid stationary phase. Solutes are transported through the column by a gaseous mobile phase and are detected as they are eluted.

Mobile phase — The mobile phase is an inert gas, generally nitrogen or helium, supplied from a cylinder via pressure and flow controls, and passing through purification cartridges before entering the column.

Sample injection — Gaseous, liquid and solid samples are introduced into the flowing mobile phase at the top of the column through an injection port using a microsyringe, valve or other device.

Column and stationary phase — Columns are either long, narrow, capillary tubes with the stationary phase coated onto the inside wall, or shorter, larger diameter tubes packed with a particulate stationary phase. Stationary phases are high-boiling liquids, waxes or solid sorbents.

Temperature control — The column is enclosed in a thermostatically-controlled oven that is maintained at a steady temperature or programmed to increase progressively during a separation.

Solute detection — Solutes are detected in the mobile phase as they are eluted from the end of the column. The detector generates an electrical signal that can be amplified and presented in the form of a chromatogram of solute concentration as a function of time.

Instrument control and data processing — A dedicated microcomputer is an integral part of a modern gas chromatograph. Software packages facilitate the control and monitoring of instrumental parameters, and the display and processing of data.

Related topics Principles of chromatography (D2) Gas chromatography: procedures and applications (D5)

Principles

Gas chromatography (GC) is a separation technique where **volatile, thermally stable** solutes migrate through a column containing a stationary phase at rates dependent on their distribution ratios (Topic D2). These are inversely proportional to their volatilities, which in turn are determined by their **partial vapor**

pressures and hence their **boiling points**. Solutes are therefore generally eluted in order of increasing boiling point, except where there are specific interactions with the stationary phase. The gaseous mobile phase elutes the solutes from the end of the column where they pass through a detector that responds to each one. An elevated temperature, usually in the range 50–350°C, is normally employed to ensure that the solutes have adequate volatility and are therefore eluted reasonably quickly.

There are two modes of gas chromatography:

- **Gas-liquid chromatography** (**GLC**), which employs a liquid stationary phase in which solutes can dissolve, the sorption process being **partition**. Specific interactions of solutes with the stationary phase may alter the order of elution from that of increasing boiling points. GLC is by far and away the more widely used mode of GC, the large number of alternative stationary phases enabling many types of sample to be analyzed.
- **Gas-solid chromatography** (**GSC**) employs a solid, sometimes polymeric, sorbent as the stationary phase, the sorption process being **surface adsorption**. GSC has limited specialist applications, being used mainly for analyzing mixtures of gases or solvents with relatively low relative molecular masses.

A schematic diagram of a **gas chromatograph** is shown in *Figure 1*. It consists of five major components:

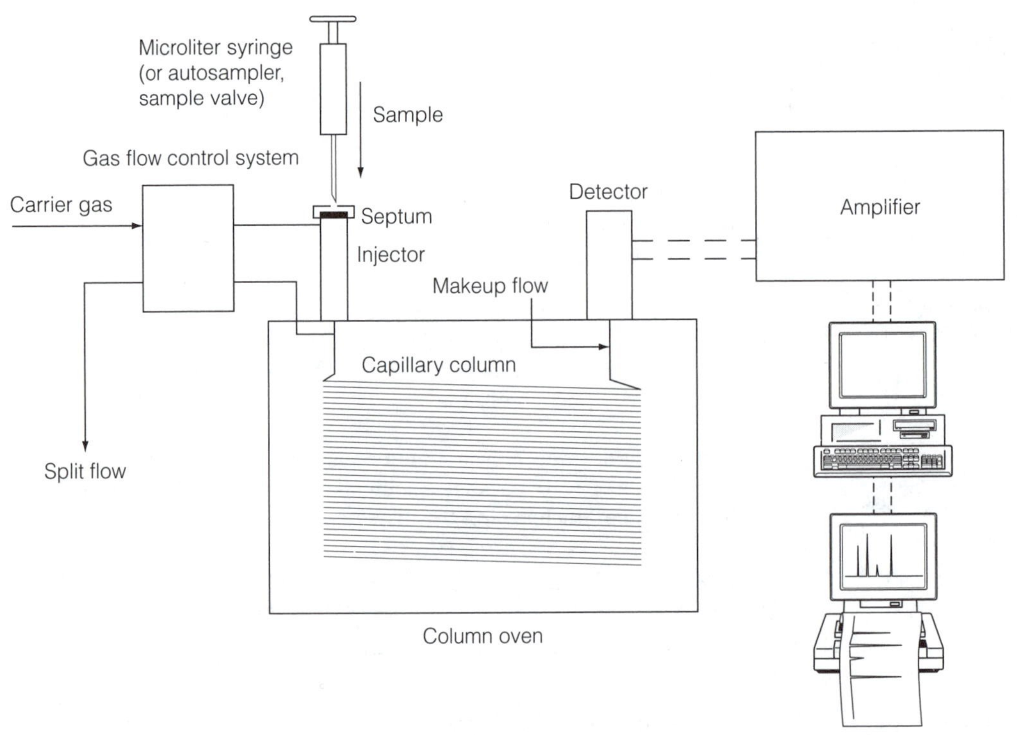

Fig. 1. Schematic diagram of a capillary column gas chromatograph. Reproduced from D.W. Grant, Capillary Gas Chromatography, *1996. © John Wiley & Sons Ltd. Reproduced with permission.*

- gas supply and controls;
- sample injection port;
- column housed in a thermostatically-controlled oven;
- detection and recording system;
- microcomputer with control and data processing software.

These are described in the following sections.

Mobile phase

The mobile phase is known as the **carrier-gas** because its sole purpose is to transport solutes through the column, thus not contributing to chromatographic selectivity. It should be inert, non toxic, non flammable and preferably cheap.

Helium and **nitrogen** are invariably used routinely, the former with **capillary** (**open tubular**) columns and the latter with **packed** columns (*vide infra*). Helium gives better chromatographic efficiency (reduced band broadening) due to faster mass transfer (Topic D2). The carrier gas must be purified by passing it through suitable adsorbents so as to avoid undesirable chemical changes to sample components and stationary phases, or adverse effects on detector performance. The most common contaminants and the means of removing them are:

- **Air or oxygen** at levels above about 10 ppm, which can oxidize sample components and liquid stationary phases, especially at high column temperatures. These can be removed by a cartridge containing **molecular sieve**.
- **Hydrocarbons**, which affect detector performance by contamination or producing a large and constant background signal. These can be removed by a cartridge containing **activated carbon**.
- **Water vapor**, which can affect some solid and bonded-liquid stationary phases, and the performance of some detectors. This can also be removed with **molecular sieve**.

The carrier gas is supplied from a cylinder via a **pressure-reducing valve** at 10–45 psi (0.7–3 bar) which provides flow rates of between 1 and 50 cm^3 min^{-1} depending on the type of column in use. A **mass-flow controller** ensures constant flow rates regardless of back-pressure and temperature (the viscosity of gases increases with temperature).

Sample injection

Samples should preferably be injected into the flowing mobile phase rapidly so as to occupy the smallest possible volume when vaporized. This ensures a narrow initial sample band that maximizes column efficiency and resolution.

There are a number of methods of injection and designs of injection port available, and the choice is determined by the type of column in use and the nature of the sample. Small volumes of liquids or solutions (0.1–10 μl) are generally injected into a heated injection port, through which the carrier gas continuously flows, from a calibrated microsyringe used to pierce a self-sealing silicone-rubber septum. Gases are introduced via a gas sampling valve or gas syringe, and solids with volatile components as solutions. Septa must be replaced regularly to avoid leakage and can be a source of contamination by previously injected samples or the bleeding of plasticizers into the gas stream, especially when operating at very high temperatures. A separate septum-purge gas stream vented to the atmosphere, or septumless valves can overcome the latter problem. Capillary (open tubular) columns require specially designed injection ports to prevent overloading them with samples, which can severely impair efficiency and resolution.

Alternative methods of sample injection are summarized below and in *Table 1,* with schematic diagrams in *Figures 2* and *3*.

Table 1. Gas chromatography sample injection systems

Type of column	Injection system	Method of injection	Advantages or disadvantages
Capillary	Split	Carrier-gas split 1:10 to 1:500 to reduce the amount of sample entering the column by 90% or more	Overloading of column avoided, but sensitivity reduced and may result in discrimination between solutes with different boiling points
	Splitless	Whole of injected sample condensed on cooled top of column, then released by heating	Increased sensitivity, but limited to low levels of solutes in sample, and may broaden bands
	On-column	Sample condensed in cooled zone at top of column, then volatilized by programmed heating	Increased sensitivity, minimal thermal degradation of solutes, no discrimination effects
Packed	Flash-vaporization	Sample injected into zone heated to 20–50°C above column temperature	Rapid volatilization of sample, but thermal degradation of some solutes may occur
	On-column	Sample injected onto top of packed bed	Increased sensitivity, minimal thermal degradation of solutes, no discrimination effects

- **Split injection** is used with capillary columns to prevent overloading the stationary phase with sample. The **split-point** is a centrally positioned hollow needle that allows a small part of the injected sample (2% or less) to reach the column and vents the remainder to the atmosphere via a control valve (*Fig. 2*). Sensitivity is less than with splitless injection because of the

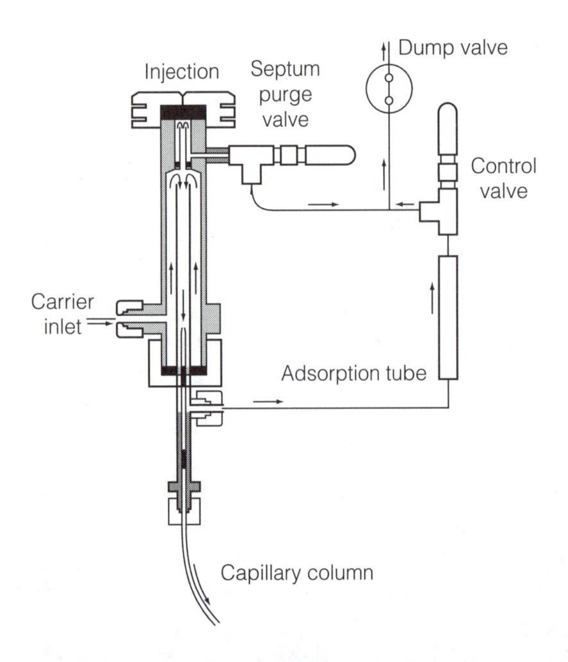

Fig. 2. Schematic diagram of a split/splitless injection port. Reproduced from A. Braithwaite & F.J. Smith, Chromatographic Methods, 5th edn, *1996, first published by Blackie Academic & Professional.*

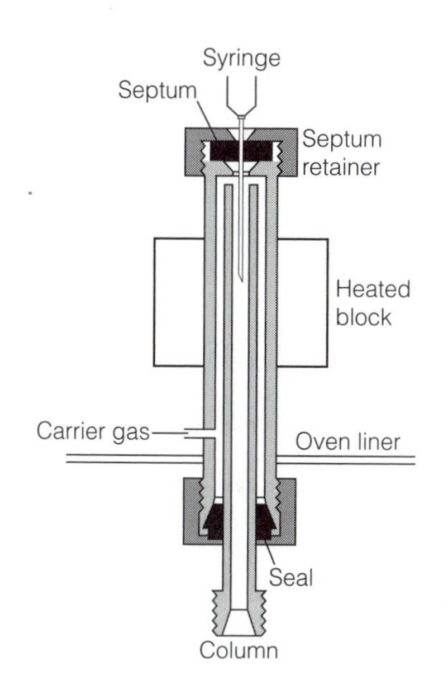

Fig. 3. Schematic diagram of a flash vaporization injection port. Reproduced from Instrumental Methods of Analysis, *2nd edn, by H.H. Willard, L.L. Merritt, J.A. Dean, F.A. Settle © 1988. Reprinted with permission of Brooks/Cole, an imprint of the Wadsworth Group, a division of Thomson Learning.*

very small fraction of injected sample reaching the column. A slow purge stream of gas prevents substances bleeding from the septum accumulating in the injection port.

- **Splitless injection** is used with capillary columns when samples contain low levels of some components, and maximum sensitivity is required. The **control valve** is kept shut as a 0.5–5 μl volume of sample in a volatile solvent is injected. By cooling the top of the column to just above the boiling point of the solvent, the sample components are trapped while the solvent travels on down the column. The control valve is then opened to purge any remaining sample vapors from the injection port, and the column temperature is raised, releasing the solutes into the gas stream.
- **On-column injection** is used with both capillary and packed columns to minimize the possibility of the decomposition of thermally labile solutes as well as providing increased sensitivity. It is also ideal for quantitative analysis.
- **Flash-vaporization** is primarily used with packed columns and is a means of rapidly volatilizing samples in a zone heated 20–50°C above the column temperature to provide as narrow a band of vapor as possible (*Fig. 3*). However, there is a risk of thermally degrading labile solutes, although a glass liner inserted into the metal block in which the samples are injected minimizes the risk.
- **Automated sample injection** is advantageous for improving precision and for processing large numbers of samples loaded into autosampler trays. Most injection ports can be adapted for this purpose, which is often under computer control.
- **Special injection techniques** are employed for **headspace analysis, pyrolysis gas chromatography** and **thermal desorption** to concentrate samples (Topic D5).

Column and stationary phase

The column is where the separation process occurs and it is, therefore, the central component of a gas chromatograph. There are two types of GC column, and a comparative summary is given in *Table 2*.

Table 2. A comparison of capillary and packed GC columns

Capillary columns (open tubular)	Packed columns
Tubing Fused quartz (SiO_2) Very high purity (<1 ppm metals) Length 10–100 m, coiled Internal diameter 0.1–0.7 mm External protective coating of a polyimide or aluminum	*Tubing* Stainless steel or glass Length 1–3 m, coiled Internal diameter 2–3 mm *Packing* Granular (0.125–0.25 mm), inert, silaceous solid support for liquid stationary phase or porous solid adsorbent
Stationary phase Very thin liquid layer (0.1–5 μm) Coated or chemically bonded to inside wall (WCOT) For GLC, bonded phases reduce column-bleed considerably Very thin finely-divided porous solid (PLOT) for GSC	*Stationary phase* Thin, 1–10 percent w/w coating of liquid for GLC; must cover solid support completely Porous solid adsorbent or polymer for GSC
Injection systems Split, splitless or on-column	*Injection systems* Flash-vaporization or on-column
Sample capacity Narrow bore <<0.1 μl Megabore 0.1 to 10 μl	*Sample capacity* 0.1–20 μl
Preferred carrier gas Helium or hydrogen	*Preferred carrier gas* Nitrogen
Performance (Fig. 4) Very high efficiencies and resolving power for complex mixtures of up to 100 or more components, especially for narrow bore columns, but sample capacities limited; solutes elute at lower temperatures than with corresponding packed column	*Performance (Fig. 4)* Limited efficiency and resolving power for up to about 20 components
Price and source Expensive, from specialist supplier	*Price and source* Inexpensive, can be packed in the laboratory

- **Capillary** (**open tubular**) columns have become the most widely-used in recent years. They consist of long, narrow bore, high-purity quartz tubing with a very thin layer of a liquid or solid stationary phase (*vide infra*) coated or chemically bonded to the surface of the inner wall. This minimizes band spreading because of rapid **mass transfer** (Topic D2). Also, as there is an unrestricted flow of carrier gas through the center of the column generating little back-pressure, very long lengths (up to 100 m) can be used, resulting in extremely high efficiencies and resolving power for the separation of complex mixtures.

  Capillary columns are available with a range of internal diameters and thicknesses of stationary phase. The narrower the bore and the thinner the coating, the greater the efficiency, but the lower the sample capacity before overloading causes peaks to tail and resolution to deteriorate. Special injection systems are therefore required, as described in the previous section, to reduce the amount of sample injected with a conventional microsyringe. The widest bore columns (**megabore**), which exceed 0.5 mm internal diameter, are the least efficient, but have the highest sample capacities, so special injection techniques are not needed.

- **Packed columns** are much shorter than capillary columns, rarely exceeding 2 m, the length being limited by the back-pressure generated by the gas flowing through a packed bed. The stainless steel or glass tubing has an internal diameter of 2 or 3 mm and is filled with a granular material that acts as a **solid support** for a thin coating of a liquid stationary phase for GLC, or as an adsorbent for GSC. Solid supports are inert, porous silaceous materials such as diatomaceous earths (kieselguhrs) with a large surface area. Their particle sizes vary between 0.125 mm (US sieve mesh 120) and 0.25 mm (US sieve mesh 40), individual columns being packed with particles having a narrow range between these limits to improve packing characteristics and chromatographic efficiency. The smaller the **particle size**, and the thinner the coating of stationary phase, the less solute bands spread by the **multiple path** and **mass transfer effects** (Topic D2). Packed columns are much cheaper than capillary columns but their overall efficiencies and resolving power are limited. They are best suited to the separation of mixtures of up to ten or twenty components.

Examples of separations on capillary and packed columns are shown in *Figure 4*.

GLC stationary phases are thin coatings of very high boiling liquids, oils or waxes, some with a polymeric structure, e.g. polysiloxanes and polyethylene glycol. Those for **GSC** are solid adsorbents and polymers. They can be classified according to their polarity, varying from nonpolar hydrocarbons to polar polyesters, cyanopropyl silicones and alumina. There are special phases that show particular selectivities for specific types of solute such as fatty acids, bases and enantiomers, and high-temperature phases based on silicone-carborane co-polymers. Hundreds of stationary phases have been investigated, many having very similar characteristics, but most laboratories use only a few for routine work.

The most important features of stationary phases are:

- They should be nonvolatile, chemically and thermally stable over a wide temperature range, and nonreactive towards the separating solutes;
- Most liquids have a recommended operating temperature range; beneath the lower limit, peak shapes become badly distorted because the phases solidify. At temperatures close to or exceeding the upper limit, the liquid gradually **bleeds** from the column and/or degrades, which changes the chromatographic characteristics and leads to an unstable detector signal;
- Liquids **chemically bonded** to the walls of capillary columns bleed much less, and the columns can be washed through with solvents to remove strongly retained sample residues contaminating the stationary phase;
- The thinnest coatings of stationary phase give the highest efficiencies and resolving power, but the lowest sample capacities;
- The choice of stationary phase is determined by the sample; generally, they should have similar polarities otherwise peaks may be distorted, but compromise choices must be made where solutes in a mixture have a wide range of polarities;
- Elution order can be altered by changing the stationary phase where there

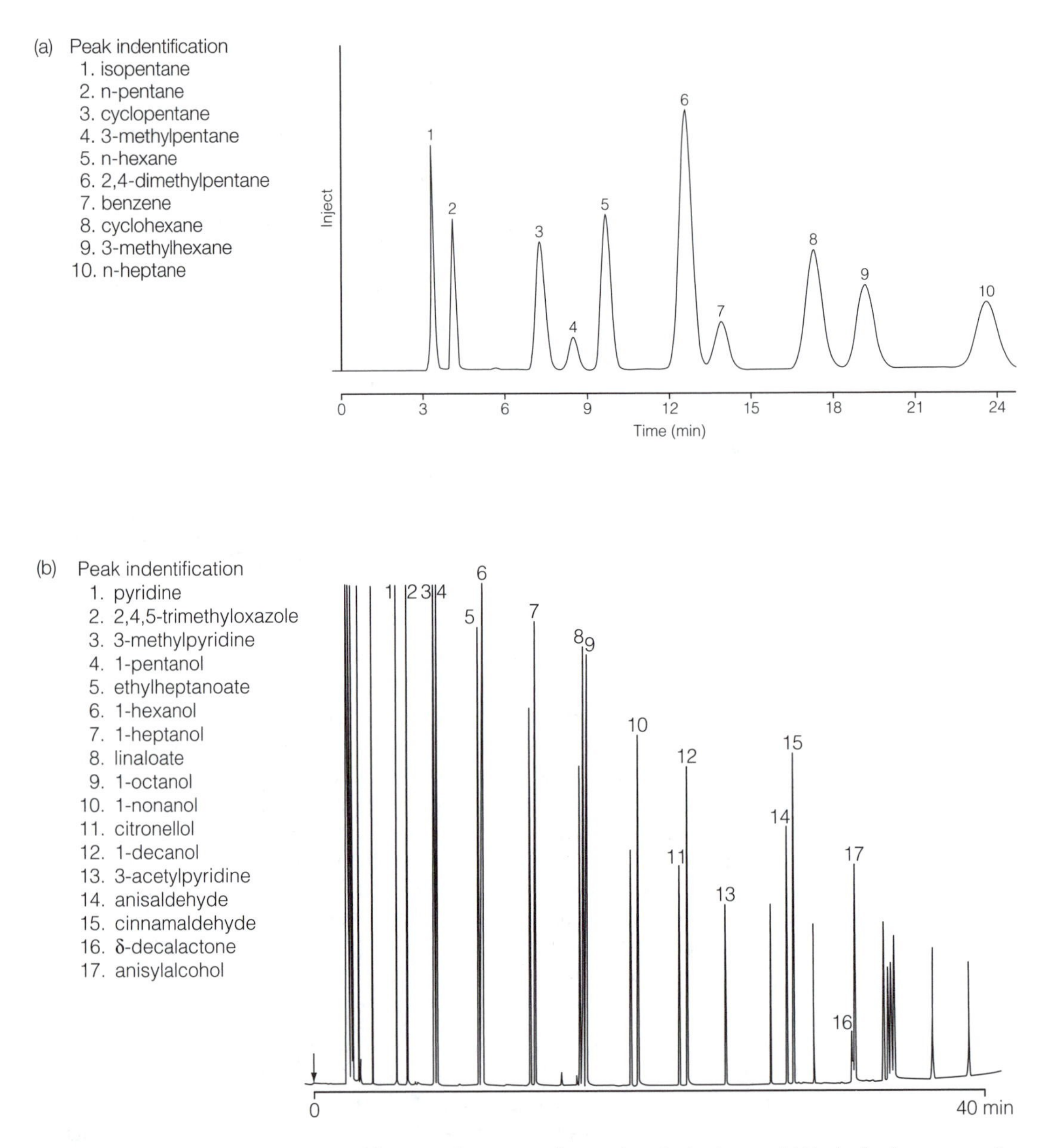

Fig. 4. Examples of gas chromatographic separations on capillary and packed columns. (a) Packed column separation of mixed hydrocarbons; (b) capillary column separation of some flavor compounds. Reproduced from D.W. Grant, Capillary Gas Chromatography, *1996. © John Wiley & Sons Ltd. Reproduced with permission.*

are specific interactions with a solute, e.g. on a nonpolar stationary phase, *t*-butyl alcohol (bp 82.6°C) elutes before cyclohexane (bp 80.8°C) because the latter, being nonpolar itself, dissolves better in the stationary phase. However, on a polar stationary phase with hydroxy groups, the elution order is reversed because the alcohol can H-bond to it.

Selections of stationary phases are given in *Tables 3 and 4*.

Table 3. A selection of stationary phases for packed columns

Stationary phase	Chemical type	Polarity
Apiezon L	Branched-chain alkane grease	Nonpolar
OV101	Dimethyl silicone	Nonpolar
OV17	50% Phenyldimethyl silicone	Medium polarity
Carbowax 20M	Polyethylene glycol	Very polar

Table 4. A selection of stationary phases for capillary columns

Stationary phase	Chemical type		Polarity	Applications
BP1	100% methyl	polysiloxanes	Nonpolar	Solvents, VOCs, petroleum products
BP5	5% phenyl 95% methyl	polysiloxanes	Nonpolar	Aromatics, PAHs, drugs, perfumes
OV1701 BP10	14% cyanopropyl 86% methyl	polysiloxanes	Medium polarity	Alcohols, phenols, esters, ketones, pesticides
DB17 RT50	50% phenyl 50% methyl	polysiloxanes	Medium polarity	Esters, ketones, plasticizers
CP-Wax DB-Wax	Polyethylene glycol		Very polar	Alcohols, esters, acids, amines, solvents

VOCs = volatile organic compounds; PAHs = polyaromatic hydrocarbons.

Temperature control

Temperature control is essential in ensuring reproducible separations by GC. The column is enclosed in an insulated and thermostatically-controlled oven with a heater and circulating fan to maintain a uniform temperature from ambient to about 400°C. For **isothermal** (constant temperature) chromatography, the selected temperature must be maintained to ±0.1°C, as solute distribution ratios are highly temperature sensitive, e.g. a 20° increase in column temperature results in about a two-fold decrease in distribution ratio and a corresponding decrease in retention time, t_R. Temperature programming is a procedure used to optimize the separation of complex mixtures (Topic D5).

Solute detection

The carrier-gas flows through a **detector** that responds to changes in a **bulk physical property**, such as its **thermal conductivity**, in the presence of a solute vapor, or to a specific property of the eluting solutes themselves, such as their ability to be **ionized**. Detectors may be **universal**, responding to practically all solutes, or **selective**, where they respond to solutes with particular characteristics, such as specific elements or structural features. Ideally, detectors should have the following characteristics:

- a rapid and reproducible reponse to the presence of solute vapors in the carrier gas;
- high sensitivity, i.e. able to detect very low levels of solutes;
- stablity in operation;
- a signal directly proportional to solute concentration or mass over a wide range (wide **linear dynamic range**).

Although many types of GC detector have been investigated, only four are in widespread use. Details of these are summarized below and in *Table 5*.

Table 5. Characteristics of GC detectors

Detector	Sensitivity ($g\ s^{-1}$)	Linear range	Characteristics
Thermal conductivity (TCD)	10^{-9}	10^4	Robust, non-destructive, flow and temperature sensitive, poor linear dynamic range, insensitive to inorganic solutes
Flame ionization (FID)	10^{-12}	10^7	Excellent sensitivity and linear dynamic range, best universal GC detector
Nitrogen-phosphorus (NPD)	10^{-14} (N) 10^{-15} (P)	10^5 10^5	Similar to FID, but selective for N and P containing solutes, limited linear dynamic range
Electron capture (ECD)	10^{-13}	10^3	Excellent sensitivity for solutes with electronegative elements, temperature sensitive, easily contaminated, limited linear dynamic range

(i) **Thermal conductivity detector (TCD).** This is one of the oldest types, and is known also as a **katharometer** or **hot-wire detector**. It is a **universal** detector consisting of a heated metal block containing a **reference cell** through which pure carrier gas constantly flows, and a **sample cell** through which carrier gas flows after emerging from the end of the column (*Fig. 5*). The cells contain identical, heated platinum filaments whose **resistances** depend on their **temperatures**, which in turn depend on the rates of heat loss from their surfaces, these being a function of the **thermal conductivity** of the surrounding gas. When pure carrier gas flows through both cells, the resistances of the two filaments are the same, and a **Wheatstone bridge circuit** into which they are both incorporated can be balanced to give a stable baseline or background signal. When a solute is eluted from the column and passes through the sample cell, its presence alters the thermal conductivity of the carrier gas. The temperature of the filament and hence its resistance changes, and an out-of-balance signal proportional to the solute concentration is created in the bridge circuit. When the solute has passed through the cell, the signal returns to the baseline value.

The **TCD** is robust and reliable, but has only moderate sensitivity and a limited dynamic range making it more suitable for qualitative than quantitative work.

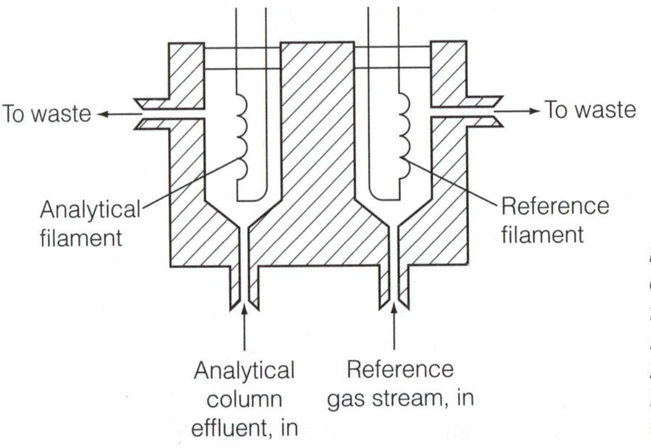

Fig. 5. Schematic diagram of a thermal conductivity detector, TCD. Reproduced from I.A. Fowlis, Gas Chromatography: Analytical Chemistry by Open Learning, *2nd edn, 1995, with permission from Her Majesty's Stationery Office. Crown Copyright.*

(ii) **Flame ionization detector (FID).** This is the most important of a group of detectors where the signal is related to the **ionization** of eluting solutes. Carrier gas emerging from the column is mixed with **air** and **hydrogen** and burnt at a small metal jet (*Fig. 6*). A 150–200 V DC potential is applied between the burner jet (positive) and a collector electrode (negative) positioned just above the micro-flame. The electrodes are connected to an external circuit where the signal can be amplified and recorded. Eluted solutes are **combusted** to yield **ions** which increase the electrical conductivity of the flame and are collected by the negative electrode, thus allowing a current proportional to the concentration of ions derived from the solute to flow around the external circuit.

The **FID** is the most widely used **universal** detector, being extremely sensitive and responding to all organic solutes except formaldehyde, formic acid and fully halogenated compounds. It has the widest linear dynamic range of all GC detectors, making it ideal for quantitative analysis, and its only disadvantage is the lack of response to inorganic solutes.

(iii) **Nitrogen-phosphorus detector (NPD).** This is a **selective** detector that is basically a flame ionization detector, modified by positioning a **ceramic bead** containing a **rubidium** or **caesium** salt, electrically heated to 800°C, between the burner jet and the collector electrode. The response to nitrogen-containing compounds is enhanced by a factor of about 50 over that of an unmodified **FID**, and the response to phosphorus-containing compounds is enhanced by a factor of about 500. The linear dynamic range is intermediate between those of the **FID** and **TCD**.

(iv) **Electron capture detector (ECD).** This is another form of ionization detector, and shows a selective response to solutes containing halogens, sulfur and unsaturated structures, all of which have high electron affinities.

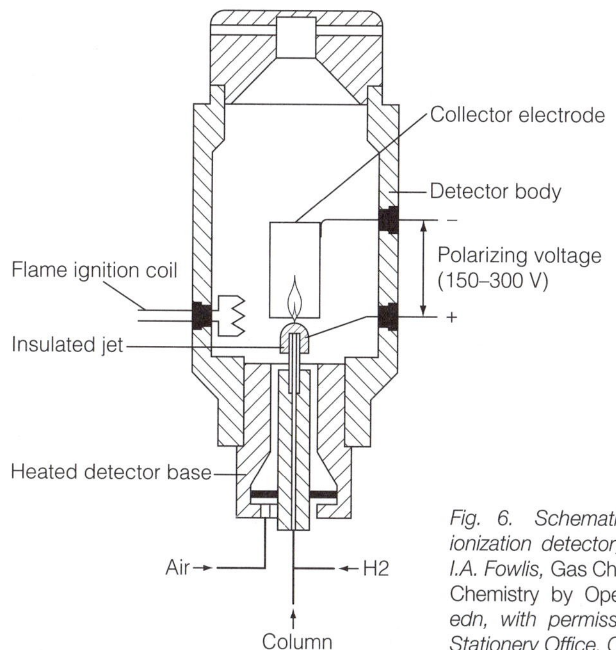

Fig. 6. Schematic diagram of a flame ionization detector, FID. Reproduced from I.A. Fowlis, Gas Chromatography: Analytical Chemistry by Open Learning, *1995, 2nd edn, with permission from Her Majesty's Stationery Office. Crown Copyright.*

The **ECD** is one of the most sensitive of GC detectors, especially if argon replaces nitrogen as the carrier gas, but traces of air, oxygen or water in the gas, liquid stationary phase bleeding from the column or residues from halogenated solvents used in sample preparation are detrimental to its performance, and its linear range is very limited. Quantitative analysis is difficult because response is dependent on solute structure.

Instrument control and data processing

Components of a modern gas chromatograph include a dedicated microcomputer with **analogue-to-digital convertor** (**ADC**) to digitize the detector signal, and software packages that perform the following functions (see also Section H):

- Facilitate the setting and monitoring of instrument parameters, i.e.:
  (i) carrier gas flow;
  (ii) oven temperature and temperature program;
  (iii) automated sample injection;
  (iv) detector gases, mode and sensitivity.
- Display of chromatograms and other information in real time with high-resolution color graphics, and electronic integration of peak areas (instrumentation may include a separate **computing integrator**).
- Recording and processing of retention and calibration data, calculations and statistical assessment of results.
- Storage and retrieval of method parameters for specific analyses.
- Diagnostic testing of the condition and performance of instrumental components.
- Communication with the database of a **laboratory information and management system** (**LIMS**) for further data-processing and archiving of results.

D5 Gas chromatography: procedures and applications

Key Notes

Temperature programming This is a form of gradient elution whereby the temperature of the column is progressively increased during a separation to optimize chromatographic performance.

Special procedures used in GC The analysis of nonvolatile materials, volatile components of complex mixtures, trace levels of solutes and multicomponent samples with very large numbers of solutes require special sampling procedures.

Qualitative analysis Unknown solutes can be identified by comparisons of retention times, spiking samples with known substances, or using retention indices. These may be ambiguous, and more reliable information can be provided by interfacing GC with a spectrometric technique.

Quantitative analysis Quantitative information is obtained from peak area measurements and calibration graphs using internal or external standards, or by standard addition or internal normalization.

Related topics Principles of chromatography (D2) Gas chromatography: principles and instrumentation (D4)

Temperature programming

The effect of column temperature on chromatographic retention is pronounced in that there is an inverse exponential relation with the distribution ratio, D, which results in a shortening of retention times as the temperature is increased. **Temperature programming**, which is a form of **gradient elution**, involves raising the oven temperature progressively during a separation to improve the resolution of mixtures where the components have a wide range of boiling points, and to shorten the overall analysis time by speeding up the elution of the higher boiling compounds. **Isothermal** (constant temperature) conditions may be unsatisfactory for the following reasons:

- if the isothermal temperature is too high, early eluting peaks may not be fully resolved;
- if the isothermal temperature is too low, later eluting peaks may have unacceptably long retention times, poor detection limits and small, broad and/or skewed peaks;
- intermediate isothermal temperatures may result in part of the chromatogram having acceptable resolution and detection limits whilst other parts do not.

The separation of a series of n-alkanes isothermally at 150°C on a packed column is shown in *Figure 1(a)*, and illustrates the first two problems. The first four alkanes, n-hexane (C6) to n-decane (C10), are incompletely resolved, C6 and C7 co-eluting, whilst later eluting peaks are smaller, broader and are **fronting** (Topic D2). In the temperature programmed chromatogram (*Fig. 1(b)*), a complete separation of all the n-alkanes up to C21 has been achieved and in about a third of the time taken to separate only up to C15 isothermally. C6–C8 are fully resolved and the later peaks are sharper and more symmetrical. For mixtures where the individual components are not, as in the case of the n-alkanes, members of a homologous series, temperature programs are often more complex. They may involve initial, intermediate and final isothermal periods separated by temperature ramps of rates varying between 2°C and 30°C per minute. The optimum conditions for a particular sample are generally established by trial and error.

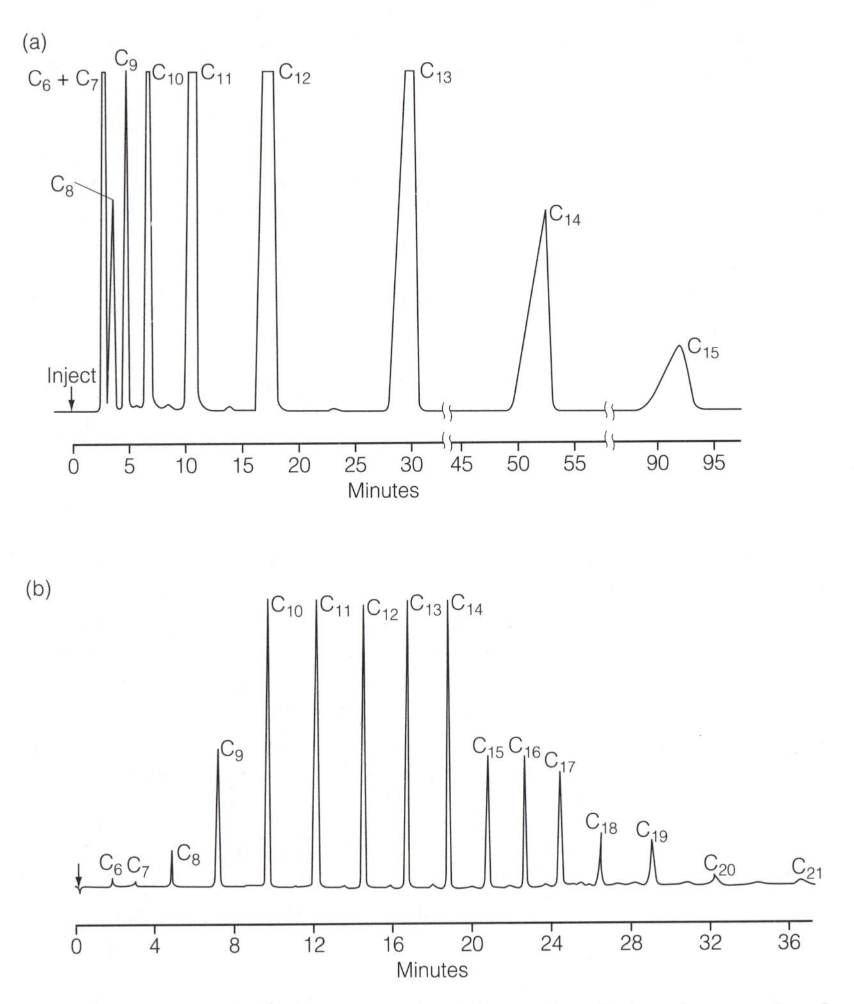

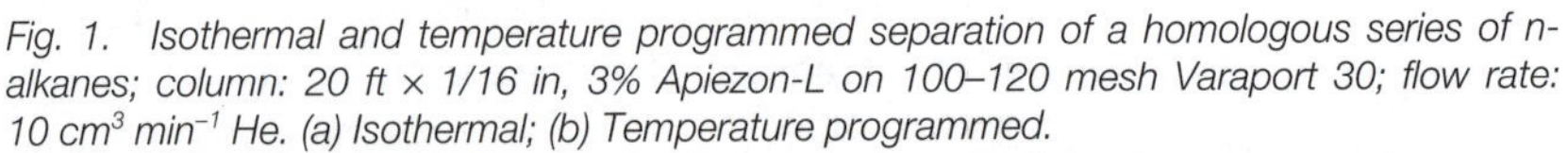

Fig. 1. Isothermal and temperature programmed separation of a homologous series of n-alkanes; column: 20 ft × 1/16 in, 3% Apiezon-L on 100–120 mesh Varaport 30; flow rate: 10 cm^3 min^{-1} He. (a) Isothermal; (b) Temperature programmed.

Two potential disadvantages of temperature programming are the inevitable delay between consecutive chromatographic runs while the oven is cooled down and a stable starting temperature re-established, and the possible decomposition of thermally-labile compounds at the higher temperatures. Computer-controlled systems improve the reproducibility of temperature programming, and the oven can be automatically force-cooled between runs to save time.

Special procedures used in GC

Although **GC** is primarily a technique for analyzing mixtures of volatile solutes, there are many compounds and materials that are either nonvolatile, have volatile components in a nonvolatile matrix, or are thermally labile. In some instances, samples may contain solutes at such low levels that they must be pre-concentrated prior to analysis.

Special procedures have been developed for handling such samples, e.g.

- **Nonvolatile and thermally labile materials** can either be **pyrolyzed** or chemically **derivatized** to yield volatile products that can be successfully chromatographed.

  Small samples of paints, plastics, polymers and many ionic compounds can be pyrolyzed (thermally decomposed) in a modified injection port under controlled conditions to yield characteristic lower molecular mass and volatile products that are swept onto the column. The resulting **pyrograms** can be used as **fingerprints** of the original materials for identification purposes (*Fig. 2*). Compounds of very limited volatility and/or thermal sensitivity containing hydroxyl, carboxyl and amino functional groups can be readily reacted with appropriate reagents to convert these into much less polar methyl, trimethylsilyl or trifluoroacetyl funtionalities. Fatty acid, carbohydrate, phenol and aminoacid derivatives can be chromatographed, but often **HPLC** (Topics D6 and D7) is the preferred technique.
- **Headspace analysis** involves chromatographing the vapors derived from a sample by warming it in a partially filled vial sealed with a septum cap. After equilibration under controlled conditions, the proportions of volatile sample components in the **headspace** above the sample are representative of those in the bulk sample. The **headspace vapors**, which are under slight positive pressure, are sampled by a modified and automated injection system or gas syringe, and injected onto the column (*Fig. 3(a)*). The procedure is useful for mixtures of volatile and nonvolatile components, such as residual monomers in polymers, alcohol or solvents in blood samples (*Fig. 3(b)*), and flavors and perfumes in manufactured products, as it simplifies the chromatograms and protects the column from contamination by nonvolatile substances.
- **Thermal desorption** is a procedure where solutes can be collected on a solid sorbent in a pre-concentration step, then **thermally desorbed** by rapid heating in a unit linked to a modified injection port and through which the carrier gas is flowing. Sorbents, such as activated charcoal or one of the granular packings used in packed column **GC**, are normally contained in a small tube with which polluted industrial or urban atmospheres can be sampled by allowing passive diffusion through the tube over a prolonged period or drawing the air through over shorter periods. Thermal desorption can also be used in conjunction with **headspace analysis** to pre-concentrate volatile solutes, and to **purge and trap** volatile solutes in liquid samples using a stream of gas (*Fig. 3(a)*).

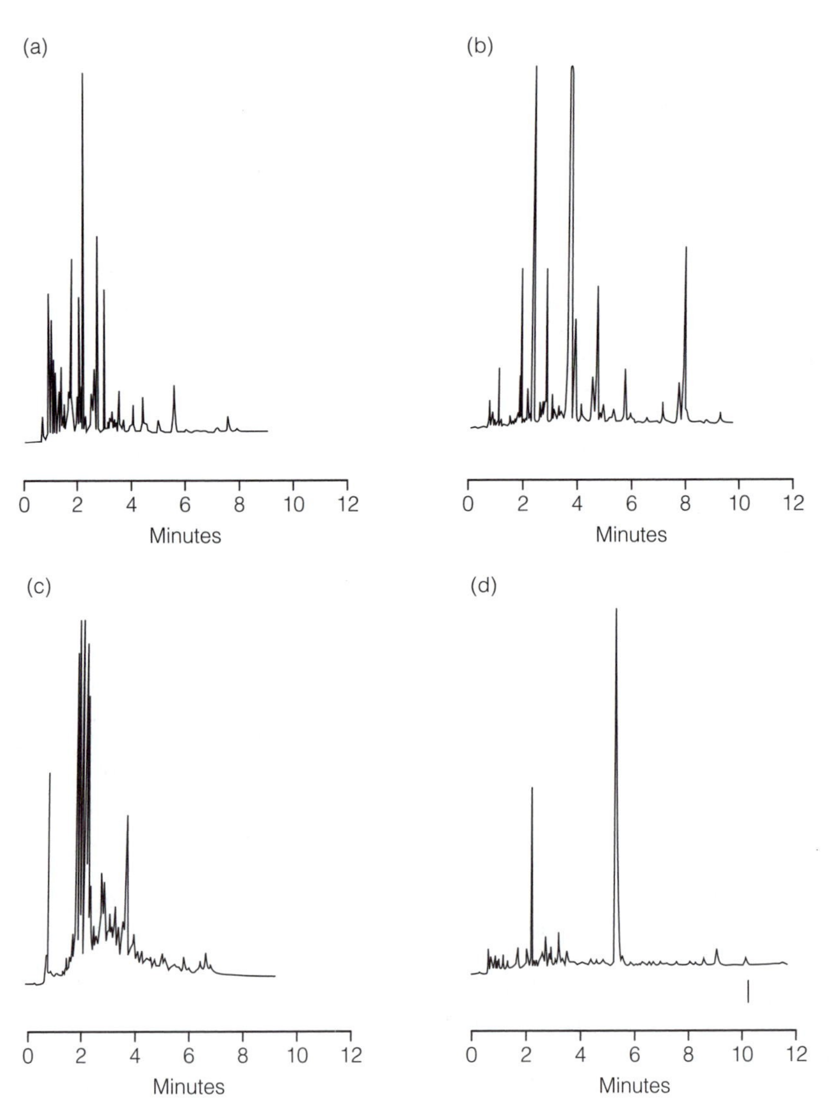

Fig. 2. Pyrograms of four common polymers. (a) Polyethylene; (b) A polyester; (c) A fluorocarbon copolymer; (d) Nylon.

Qualitative analysis

Methods of identifying unknown solutes separated by chromatographic techniques are described in Topic D2. In the case of gas chromatography, there are four alternatives:

- **Comparisons of retention times** (t_R) with those of known solutes under identical conditions, preferably on two columns of differing polarity to reduce the chances of ambiguous identifications;
- **Comparisons of chromatograms** of samples **spiked** with known solutes with the chromatogram of the unspiked sample;
- Calculation of a **retention index** based on a set of standards that can be compared with published or in-house library data. For a homologous series of compounds, the **logarithm of the retention time** is directly proportional to the number of carbon atoms. The **Kováts** system, based on the homologous

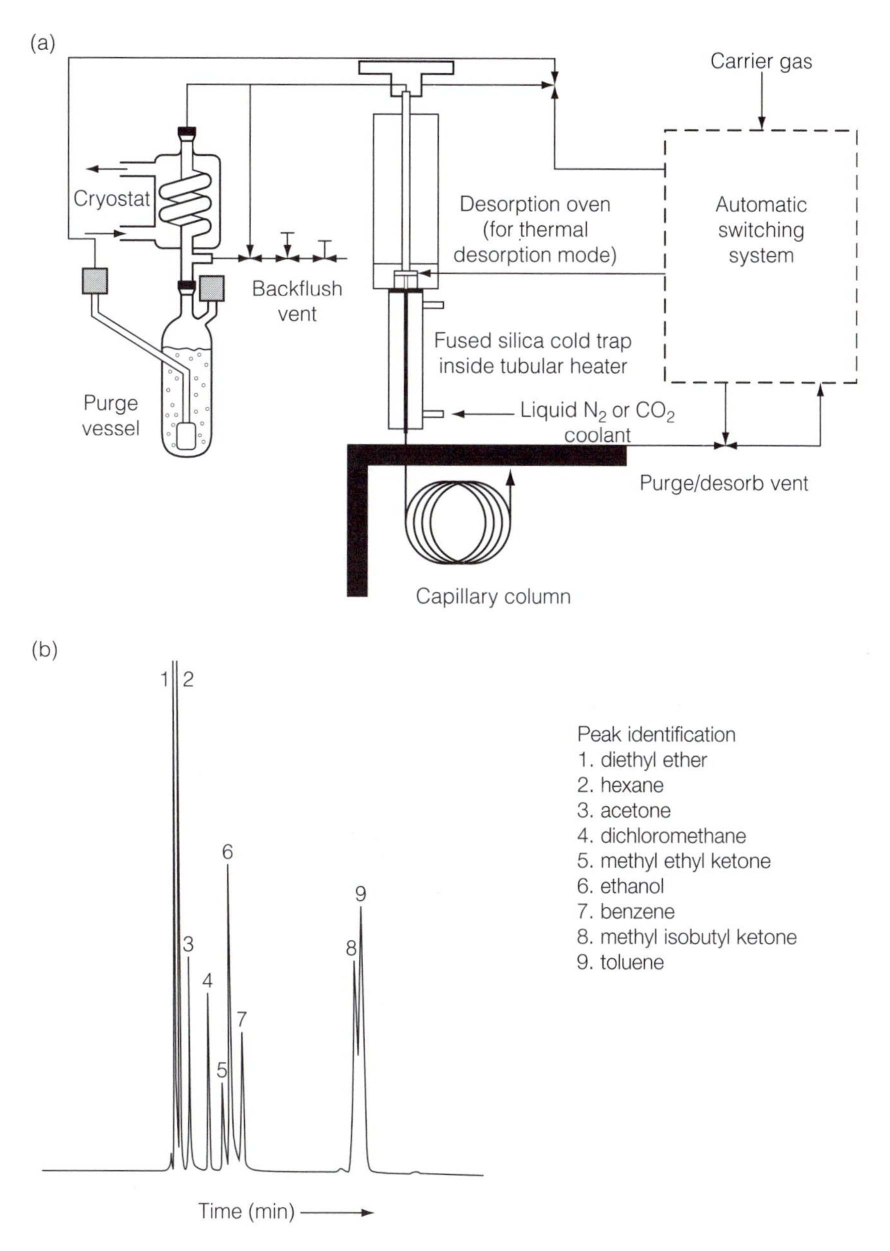

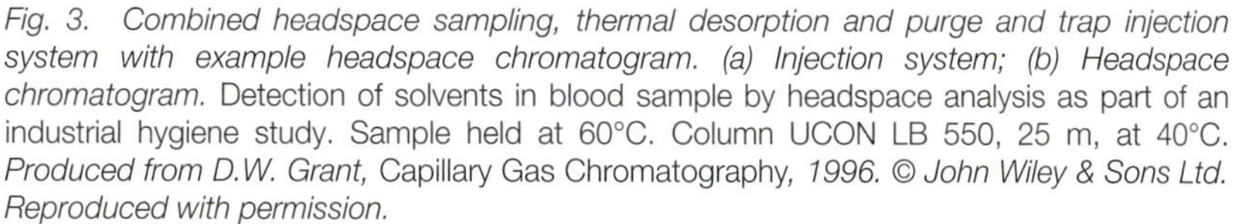

Fig. 3. Combined headspace sampling, thermal desorption and purge and trap injection system with example headspace chromatogram. (a) Injection system; (b) Headspace chromatogram. Detection of solvents in blood sample by headspace analysis as part of an industrial hygiene study. Sample held at 60°C. Column UCON LB 550, 25 m, at 40°C. *Produced from D.W. Grant,* Capillary Gas Chromatography, *1996. © John Wiley & Sons Ltd. Reproduced with permission.*

series of *n*-alkanes, defines an index for each alkane as **100** × the **number of carbon atoms** for any column at any temperature, i.e. ***n*-pentane** is **500**, ***n*-octane** is **800** etc. The retention index of any other solute, relative to the *n*-alkane scale, is calculated or read from a graph of log(retention time) against

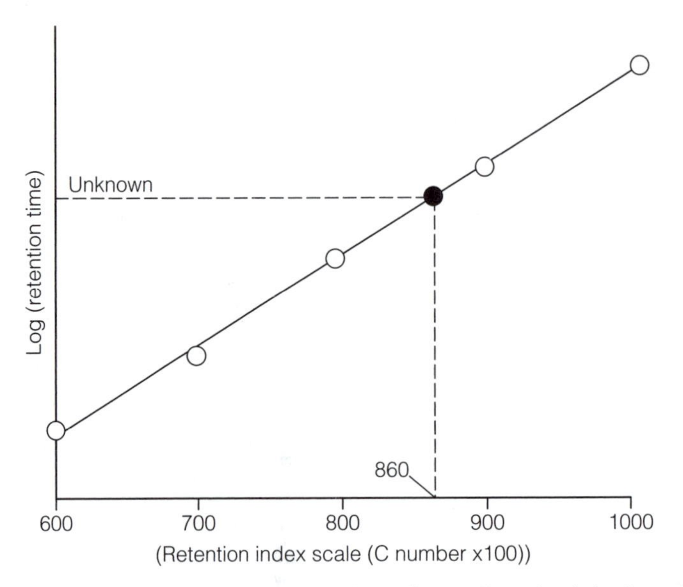

Fig. 4. Kováts retention index plot for n-alkanes. Retention index for unknown interpolated from its log(retention time) as 860.

carbon number by interpolation (*Fig. 4*). For an unknown solute, its Kováts index can then be checked against data bases by computer searching;

- **Interfacing** a gas chromatograph with a mass or infrared spectrometer. This enables spectral information for an unknown solute to be recorded and interpreted. Identifications are facilitated by searching libraries of computerized spectra (Topics F3 and F4).

Quantitative analysis

Methods used in quantitative chromatography are decribed in Topic D2 and alternative calibration procedures are described in Topics A5 and B4. Detector response factors must be established for each analyte as these can vary considerably, especially where selective detectors such as the **ECD** or **NPD** are used. Calibration can be with **external standards** chromatographed separately from the samples, by **internal standardization**, **standard addition** or **internal normalization**. An internal standard should have similar chromatographic characteristics to the analyte(s); homologues or isomers are often the most suitable. The use of **internal standards**, where peak/area ratios of analyte to internal standard are calculated, is preferable because a major source of variability arises from the very small volumes injected by microsyringe, and peak area ratios are independent of the volume injected. Autoinjectors minimize this source of error, and quantitative analysis by **GC** can be expected to have an overall relative precision of between 1 and 5%.

D6 HIGH-PERFORMANCE LIQUID CHROMATOGRAPHY: PRINCIPLES AND INSTRUMENTATION

Key Notes

Principles

High-performance liquid chromatography (HPLC) is a technique for the separation of components of mixtures by differential migration through a column containing a microparticulate solid stationary phase. Solutes are transported through the column by a pressurized flow of liquid mobile phase, and are detected as they are eluted.

Mobile phase

The mobile phase is either a single solvent or a blend of two or more having the appropriate eluting power for the sample components. It ranges from a nonpolar liquid to aqueous buffers mixed with an organic solvent.

Solvent delivery system

The solvent delivery system comprises a means of degassing, filtering and blending up to four solvents which are then delivered to the top of the column under pressure by a constant flow pump.

Sample injection

Liquid samples or solutions are introduced into the flowing mobile phase at the top of the column through a constant or variable volume loop and valve injector that is loaded with a syringe.

Column and stationary phase

Columns are straight lengths of stainless steel tubing tightly packed with a microparticulate stationary phase. The column packings are chemically-modified silicas, unmodified silica or polymeric resins or gels.

Solute detection

Solutes are detected in the mobile phase as they are eluted from the end of the column. The detector generates an electrical signal that can be amplified and presented in the form of a chromatogram of solute concentration as a function of time.

Instrument control and data processing

A dedicated microcomputer is an integral part of a modern high-performance liquid chromatograph. Software packages facilitate the control and monitoring of instrumental parameters, and the display and processing of data.

Related topics

Principles of chromatography (D2)

High-performance liquid chromatography: modes, procedures and applications (D7)

Principles

High-performance liquid chromatography (**HPLC**) is a separation technique where solutes migrate through a column containing a microparticulate stationary phase at rates dependent on their distribution ratios (Topic D2). These are functions of the relative affinities of the solutes for the mobile and stationary phases, the elution order depending on the chemical nature of the solutes and the overall polarity of the two phases. Very small particles of stationary phase are essential for satisfactory chromatographic efficiency and resolution, and the mobile phase must consequently be pumped through the column, resulting in the generation of a considerable back-pressure. The composition of the mobile phase is adjusted to elute all the sample components reasonably quickly. Solutes eluted from the end of the column pass through a detector that responds to each one. There are a number of **modes of HPLC** enabling an extremely wide range of solute mixtures to be separated. The modes (Topic D7) are defined by the type of stationary phase and associated sorption mechanism.

A schematic diagram of a **high-performance liquid chromatograph** is shown in *Figure 1*. It consists of five major components:

- solvent delivery system;
- sample injection valve;
- column;
- detection and recording system;
- microcomputer with control and data-processing software.

These are described in the following sections.

Mobile phase

The mobile phase, or **eluent**, is most frequently a blend of two miscible solvents that together provide adequate **eluting power** and **resolution**. These are deter-

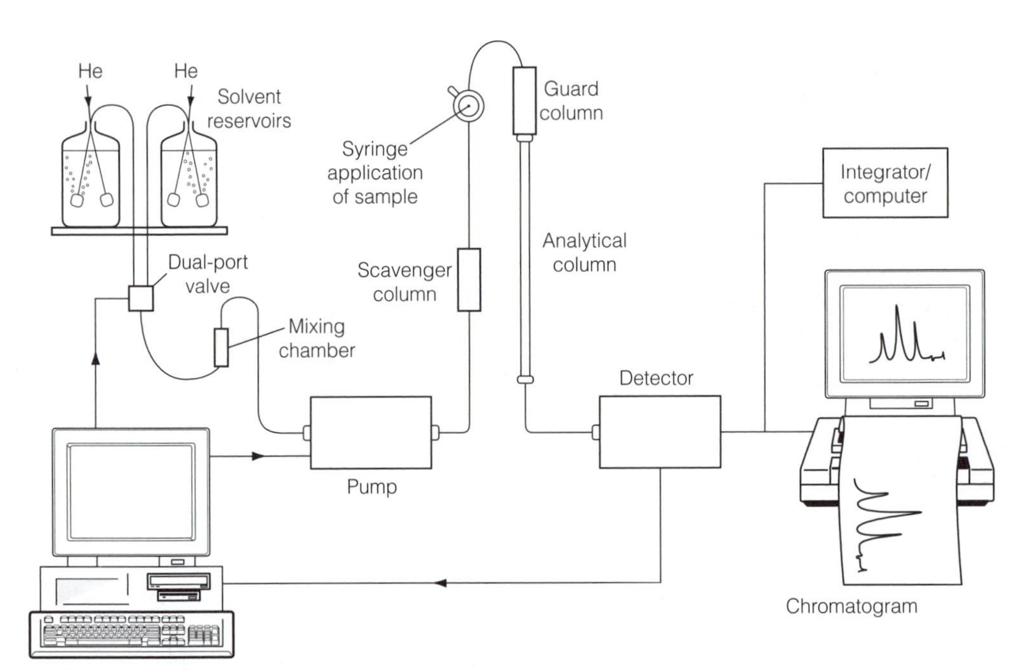

Fig. 1. Schematic diagram of a high-performance liquid chromatograph. Reproduced from A. Braithwaite & F.J. Smith, Chromatographic Methods, 5th edn, *1996, first published by Blackie Academic & Professional.*

mined by its overall polarity, the polarity of the stationary phase and the nature of the sample components. Unlike a **GC** carrier gas, which plays no part in chromatographic retention and selectivity, the composition of an **HPLC** mobile phase is crucial in both respects. For **normal-phase** separations (stationary phase **more** polar than mobile phase), eluting power **increases** with increasing solvent polarity, whilst for **reversed-phase** separations (stationary phase **less** polar than mobile phase), eluting power **decreases** with increasing solvent polarity. An **eluotropic series** of solvents, which lists them in order of increasing polarity, is a useful guide to solvent selection for HPLC separations. *Table 1* is an example that also includes **UV cut-off wavelengths** as **UV absorbance** detectors are the most widely used (*vide infra*). Elution can be under **isocratic** conditions (constant mobile phase composition) or a composition **gradient** can be generated by a **gradient former** to improve the resolution of complex mixtures, especially if the sample components have a wide range of polarities. The most widely used mobile phases for reversed-phase separations are mixtures of aqueous buffers with methanol, or water with acetonitrile. For normal-phase separations, which are less common, hydrocarbons blended with chlorinated solvents or alcohols are typical.

Table 1. An eluotropic series of solvents for HPLC

Solvent	Solvent strength parameter, ε° (adsorption)	Solvent strength parameter, p' (partition)	UV cut-off (nm)
n-Hexane	0.01	0.1	195
Cyclohexane	0.04	−0.2	200
Tetrachloromethane	0.18	1.6	265
Methylbenzene	0.29	2.4	285
Trichloromethane	0.40	4.1	245
Dichloromethane	0.42	3.1	230
Tetrahydrofuran	0.56	4.0	212
Propanone	0.56	3.9	330
Acetonitrile	0.65	5.8	190
iso-Propanol	0.82	3.9	205
Ethanol	0.88	4.3	205
Methanol	0.95	5.1	205
Ethanoic acid	>1	4.4	255
Water	>1	10.2	170

Solvent delivery system

The mobile phase is either a single solvent or a blend of two to four solvents delivered at pressures of up to about 5000 psi (350 bar) with a constant and reproducible flow rate of <0.01–5 $cm^3\ min^{-1}$. The solvent delivery system comprises the following components:

- A **mechanical pump** designed to deliver a **pulse-free** flow of mobile phase. Most are single or dual piston **reciprocating pumps** (*Fig. 2*) with specially designed cams and pulse dampers to minimize or eliminate inherent flow variations, or one-shot pulseless **syringe pumps** used primarily with micro-bore columns (*vide infra*) requiring low flow rates. The wetted parts of the pump should be inert to all solvents (stainless steel, titanium, industrial sapphire, ruby, and Teflon being the principal choices) with minimal volume pumping chambers to facilitate rapid changes of mobile phase composition.

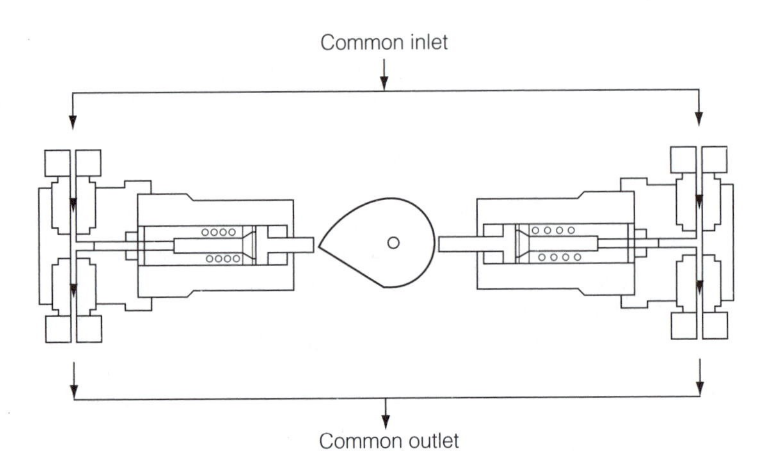

Fig. 2. A typical twin-headed reciprocating pump. Reproduced from W.J. Lough & I.W. Wainer, High Performance Liquid Chromatography, *1996, first published by Blackie Academic & Professional.*

- **Solvent reservoirs** with **in-line filters** (2 μm porosity or less) for each solvent to remove dust and other particulate material. This reduces pump wear and protects the column from becoming clogged which results in increased back-pressures.
- A means of **de-gassing** the solvents to remove dissolved air. Air interferes with the detector response by forming bubbles in the flow-through cell as the pressure reduces to atmospheric at the end of the column. De-gassing is normally accomplished by bubbling helium through each solvent to displace the air, or by passing them through a commercial permeable-membrane de-gassing unit.
- A **gradient former** to generate binary, ternary or quaternary mixtures of solvents with a pre-programmed composition profile during a separation (**gradient elution**).

Sample injection

Liquid samples and solutions are injected directly into the pressurized flowing mobile phase just ahead of the column using a stainless steel and Teflon **valve** fitted with an internal or external **sample loop** (*Fig. 3*). The loop, generally of between 0.5 and 20 μl capacity, is first filled or partially filled with sample from a syringe while the mobile phase flows directly to the column. By turning a handle to rotate the body of the valve, the mobile phase is diverted through the loop thus injecting the sample onto the top of the column without stopping the flow. A disposable **guard column** is sometimes positioned between the injector and the analytical column to protect the latter from a buildup of particulate matter and strongly retained matrix components from injected samples. It consists of a short length of column tubing, or a cartridge, packed with the same stationary phase as is in the analytical column.

Valve injection can easily be automated, controlled by computer software and used with autosamplers. For quantitative analysis, filled-loop injection has a relative precision of about 0.5%.

Column and stationary phase

The column is where the separation process occurs and it is, therefore, the central component of a high-performance liquid chromatograph. There are two

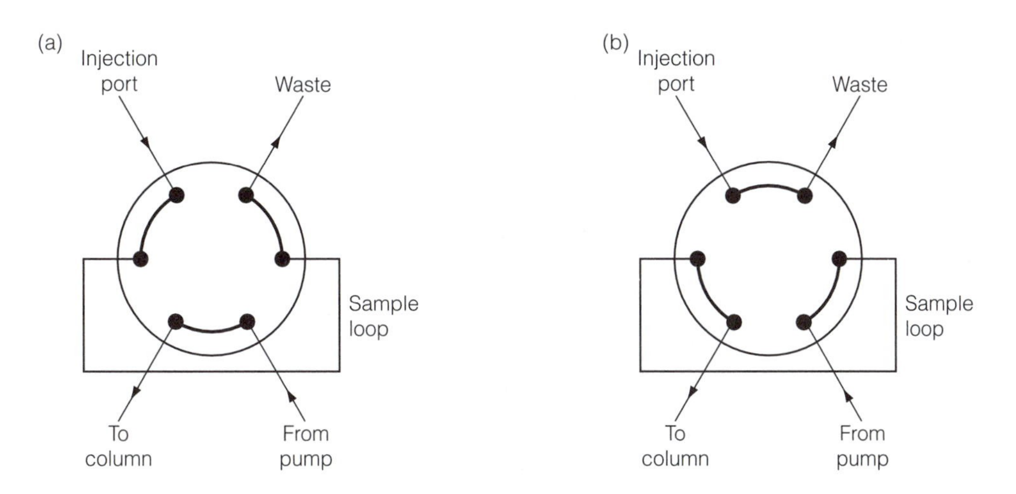

Fig. 3. Sample-injection valve and loop. (a) Sample-loading position; (b) sample-injection position.

types of **HPLC** column, **conventional** and **microbore**, and a comparative summary is given in *Table 2*.

Microbore columns have three principal advantages over conventional columns, i.e.:

- solvent consumption is about 80% less because of the much lower mobile phase flow rate (10–100 μl min^{-1})
- the low volume flow rate makes them ideal for interfacing with a mass spectrometer (Topics F3 and F4)
- sensitivity is increased because solutes are more concentrated, which is especially useful if sample size is limited, e.g. for clinical specimens.

However, in practice, they are not as robust as conventional columns and are not necessary for many routine applications.

Columns are connected to the sample injection valve and the detector using short lengths of very narrow bore (~0.15 mm internal diameter) stainless steel or PEEK (polyether ether ketone) tubing to minimize dead-volume which contributes to band spreading in the mobile phase by diffusion.

HPLC stationary phases are predominantly chemically-modified silicas, unmodified silica or cross-linked co-polymers of styrene and divinyl benzene. The surface of silica is polar and slightly acidic due to the presence of silanol (Si-OH) groups. It can be chemically modified with reagents, such as chlorosilanes, which react with the silanol groups replacing them with a range of other functionalities (*Fig. 4(a)*). The resulting **bonded phases**, which are hydrolytically stable through the formation of siloxane (Si—O—Si—C) bonds, have different chromatographic characteristics and selectivities to unmodified silica.

Octadecyl silica (**ODS** or **C18**) is the most widely used of all the stationary phases, being able to separate solutes of low, intermediate and high polarities. **Octyl** and shorter alkyl chains are considered to be more suitable for polar solutes. **Aminopropyl** and **cyanopropyl** (**nitrile**) silicas are good replacements for unmodified silica, which can give variable retention times due to traces of water in the solvents. Polar, and especially basic solutes, tend to give tailing

Table 2. A comparison of conventional and microbore HPLC columns

Conventional columns	Microbore columns
Tubing Stainless steel Lengths 3, 10, 15, 20 and 25 cm ¼″ outside diameter Internal diameter 4.6 mm	**Tubing** Stainless steel Lengths 25 and 50 cm Coupled lengths 1 m or more ¼″ outside diameter Internal diameter 1 or 2 mm
Stationary phase (packing) Porous, microparticulate silica, chemically-modified silica (bonded phases) or styrene/divinyl benzene co-polymers Mean particle diameters 3, 5 or 10 μm with a narrow range of particle sizes	**Stationary phase (packing)** Porous, microparticulate silica, chemically-modified silica (bonded phases) or styrene/divinyl benzene co-polymers Mean particle diameters 3, 5 or 10 μm with a narrow range of particle sizes
Operating pressures 500–3000 psi (35–215 bar)	**Operating pressures** 1000–5000 psi (70–350 bar)
Typical mobile phases Hydrocarbons + chlorinated solvents or alcohols for normal-phase; methanol or acetonitrile + water or aqueous buffers for reversed-phase Flow rate 1–3 $cm^3 min^{-1}$	**Typical mobile phases** Hydrocarbons + chlorinated solvents or alcohols for normal-phase; methanol or acetonitrile + water or aqueous buffers for reversed-phase Flow rate 10–100 $\mu l\ min^{-1}$
	Modified instrumentation Solvent delivery system capable of accurate flow control down to 10 $\mu l\ min^{-1}$ or less Small volume sample injection valves Small volume detector cells
Performance Efficiency increases with diminishing particle size, but column life for 3 μm particles is shorter Separations on 3 cm *fast columns* in less than 1 minute	**Performance** Very efficient and sensitive, but slow Solvent consumption only a quarter that of conventional columns

peaks on bonded phase silicas because of adsorptive interactions with residual silanols and metallic impurities in the silica. The problem is reduced by **end-capping**, a process of blocking the sites with trimethylsilyl ($(CH_3)_3$-Si-) groups (*Fig. 4(b)*), and by using highly purified silica (< 1 ppm metals).

Size exclusion and **ion-exchange** stationary phases are either silica or polymer based. Sulphonic acid or quaternary ammonium groups provide cation and anion-exchange capabilities respectively, but slow rates of exchange leading to poor efficiencies and low sample capacities have limited their use, except for **ion chromatography** (Topic D7).

Chiral stationary phases have been developed for the separation of mixtures of enantiomers but are expensive and have a very limited working life.

The availability of a wide range of bonded phases together with polymeric materials, has resulted in the development of a number of **modes** of HPLC (Topic D7). The more important stationary phases and their characteristics are summarized in *Table 3*.

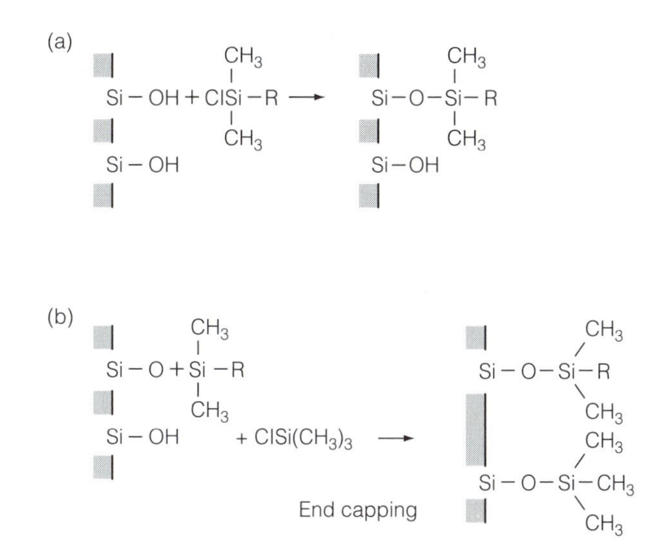

Fig. 4. Formation of bonded-phase silicas. (a) Monomeric bonded phases; (b) End-capping of residual silanols. R = alkyl, aminoalkyl, ion-exchange groups.

Table 3. Stationary phases for HPLC

Stationary phase	Sorption mechanism	Characteristics
Unmodified silica, SiO_2	Adsorption, normal-phase	Polar, retention times variable due to adsorbed water
Bonded phases		
Octadecyl silica, $-C_{18}H_{37}$ (ODS or C18) Octyl silica, $-C_8H_{17}$ Propyl silica, $-C_3H_7$	Modified partition, reversed-phase	Nonpolar, but unreacted silanol groups cause polar solutes, especially bases, to tail, pH range limited to 2.5–7.5
		All separate a very wide range of solutes
Aminopropyl, $-C_3H_6NH_2$	Modified partition, normal or reversed phase	Polar, separates carbohydrates pH range limited to 2.5–7.5
Sulphonic acid, $-(CH_2)_nSO_3H$	Cation-exchange	Slow mass transfer broadens peaks, limited sample capacity, pH range limited to 2.5–7.5 for silica-based materials
Quaternary amine, $-(CH_2)_nNR_3OH$	Anion-exchange	
Controlled-porosity silicas (some with $-Si(CH_3)_3$ groups)	Size exclusion	Compatible with both organic and aqueous solvents, pH range limited to 2.5–10
α-, β-, γ-cyclodextrin silicas	Chiral selectivity based on adsorptive interactions	Expensive, limited life, resolution sensitive to mobile phase composition
Polymer phases		
Cross-linked styrene/divinyl benzene co-polymers, unmodified or with ion-exchange groups	Modified partition, exclusion or ion-exchange	Nonpolar if unmodified, stable over pH range 1–13

Solute detection

Detectors are based on a **selective response** for the solute, such as **UV-absorbance** or **fluorescence**, or on a **bulk property** of the mobile phase which is modified by the solute, such as **refractive index**. Ideally, detectors should have the following characteristics:

- a rapid and reproducible response to solutes;
- high sensitivity, i.e. able to detect very low levels of solutes;
- stability in operation;
- a small volume cell to minimize band broadening, i.e. 8 μl or less for a conventional column, 1 μl or less for a microbore column;
- a signal directly proportional to solute concentration or mass over a wide range (**linear dynamic range**);
- insensitivity to changes in temperature and flow rate;
- a cell design that does not entrap air bubbles that outgas from the mobile phase at the end of the column.

Many types of detector have been investigated, and the most widely used are summarized below and in *Table 4*.

- **UV-visible absorbance detector.** This type, which is the most widely used, is based on the absorbance of **UV** or **visible** radiation in the range 190–800 nm by solute species containing **chromophoric groups** or structures (Topics E8 and E9). Detector cells are generally 1 mm diameter tubes with a 10 mm optical path length and designed so as to eliminate refractive index effects which can alter the measured absorbance. There are three types of **UV-visible absorbance detector**:
  (i) **Fixed-wavelength filter-photometers**, which are the simplest, employing mercury-vapor lamp sources and optical filters to select a limited number of wavelengths, e.g. 254, 280, 334 and 436 nm, and a phototube detector. They have a limited use, lacking versatility, but they are cheap.
  (ii) **Variable-wavelength spectrophotometers** (*Fig. 5)* are much more

Table 4. Characteristics of HPLC detectors

Detector	Sensitivity g cm^{-3}	Linear range	Characteristics
UV-visible absorbance			Good sensitivity, most widely used, selective for unsaturated groups and structures. Not significantly flow or temperature sensitive. Can be used with gradient elution.
Filter-photometer	5×10^{-10}	10^4	
Spectrophotometer	5×10^{-10}	10^5	
Diode-array Spectrometer	$>2 \times 10^{-10}$	10^5	
Fluorescence	10^{-12}	10^4	Excellent sensitivity, selective, including fluorescent derivatives. Not flow or temperature sensitive.
Refractive index	5×10^{-7}	10^4	Almost universal, but only moderate sensitivity. Very temperature sensitive (control to ±0.001°C). Cannot be used with gradient elution.
Electrochemical			
Conductimetric	10^{-8}	10^4	Flow and moderately temperature sensitive. Cannot be used with gradient elution. Detects only ionic solutes.
Amperometric	10^{-12}	10^5	Excellent sensitivity, selective but problems with electrode contamination.

versatile as they allow monitoring at any wavelength within the working range of the detector to give the optimum response for each solute. They employ deuterium and tungsten lamp sources for the **UV** and **visible** regions, respectively, a diffraction grating monochromator for wavelength selection and a photomultiplier detector. Many are computer-controlled for programmable wavelength switching during a separation to optimize sensitivity and selectivity.

(iii) **Photodiode-array detectors** are spectrometers with fixed optics and a detection system consisting of one or two **arrays** of **photodiodes** on a silicon chip positioned to receive radiation dispersed by a diffraction grating (*Fig. 6*). Electronic scanning, digitizing and processing of the signals by a microcomputer enables 'snapshots' of the complete spectrum of the flowing eluent to be collected and stored every 0.1 s. The spectra and the developing chromatogram at any wavelength can be displayed on a VDU screen in real time and subsequently shown as a 3-D color plot of absorbance, wavelength and time (*Fig. 7*). The data can be manipulated and re-plotted on the screen, and comparisons made with library spectra for identification purposes.

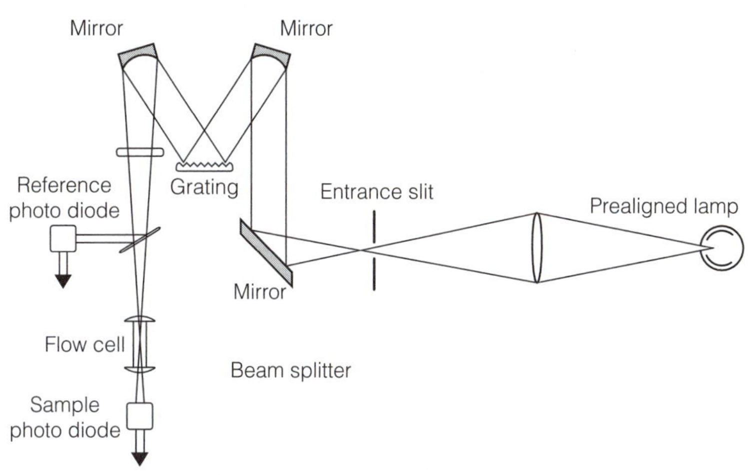

Fig. 5. UV-visible variable-wavelength spectrophotometric detector.

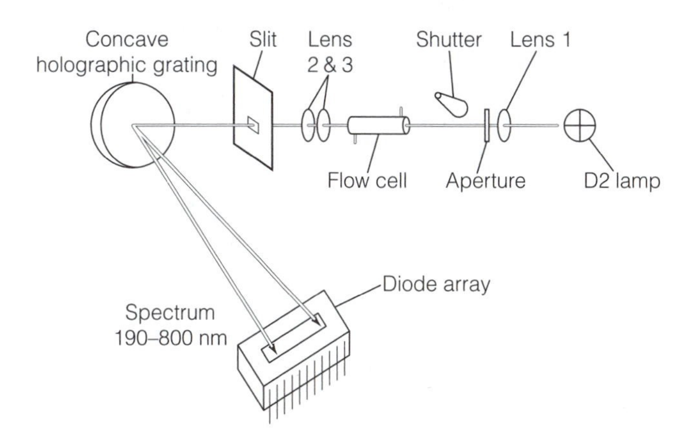

Fig. 6. Diode-array detector (DAD).

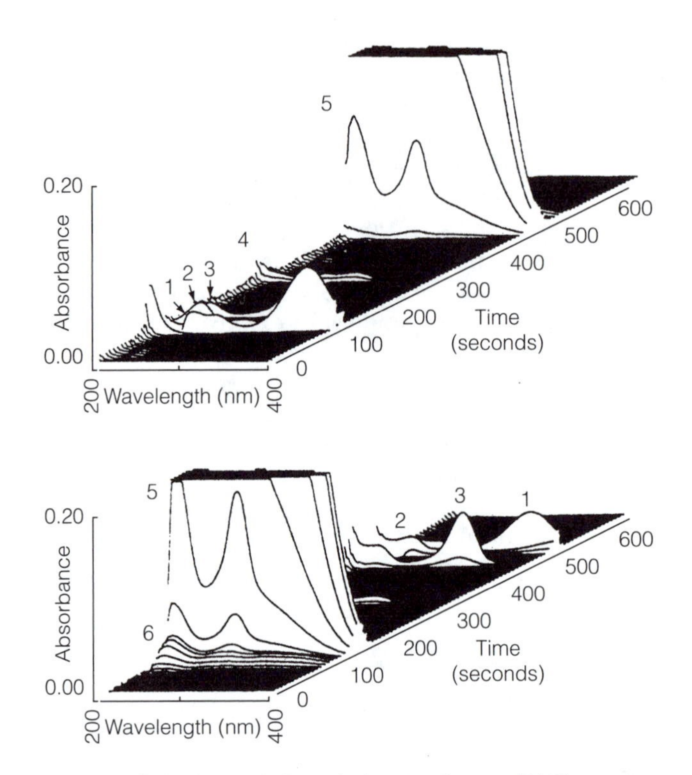

Fig. 7. 3-D display mode for a diode-array detector (DAD).

- **Fluorescence detectors** are based on filter-fluorimeters or spectrofluorimeters. They are more selective and can be up to three orders of magnitude more sensitive than **UV absorbance** detectors. The detector responds selectively to naturally fluorescing solutes such as polynuclear aromatics, quinolines, steroids and alkaloids, and to fluorescing derivatives of amines, amino acids and phenols with fluorogenic reagents such as dansyl chloride (5-(dimethylamino)-1-naphthalene sulfonic acid).
- **Refractive index** (**RI**) monitors are the closest to being universal **HPLC** detectors, as nearly all dissolved solutes alter the refractive index of the mobile phase. They are **differential detectors**, generating a signal that depends on the difference between the **RI** of the pure mobile phase and the modified value caused by the dissolved solute, which can, therefore, be positive or negative.

  They are several orders of magnitude less sensitive than **UV absorbance** detectors, but are invaluable in the separation of saturated solutes such as carbohydrates, sugars and alkanes. They are highly temperature sensitive and are very difficult to use with gradient elution because the sample and reference cells cannot be continuously matched.
- **Electrochemical detectors** are based on measuring either the **conductance** of an aqueous mobile phase containing ionic solutes, or the **current** generated by the electrochemical reduction or oxidation of solutes at a fixed applied potential (**amperometry**) (Topic C9).

Conductance monitors are used in **ion chromatography**, a mode of **HPLC** useful for separating low levels of inorganic and organic anions and cations by ion-exchange (Topic D7).

Instrument control and data processing

These aspects of HPLC closely parallel those described for GC (Topic D4). Additional instrument parameters to be set and monitored include:

- solvent composition, flow rate and pressure limit;
- gradient elution programs;
- wavelength(s) and wavelength-switching for UV-visible absorbance detectors;
- wavelength range, sampling frequency and mode of display for a DAD.

D7 HIGH-PERFORMANCE LIQUID CHROMATOGRAPHY: MODES, PROCEDURES AND APPLICATIONS

Key Notes

Modes of HPLC

Modes of HPLC are defined by the nature of the stationary phase, the mechanism of interaction with solutes, and the relative polarities of the stationary and mobile phases.

Optimization of separations

After selection of an appropriate mode, column and detector for the solutes to be separated, the composition of the mobile phase must be optimized to achieve the required separation. A trial and error approach or a computer aided investigation can be adopted.

Qualitative analysis

Unknown solutes can be identified by comparisons of retention factors or times, spiking samples with known substances or through spectrometric data.

Quantitative analysis

Quantitative information is obtained from peak area or peak height measurements and calibration graphs using internal or external standards, or by standard addition or internal normalization.

Related topics

Principles of chromatography (D2)

High-performance liquid chromatography: principles and instrumentation (D6)

Modes of HPLC

Almost any type of solute mixture can be separated by **HPLC** because of the wide range of stationary phases available, and the additional selectivity provided by varying the mobile phase composition. Both **normal-** and **reversed-phase** separations are possible, depending on the relative polarities of the two phases. Although these are sometimes referred to as **modes of HPLC**, the nature of the stationary phase and/or the solute sorption mechanism provide a more specific means of classification, and modes based on these and the types of solutes to which they are best suited are summarized below.

- **Adsorption chromatography.** Separations are usually **normal-phase** with a silica gel stationary phase and a mobile phase of a nonpolar solvent blended with additions of a more polar solvent to adjust the overall polarity or eluting power, e.g. *n*-hexane + dichloromethane or di-ethyl ether. The choice of solvent is limited if **a UV absorbance** detector is to be used. Traces of water in the solvents must be controlled, otherwise solute retention will not be reproducible. Solutes are retained by **surface adsorption**; they compete with solvent molecules for active silanol sites (Si-OH), and are eluted in order of increasing polarity. This mode is not used extensively, but is suitable for mixtures of structural isomers and solutes with differing functional groups. Members of a homologous series can not be separated by adsorption chromatography because the nonpolar parts of a solute do not interact with the polar adsorbent surface.

- **Modified partition** or **bonded-phase chromatography (BPC).** Most **HPLC** stationary phases are **chemically-modified silicas**, or **bonded phases**, by far the most widely used being those modified with nonpolar hydrocarbons. The solute sorption mechanism is described as **modified partition**, because, although the bonded hydrocarbons are not true liquids, organic solvent molecules from the mobile phase form a liquid layer on the surface.

  The most popular phase is **octadecyl (C18** or **ODS**), and most separations are **reversed-phase**, the mobile phase being a blend of methanol or acetonitrile with water or an aqueous buffer. For weakly acidic or basic solutes, the role of pH is crucial because the ionized or protonated forms have a much lower affinity for the **ODS** than the corresponding neutral species, and are therefore eluted more quickly. The dissociation of weak acids and the protonation of weak bases are shown by the following equations

$$RCOOH \rightarrow RCOO^- + H^+$$

$$RNH_2 + H^+ \rightarrow RNH_3^+$$

  Thus, **at low pH, bases are eluted more quickly than at high pH**, whilst the opposite holds for **weak acids** (*Fig. 1*).

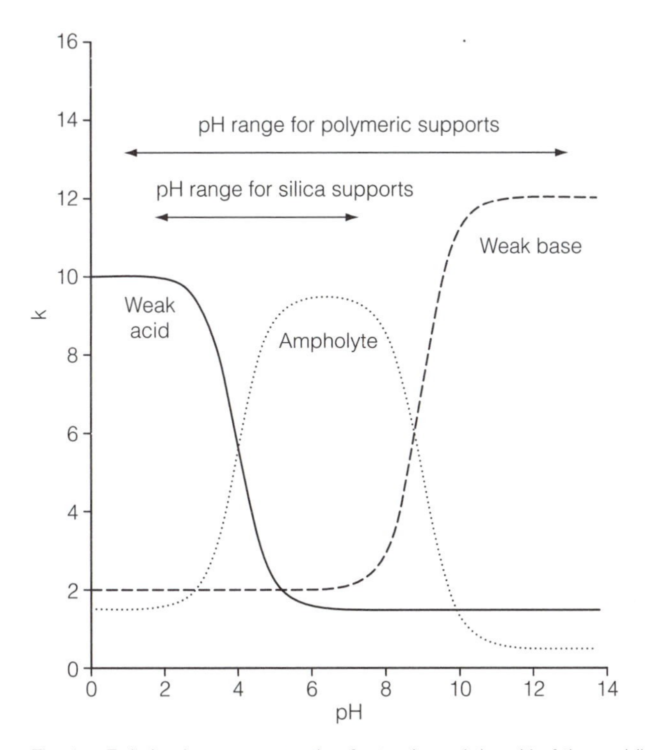

Fig. 1. Relation between retention factor, k, and the pH of the mobile phase for weak acids, bases and ampholytes in reversed-phase separations.

ODS and other **hydrocarbon** stationary phases will separate many mixtures, and are invariably a first choice in developing new **HPLC** methods. They are particularly suited to the separation of moderately polar to polar solutes (*Fig. 2(a)*).

Aminoalkyl and **cyanoalkyl** (nitrile) bonded phases (the alkyl group is usually propyl) are moderately polar. The former is particularly useful in separating mixtures of sugars and other carbohydrates (*Fig. 2(b)*), whilst the latter is used as a substitute for unmodified silica, giving more reproducible retention factors and less tailing, especially with basic solutes. Both normal-phase and reversed-phase chromatography is possible by appropriate choice of eluents.

- **Ion-exchange chromatography (IEC).** Stationary phases for the separation of mixtures of ionic solutes, such as inorganic cations and anions, amino acids and proteins, are based either on microparticulate ion-exchange resins, which are crosslinked co-polymers of styrene and divinyl benzene, or on bonded phase silicas. Both types have either sulfonic acid cation-exchange sites ($-SO_3^-H^+$) or quaternary ammonium anion-exchange sites ($-N^+R_3OH^-$) incorporated into their structures.

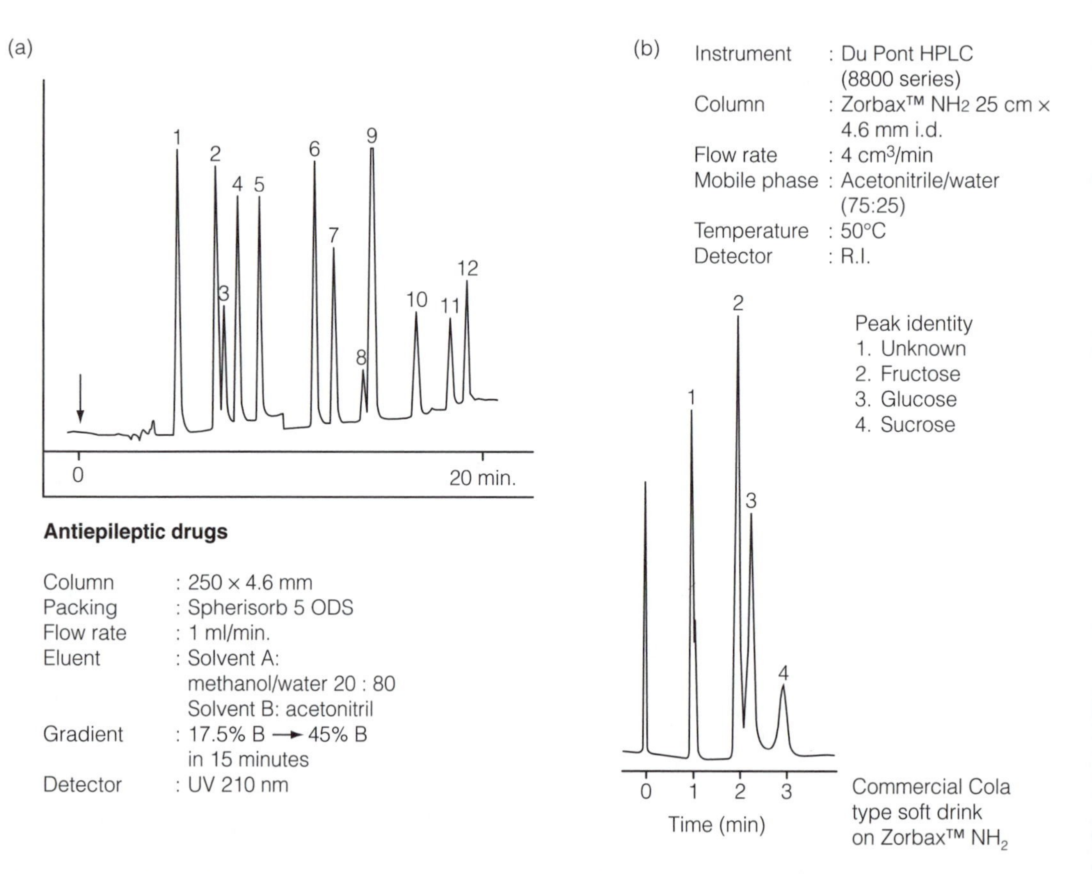

Fig. 2. Separations of solutes of different polarities on bonded-phases. (a) Pharmaceuticals separated on ODS; UV absorbance detection; (b) sugars separated on aminopropyl silica; RI detection.

Ion-exchange chromatography is not that widely used. Inorganic ions and some cations are better separated by a related mode known **as ion chromatography** (*vide infra*), whilst for organic ions, **ion-pair chromatography** is generally preferred because of its superior efficiency, resolution and selectivity.

- **Ion chromatography (IC).** This is a form of ion-exchange chromatography for the separation of inorganic and some organic cations and anions with **conductometric detection** after **suppressing** (removing) the mobile phase electrolyte (*Fig. 3(a)*).

  The stationary phase is a **pellicular** material (**porous-layer beads**), the particles consisting of an impervious central core surrounded by a thin porous outer layer (~2 μm thick) incorporating cation- or anion-exchange sites. The thin layer results in much faster rates of exchange (mass transfer) than is normally the case with ion-exchange and therefore higher efficiencies. Mobile phases are electrolytes such as NaOH, $NaCO_3$ or $NaHCO_3$ for the separation of anions, and HCl or CH_3SO_3H for the separation of cations. The detection of low levels of ionic solutes in the presence of high levels of an eluting electrolyte is not feasible unless the latter can be removed. This is accomplished by a **suppressor cartridge** that essentially converts the electrolyte into water, leaving the solute ions as the only ionic species in the mobile phase.

  The following equations summarize the reactions for the separation of inorganic anions on an anion-exchange column in the HCO_3^- form using a sodium hydrogen carbonate mobile phase:

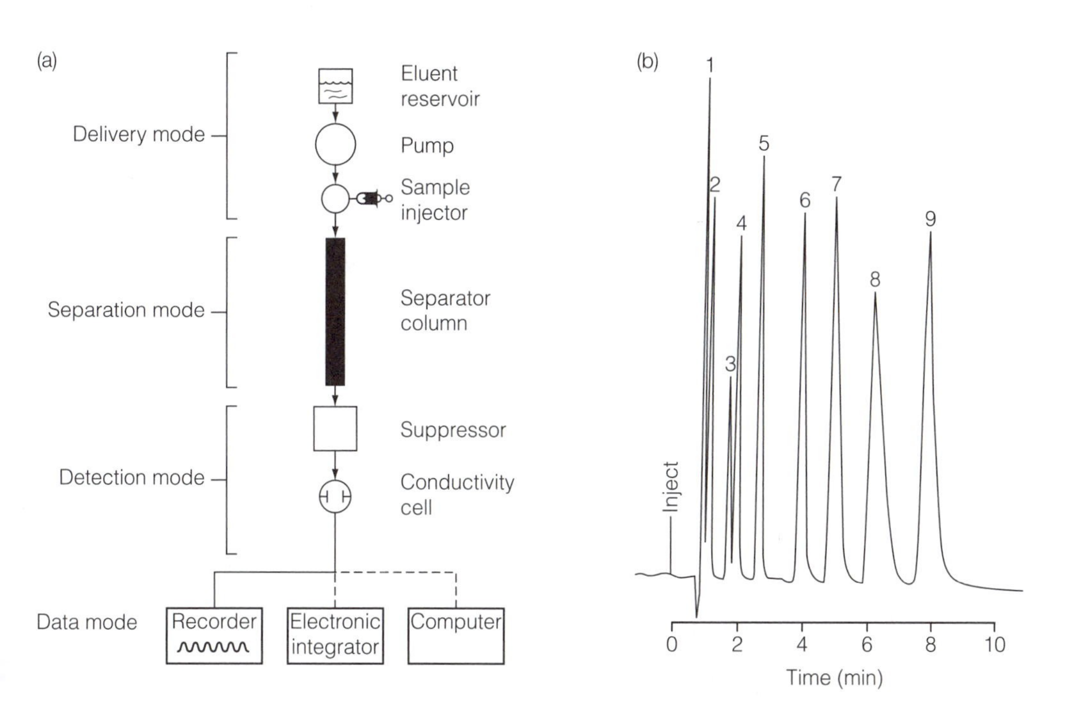

Fig. 3. Ion chromatography. (a) Schematic diagram of an ion chromatograph. (b) Anions in water separated on an anion-exchange column. Reproduced from Dionex UK Ltd with permission.

Column reaction:

$$n(\text{Resin-N}^+R_3HCO_3^-) + X^{n-} \rightarrow (\text{Resin-N}^+R_3)_nX^{n-} + nHCO_3^-$$

$$\text{where } X^{n-} = F^-, Cl^-, NO_3^-, SO_4^{2-}, PO_4^{3-} \text{ etc.}$$

Suppressor reactions:

$Na^+HCO_3^-$ +	H^+	$\rightarrow H_2O + CO_2$ +	Na^+
	introduced via a membrane		removed via a membrane

$Na_n{}^+X^{n-}$	+	nH^+	$\rightarrow$	$H_n{}^+X^{n-}$	+	nNa^+
separated by the column		introduced via a membrane		detected by conductance		removed via a membrane

Similar reactions form the basis of the separation of cations. An example of the separation of inorganic anions at the ppm level is shown in *Figure 3(b)*.

- **Size exclusion chromatography (SEC).** This is suitable for mixtures of solutes with relative molecular masses (RMM) in the range 10^2–10^8 Da. Stationary phases are either microparticulate cross-linked co-polymers of styrene and divinyl benzene with a narrow distribution of pore sizes, or controlled-porosity silica gels, usually end-capped with a short alkyl chain reagent to prevent adsorptive interactions with solutes. **Exclusion** is not a true sorption mechanism because solutes do not interact with the stationary phase (Topic D2). They can be divided into three groups:
  (i) Those larger than the largest pores are excluded completely, and are eluted in the same volume as the interstitial space in the column, V_o.
  (ii) Those smaller than the smallest pores, can diffuse throughout the entire network and are eluted in a total volume, V_{tot}.
  (iii) Those of an intermediate size separate according to the extent to which they diffuse through the network of pores, of volume V_p and are eluted in volumes between V_o and V_{tot}.

  Only those solutes in the third group will be separated from one another, and their retention volumes are directly proportional to the logarithm of their **relative molecular mass** (**RMM**; molecular weight). Columns can be calibrated with standards of known RMM before analyzing unknowns.

  Figure 4 shows a typical plot of **elution volume** against **log (RMM)** and a chromatogram of a mixture with a range of solutes of differing molecular mass. **SEC** is of particular value in characterizing polymer mixtures and in separating biological macromolecules such as peptides and proteins. It is also used for preliminary separations prior to further analysis by other more efficient modes of **HPLC**.
- **Chiral chromatography.** Chiral stationary phases (CSP) enable **enantiomers** (mirror image forms) of a solute to be separated. Several types of these stereoselective materials have been investigated and marketed commercially, some of the most useful being **cyclodextrins** bonded to silica. The cyclodextrins are cyclic chiral carbohydrates with barrel-shaped cavities into which solutes can fit and be bound by H-bonding, π–π and dipolar interactions. Where the total adsorptive binding energies of two enantiomers differ, they will have different retention factors and can be resolved. Steric repulsion and the pH, ionic strength and temperature of the mobile phase all affect the resolution. Although of great interest to the pharmaceutical industry for the

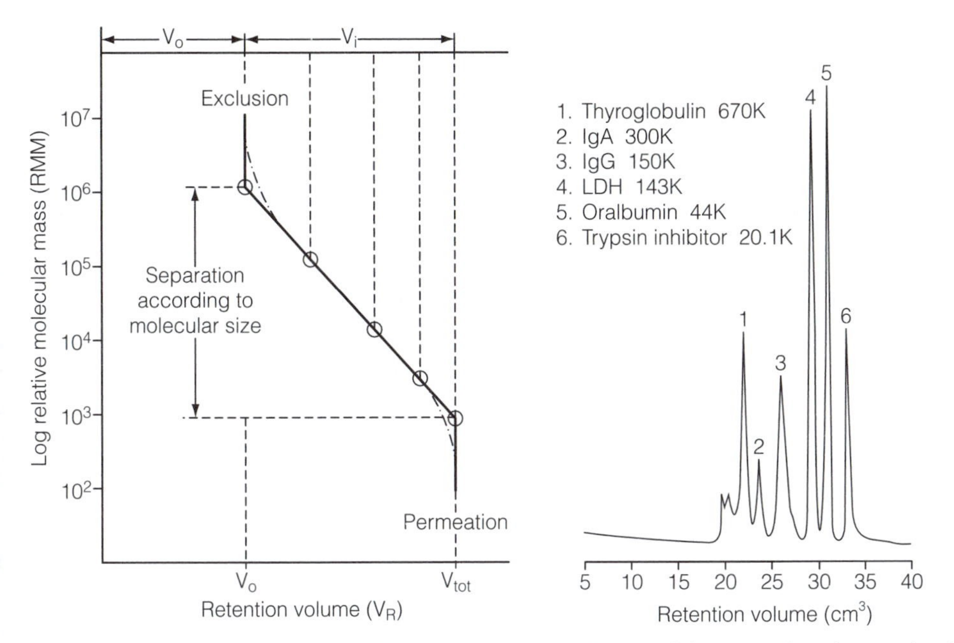

Fig. 4. A typical size exclusion calibration curve and chromatogram of the separation of a protein mixture. Column: BIOSEP-SEC-S3000. Mobile phase: pH 6.8 phosphate buffer. Detector: UV abs. at 280 nm. Reproduced from W.J. Lough & I.W. Wainer (eds), High Performance Liquid Chromatography, *1996, first published by Blackie Academic & Professional.*

separation of enantiomeric forms of drugs having different pharmacological activities, chiral columns are expensive and most have very limited working lives. **Capillary electrophoresis** (Topic D8) provides a cheaper alternative.

Optimization of separations

The optimum conditions for an **HPLC** separation are those which give the required **resolution** in the minimum **time**. A stepwise approach based on the characteristics of the solutes to be separated, and trial chromatograms with different mobile phase compositions is usually adopted. The following is an outline of the procedure for a reversed-phase separation where a hydrocarbon stationary phase, usually **C18** (**ODS**), is the first choice.

- The **mode of HPLC** most suited to the structures and properties of the solutes to be separated is selected, having regard to their relative molecular mass, polarity, ionic or ionizable character, and solubility in organic and aqueous solvents.
- The **stationary phase and column** are selected (Topic D6, *Tables 2 and 3*). The shortest column and the smallest particle size of stationary phase consistent with adequate resolution should be used.
- The **detector**, subject to availability, should match the solute characteristics. **UV-visible absorbance** detectors are suitable for many solutes except those that are fully saturated. Fluorescence and electrochemical detectors should be considered where high sensitivity is required.
- **Mobile phase composition** is optimized by obtaining and evaluating a number of trial chromatograms, often with the aid of computer optimization software packages. A typical series of chromatograms for a reversed-phase separation on a hydrocarbon bonded phase column is shown in *Figure 5*.

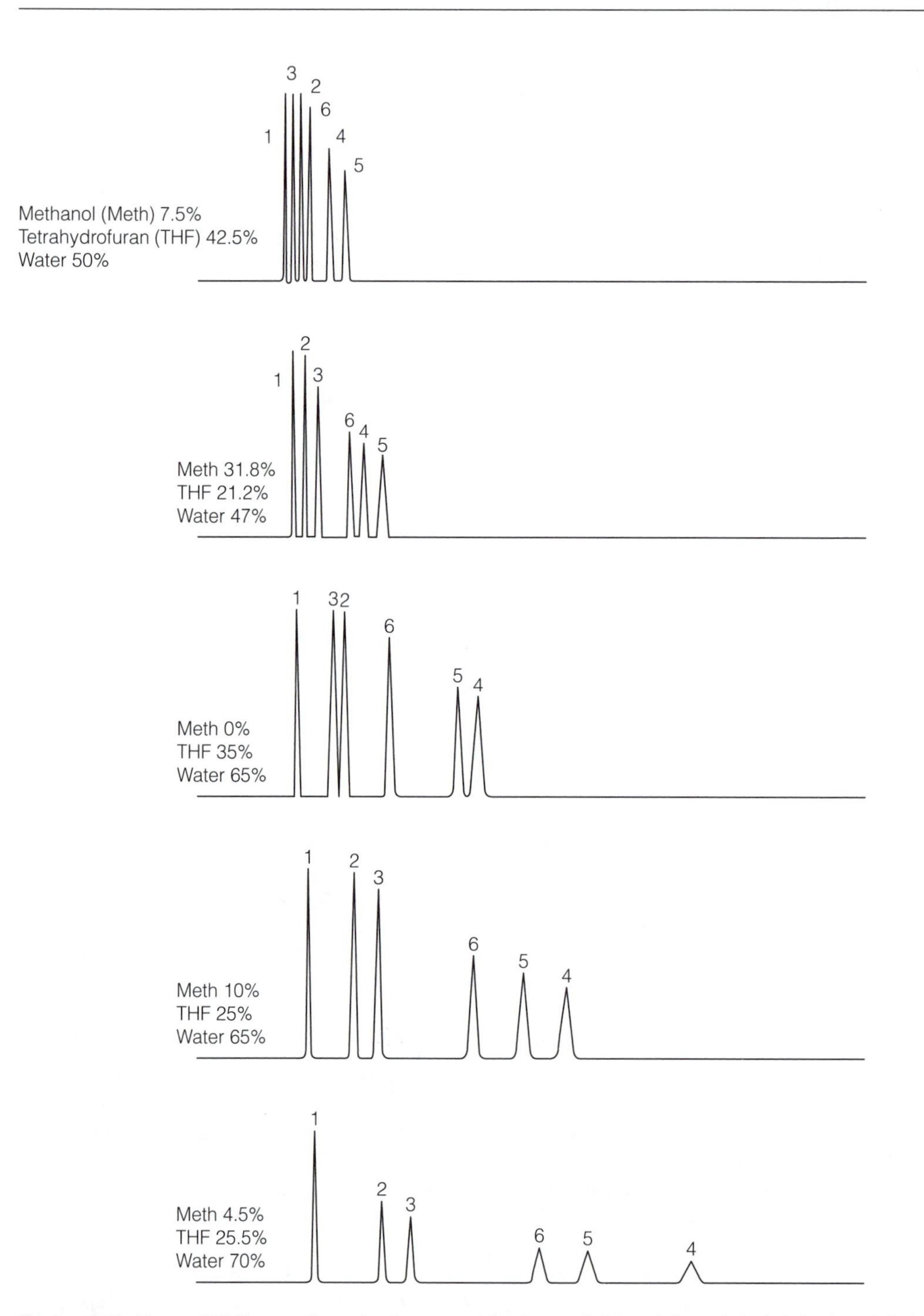

Fig. 5. Optimizing an HPLC separation using ternary mobile phases. Solutes: 1. benzyl alcohol; 2. phenol; 3. 3-phenylpropanol; 4. 2,4-dimethylphenol; 5. benzene; 6. diethyl o-phthalate.

Note how the elution order of the six components in the mixture alters with the mobile phase composition.

Qualitative analysis

Methods of identifying unknown solutes separated by chromatographic techniques are described in Topic D2. In the case of **HPLC**, there are four alternatives:

- **Comparisons of retention factors** (k) or **retention times** (t_R) with those of known solutes under identical conditions, preferably on two columns of differing selectivity to reduce the chances of ambiguous identifications.
- **Comparisons of chromatograms** of samples **spiked** with known solutes with the chromatogram of the unspiked sample.
- **Comparisons of UV-visible spectra** recorded by a diode-array detector with those of known solutes. This is of limited value because most spectra have only two or three broad peaks so many solutes have very similar spectral features.
- **Interfacing** a high-performance liquid chromatograph with a mass spectrometer. This enables spectral information for an unknown solute to be recorded and interpreted. Identifications are facilitated by searching libraries of computerized spectra (Topics F3 and F4).

Quantitative analysis

Methods used in quantitative chromatography are decribed in Topic D2, and alternative calibration procedures are described in Topic A5. **Peak areas** are more reliable than peak heights as they are directly proportional to the quantity of a solute injected when working within the linear range of the detector. Most **HPLC** detectors have a wide linear dynamic range (Topic D6, *Table* 4), but response factors must be established for each analyte as these can vary considerably. Calibration is normally with **external standards** chromatographed separately from the samples, or by **standard addition.** Constant volume loops for sample injection give very good reproducibility (about 0.5% relative precision), making internal standards unnecessary, *cf* gas chromatography (Topic D5), and auto-injectors are frequently employed for routine work. An overall relative precision of between 1 and 3% can be expected.

D8 ELECTROPHORESIS AND ELECTROCHROMATOGRAPHY: PRINCIPLES AND INSTRUMENTATION

Key Notes

Principles — Electrophoresis is a technique for the separation of components of mixtures by differential migration through a buffered medium across which an electric field is applied. Electrochromatography is a hybrid of electrophoresis and HPLC.

Running buffer — Samples are introduced into a buffer solution that provides an electrically conducting medium and pH stability throughout the separation.

Supporting medium — In some modes of electrophoresis, solutes migrate through a solid medium consisting of a polymeric gel which supports the running buffer.

Electro-osmosis — The application of a potential gradient across the running buffer causes hydrated buffer cations to move towards the cathode. The resulting bulk flow of liquid is described as an electro-osmotic flow, and is of particular significance in high-performance capillary electrophoresis.

Sample injection — Samples are placed in wells formed in the gel slab or column, or introduced into a capillary column by hydrodynamic or electrokinetic means.

Solute detection — Solutes are either all detected *in situ* after the separation is complete by treating the supporting medium with a chromogenic reagent, or sequentially in the running buffer as they migrate towards one end of a capillary tube.

Instrument control and data processing — A dedicated microcomputer is an integral part of a modern electrophoresis system. Software packages facilitate the control and monitoring of instrumental parameters, and the display and processing of data.

Related topics

Principles of chromatography (D2)

High-performance liquid chromatography: principles and instrumentation (D6)

Principles

Electrophoresis, in its **classical** form, is used to separate mixtures of **charged solute species** by differential migration through a buffered electrolyte solution supported by a thin slab or short column of a polymeric gel, such as polyacrylamide or agarose, under the influence of an **applied electric field** that creates a **potential gradient**. Two platinum electrodes (cathode and anode) make contact with the electrolyte which is contained in reservoirs at opposite ends of the supporting medium, and these are connected to an external DC power supply. Considerable amounts of heat may be generated during separations at higher operating voltages (**Joule heating**), and many systems stabilize the operating temperature by water-cooling. Buffer solutions undergo electrolysis, producing hydrogen and oxygen at the cathode and anode, respectively, and the reservoirs have to be replenished or the buffer renewed to maintain pH stability.

Cationic solute species (positively-charged) migrate towards the cathode, **anionic species** (negatively-charged) migrate towards the anode, but **neutral species** do not migrate, remaining at or close to the point at which the sample is applied. The rate of migration of each solute is determined by its **electrophoretic mobility**, μ, which is a function of its net charge, overall size and shape, and the viscosity of the electrolyte. The latter slows the migration rate by viscous drag (frictional forces) as the solute moves through the buffer solution and supporting medium.

The distance travelled, d, after the application of a potential, E, between two electrodes for time, t, is given by

$$d = \mu \times t \times \left(\frac{E}{S}\right) \tag{1}$$

where S is the distance between the two electrodes and E/S is the potential gradient. For two separating solutes with mobilities μ_1 and μ_2, their separation, Δd , after time, t, is given by

$$\Delta d = (\mu_1 - \mu_2) \cdot t \cdot \left(\frac{E}{S}\right) \tag{2}$$

Δd is maximized by the application of a large potential gradient over a long period. However, as in chromatographic separations, diffusion of the solute species in the buffer solution causes band spreading which adversely affects resolution, so excessive separation times should be avoided.

The role of buffers in electrophoresis is crucial because many solutes are weakly acidic, basic or ampholytic, and even small changes in pH can affect their mobility. Amino acids, peptides and proteins are particularly susceptible as their direction of migration is a function of pH, e.g for glycine

$$\underset{\text{cation}}{\underset{\text{low pH}}{H_3N^+{-}CH_2{-}COOH}} \underset{H^+}{\overset{OH^-}{\longleftrightarrow}} \underset{\text{neutral/zwitterion}}{\underset{H_3N^+{-}CH_2{-}COO^-}{\underset{\text{or}}{H_2N{-}CH_2{-}COOH}}} \underset{H^+}{\overset{OH^-}{\longleftrightarrow}} \underset{\text{anion}}{\underset{\text{high pH}}{H_2N{-}CH_2{-}COO^-}}$$

Typical formats for **classical electrophoresis** are shown in *Figure 1*. In the slab gel method, the supporting gel is pre-formed into thin rectangular slabs on which a number of samples and standards can be separated simultaneously. Alternatively, it can be polymerized in a set of short tubes. The whole assembly is enclosed in a protective perspex chamber for safety reasons because of the high voltages employed (500 V–2 kV DC, or up to 50 V cm^{-1}). A separation may take from about 30 minutes to several hours, after which the gels are treated with a suitable visualizing agent to reveal the separated solutes (*vide infra*). Slab

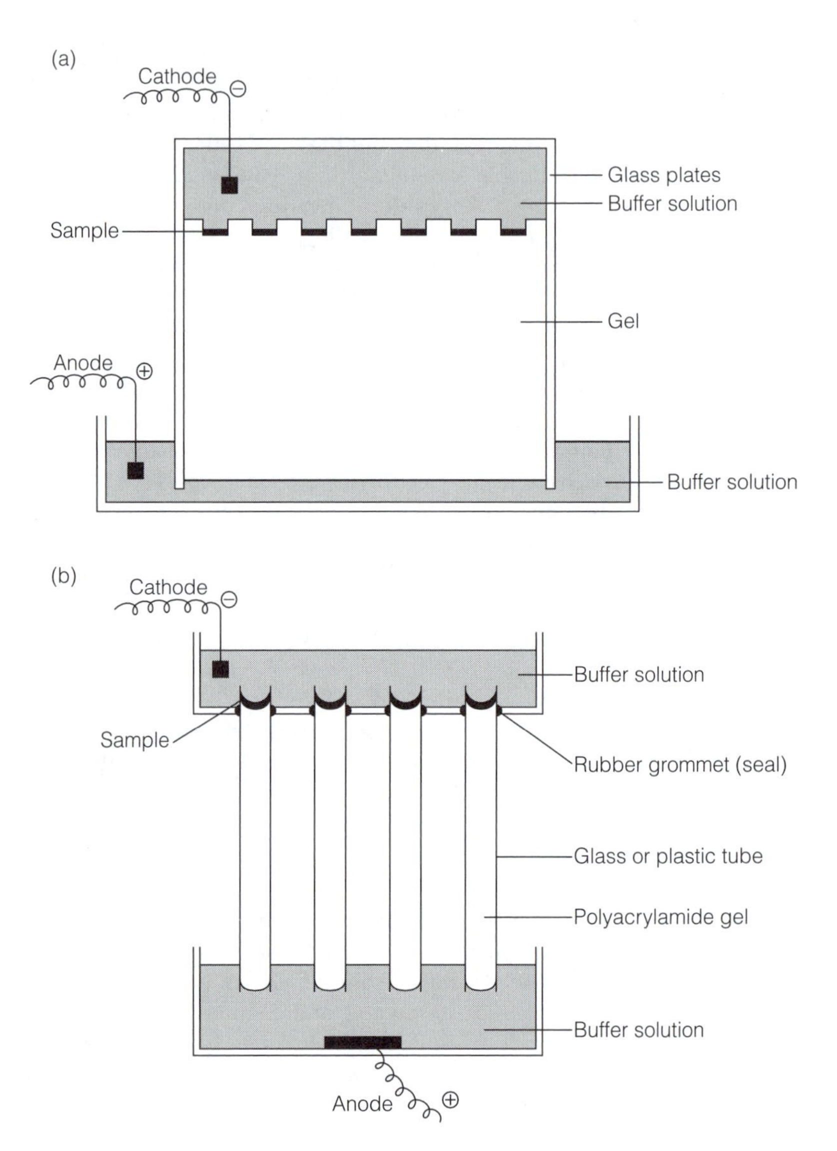

Fig. 1. Typical formats for classical gel electrophoresis. (a) Slab gel. (b) Tube gel. Reproduced from M. Melvin, Electrophoresis, *1987 with permission from Wiley-VCH and from M. Melvin.*

gels dissipate Joule heat more efficiently than column gels and have a superior resolving power.

High-performance capillary electrophoresis (**HPCE** or **CE**) is a relatively recently developed form where solutes are separated in a narrow-bore fused-quartz capillary tube, 10–75 cm in length and 25–100 μm internal diameter (*Fig. 2*). High voltages (10–50 kV DC, or up to 500 V cm^{-1}) result in rapid separations, and the heat generated is rapidly dissipated through the capillary wall. Solutes are detected sequentially by a variety of means as they travel towards one end of the capillary by migration and **electro-osmosis** (*vide infra*), but, unlike classical electrophoresis, only one sample at a time can be analyzed. A comparative summary of classical and capillary electrophoresis is given in *Table 1*.

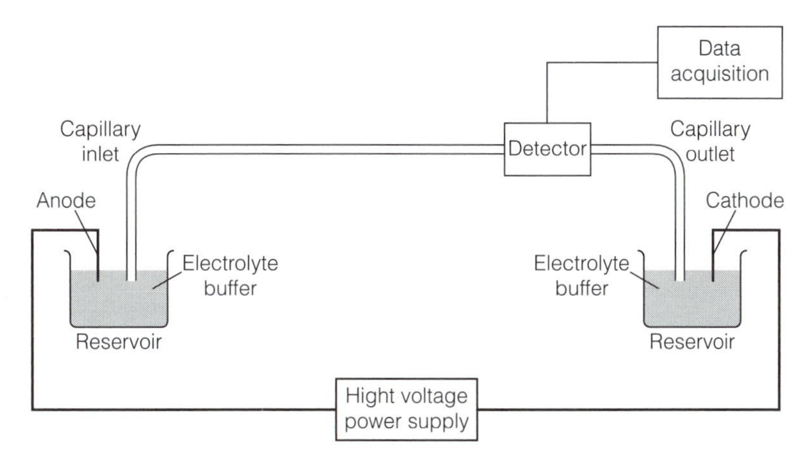

Fig. 2. Schematic diagram of a high performance capillary electrophoresis system.

Capillary electrochromatography (CEC) is an even newer technique than **CE** and uses capillaries packed with 5 μm or smaller particles of a stationary phase similar to those used in **HPLC** (Topics D6 and D7). Solutes are separated by a combination of electrophoretic migration and chromatographic sorption processes (Topic D2) giving the technique additional versatility in varying the selectivity. Efficiencies are particularly high because the packing minimizes band spreading by solute diffusion in the buffer solution.

Running buffer

The function of the running buffer is to provide an electrically conducting medium and pH stability. The latter is essential in ensuring that solutes have a constant mobility throughout the separation. Typical buffers covering a wide range of pH values are listed in *Table 2(a)*. Concentrations of 0.05–0.5 M provide an optimum ionic strength that allows rapid migration of solutes without the generation of excessive heat or losses by evaporation. Buffer additives in the form of surfactants, complexing agents and organic solvents are sometimes added to control migration rates and selectivity. Some examples are given in *Table 2(b)*.

Supporting medium

Various solid media are employed to support the running buffer in traditional electrophoresis. **Polymeric gels**, such as polyacrylamide, agarose, starch and cross-linked dextrans (Sephadex) are the most common, although paper and cellulose acetate have also been used. The gels are saturated with the running buffer, and have a restricted range of pore sizes that can be controlled during polymerization. This facilitates the separation of solutes by a **size exclusion** mechanism (Topic D2) in addition to their differential electrophoretic migration. **Polyacrylamide gels** are the most versatile, offering superior resolution and dissipating heat efficiently. **Agarose gels** are particularly effective for the separation of mixtures of large biomolecules such as DNA proteins, as the pore sizes are larger than those in polyacrylamide.

Electro-osmosis

When a potential gradient is applied across a running buffer, hydrated buffer cations tend to be drawn towards the cathode producing a bulk flow of liquid known as the **electro-osmotic flow** (EOF). Although the effect is not pronounced in classical electrophoresis, in capillary electrophoresis it is

Table 1. A comparison of classical gel and capillary electrophoresis

Classical gel	Capillary
Gels Polyacrylamide or agarose slabs: length and width 5–25 cm thickness 1–2 mm columns: length 7–10 cm internal diameter ~5 mm	*Tubing* Fused-quartz capillary length 25–75 cm internal diameter 20–100 μm
Applied field 100 V–2 kV (5–500 V cm^{-1})	*Applied field* 10–50 kV (up to 500 V cm^{-1})
Heat dissipation Slow from columns, quicker from slabs, aided by water cooling	*Heat dissipation* Rapid due to high surface area to volume ratio, aided by air cooling
Running buffer 0.05–0.5 M electrolyte providing conductivity and pH stability	*Running buffer* As for classical gel
Buffer additives Urea, surfactants, complexing agents, organic solvents	*Buffer additives* Urea, surfactants, inorganic salts, organic solvents, sulfonic acids, chiral cyclodextrins, amines
Sample injection 0.1–1 cm^3 loaded into well in slab or layered onto top of column	*Sample injection* 1–50 nl by hydrodynamic or electrokinetic method
Solute detection Chromogenic reagent or staining dye e.g. ninhydrin for amino acids, coomassie brilliant blue or ponceau S for proteins, acridine orange for nucleotides including DNA	*Solute detection* Similar to HPLC detectors, UV absorbance and fluorescence the most common, see *Table 3*
Modes Horizontal or vertical slab, disk or column SDS -PAGE Isoelectric focusing Immunoelectrophoresis	*Modes* Capillary zone electrophoresis (CZE) Micellar electrokinetic chromatography (MEKC) Capillary gel electrophoresis (CGE) Capillary isoelectric focusing (CIEF)
Performance Slow with limited resolution, but many samples run simultaneously, may require several hours	*Performance* Very high efficiency, resolution and sensitivity, moderately fast, 5 min to about 1 h

responsible for the movement of all species (cationic, anionic and neutral) along the capillary towards the detector and is the basis of the technique. The strength of the **EOF** in fused-quartz capillaries filled with a running buffer arises from the ionization of the surface silanol groups (SiOH) on the inner wall above about pH 4. Hydrated buffer cations accumulate close to the negatively charged

Table 2(a). Typical running buffers for electrophoresis

Buffer	Useful pH range
Phosphate	1.1–3.1
Ethanoate	3.8–5.8
Phosphate	6.2–8.2
Borate	8.1–10.1
Zwitterionic buffers	
MES (2-(4-morpholino)ethanesulfonic acid)	5.2–7.2
TRIS ((2,3-dibromopropyl) phosphate)	7.3–9.3

Table 2(b). Typical buffer additives for electrophoresis

Additive	Function
Inorganic salts	Change protein conformations
Organic solvents	Modify EOF, increase solute solubilities
Urea	Solubilize proteins, denature oligonucleotides
Surfactants	Form micelles, cationic ones reverse charge on capillary wall
Cyclodextrins	Provide chiral selectivity

surface to form an electrical double layer with an associated potential (zeta potential) (*Fig. 3*). Application of a voltage across the capillary initiates an **EOF**, the velocity or mobility, μ_{EOF}, of which is typically about an order of magnitude greater than those of individual solutes. Hence, the total migration rate of a solute, μ_{Tot}, is

$$\mu_{Tot} = \mu_{solute} + \mu_{EOF} \tag{3}$$

and all solutes, are carried towards the cathodic end of the capillary by the **EOF** where they pass through the detector cell.

Important characteristics of the **EOF** are:

- The velocity **increases** with **increasing pH** of the running buffer as more silanol groups become ionized.
- The velocity **increases** with the magnitude of the **applied potential**.
- The velocity **decreases** along with the **zeta potential** with **increasing ionic strength**.
- The velocity **decreases** with **increasing viscosity** of the running buffer.
- It has a **flat flow profile** across the capillary in contrast to the parabolic profile of a pumped **HPLC** mobile phase (*Fig. 3*). This minimizes **band spreading** and results in very high efficiencies (plate numbers) for the separating solutes.
- There is **no pressure drop** along the capillary, as occurs when the mobile phase is pumped through an **HPLC** column, because the electrically-generated driving force acts equally along the whole length.
- The velocity can be **controlled, reduced to zero or reversed** by additives such as surfactants, organic solvents or quaternary amine salts incorporated into the running buffer, or by modifying the inside wall of the capillary.

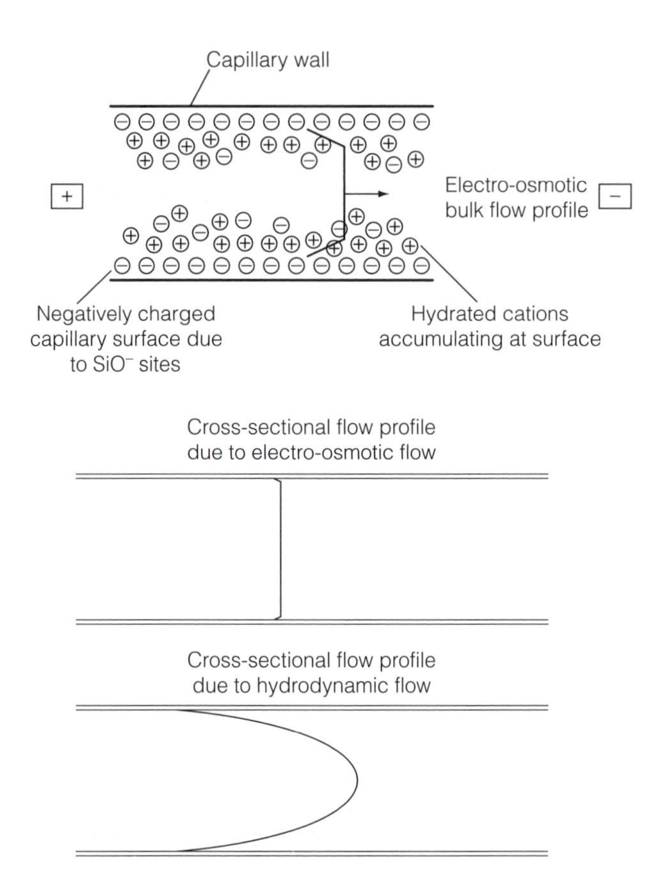

Fig. 3. EOF and comparisons of flow-profiles in a fused-quartz capillary and an HPLC column.

Sample injection

For classical electrophoresis, samples of 0.1–1 cm^3 are loaded into wells formed in gel slabs or layered onto the tops of gel columns, often with the addition of a sucrose solution to increase the density. For **CE** and **CEC**, much smaller samples (1–50 nl) are drawn into one end of the capillary (usually the anodic end) from a sample vial, either hydrodynamically using gravity, positive pressure or a vacuum, or electrokinetically by applying a voltage for a short time when the **EOF** causes the sample components to migrate into the capillary. The reproducibility of sample injection into capillaries, typically 0.5–3%, is variable, and electrokinetic methods may discriminate between components of a mixture because of differences in electrophoretic mobilities. Time, temperature, pressure drops and sample and running buffer viscosities are all sources of variability, and automated sample injection is preferable to minimize these effects.

Solute detection

For classical gel electrophoresis, solutes are detected on the gel after the separation is complete by treating it with a chromogenic reagent similar to those used in **TLC** (Section D3), or a staining dye. The gels are immersed in a reagent or dye solution, then the excess is removed by washing with a suitable solvent to reveal the solute bands.

Detectors for **CE** and **CEC** are similar to those used in **HPLC** (Topic D6), **UV-absorbance** spectrometers, especially diode array detectors, and fluorimeters being the most common. DADs are increasingly preferred because of their ability to monitor at multiple selected wavelengths and the complete spectral information provided. The detector cell is normally a portion of the capillary itself, sometimes enlarged (bubble cell) or bent so as to increase the optical path and hence the sensitivity. *Table 3* summarizes the characteristics of the more important detectors.

Table 3. Characteristics of CE and CEC detectors.

Detector	Sensitivity mass (moles)	Sensitivity concentration (molar)	Characteristics
UV-visible absorbance	10^{-13}–10^{-16}	10^{-5}–10^{-8}	Good sensitivity, most widely used. DADs are versatile and give spectral information.
Fluorescence	10^{-15}–10^{-17}	10^{-7}–10^{-9}	Sensitive, but many solutes need to be derivatized.
Laser-induced fluorescence	10^{-18}–10^{-20}	10^{-14}–10^{-16}	Extremely sensitive, but many solutes need to be derivatized. Expensive.
Electrochemical			Sensitive, require special electronics and capillary modification. Conductometric almost universal.
Amperometric	10^{-18}–10^{-19}	10^{-10}–10^{-11}	
Conductometric	10^{-15}–10^{-16}	10^{-7}–10^{-8}	

Instrument control and data processing

These aspects of CE and CEC closely parallel those described for GC and HPLC (Topics D4 and D6). Additional instrument parameters to be set and monitored or controlled include:

- applied potential and run-time;
- temperature (heat dissipation);
- hydrodynamic and electrokinetic sample-injection.

D9 Electrophoresis and electrochromatography: modes, procedures and applications

Key Notes

Modes of electrophoresis and electrochromatography

Modes of electrophoresis are defined by the nature and form of the supporting medium, the running buffer and any incorporated additives. For capillary electrochromatography (CEC), the capillary is filled with a stationary phase similar to those used in HPLC.

Qualitative analysis

For classical gel electrophoresis, unknown solutes are identified by comparisons of the distances migrated with those of standards run simultaneously. For capillary electrophoresis (CE) and CEC, migration times and spectrometric data are used.

Quantitative analysis

Classical gel electrophoresis provides only semi-quantitative information at best. For CE and CEC, quantitative information is readily obtained from peak area measurements and calibration graphs.

Related topic

Electrophoresis and electrochromatography: principles and instrumentation (D8)

Modes of electrophoresis and electrochromatography

There are several modes of classical gel electrophoresis. They are defined by the format and the nature of the gel, the running buffer and any incorporated additives.

- **Slab gel, column** and **disk electrophoresis** (Topic D8) are the principal formats. They allow multiple samples and standards to be run simultaneously for comparison purposes, which is comparable to separations by **TLC** (Topic D3). Samples are placed in wells near the centre of horizontal slabs when both cationic and anionic species are to be separated, or at one end if all solutes are expected to carry the same charge. With vertical slabs, samples are placed in wells at the top of the slabs so that only downwards migration is possible.
- **Sodium dodecyl sulfate polyacrylamide gel electrophoresis (SDS-PAGE)** is used to determine the relative molecular masses (RMM values) of individual proteins in a mixture. The proteins are first de-natured with **mercaptoethanol** before bonding to the SDS, which is an **anionic surfactant**. The negatively-charged proteins subsequently migrate at rates inversely proportional to the **logarithm** of their RMM, the system being calibrated with proteins of known RMM.
- **Isoelectric focusing** is a mode used to separate ampholytes such as amino

acids and peptides. A stable **pH gradient** is first created along the gel by polymerizing it with a mixture of polyamino-polycarboxylic acids having a range of pK_a values. Applying a potential causes them to migrate to positions in the gel where they become stationary on account of electrical neutrality by forming neutral molecules or zwitterions. The pH values at these positions define their **isoelectric points**, *pI*. Electrophoresis of samples results in ampholytic components migrating to their respective isoelectric points in the pH gradient, after which they can be located with a suitable dye.

- **Immunoelectrophoresis** depends on specific **antigen–antibody reactions** for the detection of separated proteins. Low-voltage electrophoresis of samples in an agarose gel is followed by the introduction of **antisera** that diffuse throughout the gel forming visible precipitates with the separated antigens.

Capillary electrophoresis separations can be performed in one of four principal modes, each depending on a different separation mechanism. They are primarily used in the pharmaceutical, clinical and biomedical fields for the analysis of mixtures of amino acids, peptides, proteins and other macromolecules, and drugs and their metabolites in body fluids. Analyses down to nanogram (10^{-9} g) or picogram (10^{-12} g) levels are often quicker, giving better resolution than corresponding **HPLC** procedures (*Fig. 2c*).

- **Capillary zone electrophoresis**, **CZE**, is the simplest and currently the most widely used mode of **CE**. The capillary is filled with a running buffer of the appropriate pH and ionic strength (Topic D8, *Table 1a*), and all solutes are carried towards the cathodic end of the capillary by a strong **EOF**. Cationic and anionic solutes are separated, but neutral species, which migrate together at the same velocity as the **EOF**, are not separated from one another. **Cationic** solutes migrate **faster** than the **EOF** because their overall mobilities are enhanced by their attraction to the cathode, whereas **anionic** solutes migrate **slower** than the **EOF** because they are attracted towards the anode. Solutes reach the detector in order of decreasing **total mobility** ($\mu_{Tot} = \mu_{solute} + \mu_{EOF}$), as shown diagrammatically in *Figure 1a*, the individual solute mobilities being determined by their size and charge, that is:
  (i) **cationic** species first in **increasing** order of size;
  (ii) **neutral** species next, but not separated;
  (iii) **anionic** species last in **decreasing** order of size.
  Buffer additives, such as urea, surfactants and organic solvents are sometimes used to alter the selectivity of a separation through controlling the **EOF**, solute mobilities, solubilities and other effects (Topic D8, *Table 1b*). An example of a **CZE** electropherogram, the separation of some artificial sweetners and preservatives, is shown in *Figure 1b*.
- **Micellar electrokinetic chromatography**, **MEKC** or **MECC**, is a more versatile mode than **CZE** because both neutral and ionic solutes can be separated. A **surfactant** is added to the running buffer, forming aggregates of molecules, or **micelles**, having a hydrophobic center and a positively or negatively-charged outer surface (*Fig. 2a*). The micelles act as a chromatographic **pseudo-stationary phase**, into which neutral solutes can partition, their **distribution ratios** depending on their degree of hydrophobicity. **Cationic micelles** migrate towards the cathode faster than the **EOF** and **anionic micelles** more slowly. Neutral solutes migrate at rates intermediate between the velocity of the **EOF** and that of the micelles. By analogy with chromatography, they are **eluted** with characteristic **retention times**, t_R, that depend on

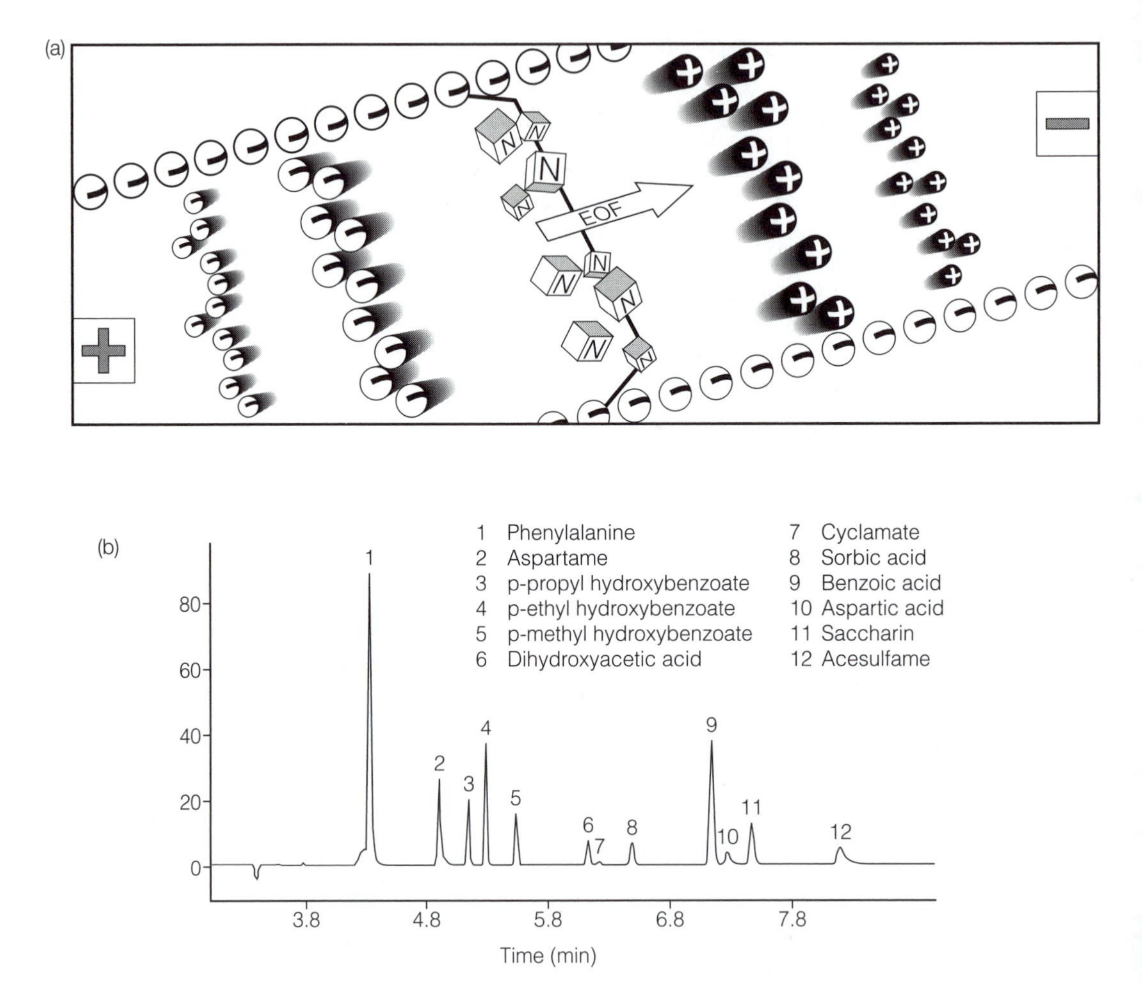

Fig. 1. Capillary zone electrophoresis (CZE). (a) EOF and order of solute migration. (b) Separation of some artificial sweeteners and preservatives by CZE capillary, 65 cm, 50 mm i.d.; buffer, 0.02 M borate, pH 9.4; temperature, 25°C; voltage, 30 kV, injection, hydrodynamic 50 mbar sec; detection, UV absorbance at 192 nm. Reproduced from D.N. Heijer, High Performance Capillary Electrophoresis, *1992, with permission from Agilent Technologies UK Ltd and D.N. Heijer.*

their distributions between the running buffer and the micelles. The general order of elution is:

(i) cationic micelles and cationic solutes;
(ii) neutral solutes partitioning into the cationic micelles;
(iii) **EOF**;
(iv) neutral solutes partitioning into anionic micelles;
(v) anionic micelles and anionic solutes.

Sodium dodecyl sulfate, SDS, is a frequently used **anionic surfactant** for **MEKC**. The most hydrophobic neutral solutes migrate at the same velocity as the SDS, which is slower than the **EOF**, and the least hydrophobic neutral solutes migrate at the same velocity as the **EOF** (*Fig. 2a*).

An example of an **MEKC** electropherogram, the separation of some components of cold relief-products, is shown in *Figure 2b*. The addition of chiral selectors such as the cyclodextrins (Topic D6), in place of a surfactant, enables mixtures of enantiomers to be separated more cheaply than by **HPLC** (*Fig. 2c*).

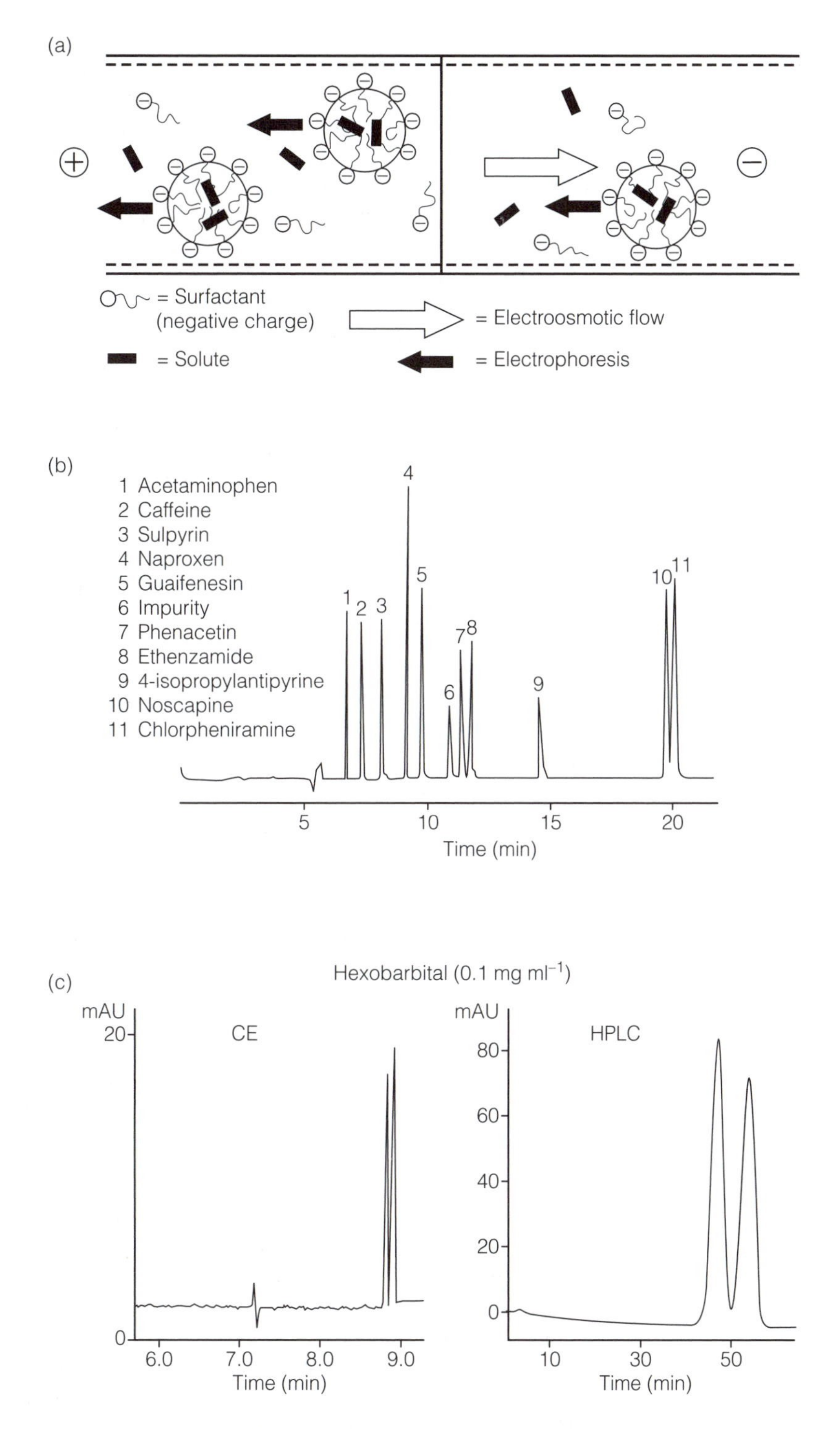

Fig. 2. Micellar electrokinetic capillary chromatography (MEKC). (a) Formation and migration of SDS micelles; (b) separation of some constituents of cold-relief products; (c) separation of the enantiomers of hexobarbital and comparison with an HPLC separation on a chiral stationary phase. b and c reproduced from D.N. Heijer, High Performance Capillary Electroporesis, *1992, with permission from Agilent Technologies UK Ltd and D.N. Heijer.*

- **capillary gel electrophoresis, CGE,** is similar to classical gel electrophoresis, the capillary being filled with a polyacrylamide or agarose gel that superimposes size exclusion selectivity onto the electrophoretic migration of ionic solutes. The larger the solute species the slower the rate of migration through the gel. Solute peaks are narrow because band spreading by diffusion in the running buffer is hindered by the gel structure. The main applications of **CGE** are in separating polymer mixtures, protein fractions and DNA sequencing.
- **capillary isoelectric focusing, CIEF,** is similar to classical isoelectric focusing, a **pH gradient** being first formed in the capillary using carrier ampholytes having *pI* values spanning the required pH range, typically 3 to 10. Sample solutes migrate and are **focused** in positions along the capillary where their isoelectric point, *pI*, is equal to the pH. Solute zones are **self-sharpening** because diffusion away from the focal points causes the solutes to aquire a charge which results in them migrating back towards their isoelectric point. After the separation is complete, pressure is applied to the anodic end of the capillary to move all the solutes sequentially through the detector cell.

Capillary electrochromatography, CEC, is a relatively new technique, and is a hybrid of capillary electrophoresis and high-performance liquid chromatography, combining elements of both. Particular features of **CEC** are:

- the capillary is packed with an **HPLC** stationary phase, usually a bonded-phase silica, and filled with a running buffer (>pH 4);
- as in **CE**, the applied potential generates a strong **EOF** with a flat flow profile, but the electrical double-layer formed is predominantly at the surface of the individual particles of packing rather than the capillary wall;
- unlike in **HPLC**, there is no pressure drop because the driving force is generated throughout the length of the column;
- even higher efficiencies are observed than in **CZE** because the column packing limits solute diffusion in the mobile phase. Very small particles of stationary phase, currently 1.5 to 3 μm nominal diameter, can be used and columns of 25 to 50 cm in length are typical. Internal diameters are generally 50 to 100 μm, but narrower columns are advantageous because the **EOF** is faster, thus speeding up the separations;
- column packings can be porous, non-porous, spherical or irregular in shape and of controlled pore size if required. In some cases, **mixed-mode separations** can be achieved by using both non-polar and polar or ionic bonded phases in the same column.

CEC separations are based on both electrophoretic migration for charged analytes and chromatographic sorption for neutral species, hence providing an additional source of selectivity over and above differences in electrophoretic mobility. The composition of the mobile phase can have dramatic effects on both the **EOF** and the selectivity of the separation. **CEC** has a number of other advantages over both **CE** and **HPLC**. Compared to **HPLC**, solvent consumption is greatly reduced, which facilitates coupling **CEC** to mass spectrometry (Section F). Furthermore, there is no need to use a micelle-forming surfactant additive as in **MEKC** when separating neutral solutes. This is also an advantage when coupling the technique to mass spectrometry because solvents containing high concentrations of surfactants such as SDS can cause difficulties with some ionization sources.

As yet, the number of applications is limited but is likely to grow as instrumentation, mostly based on existing **CE** systems, and columns are improved and the theory of **CEC** develops. Current examples include mixtures of polyaromatic hydrocarbons, peptides, proteins, DNA fragments, pharmaceuticals and dyes. Chiral separations are possible using chiral stationary phases or by the addition of cyclodextrins to the buffer. In theory, the very high efficiencies attainable in **CEC** provides high peak capacities, and therefore the possibility of separating complex mixtures of hundreds of components. A typical **CEC** separation is shown in *Figure 3*.

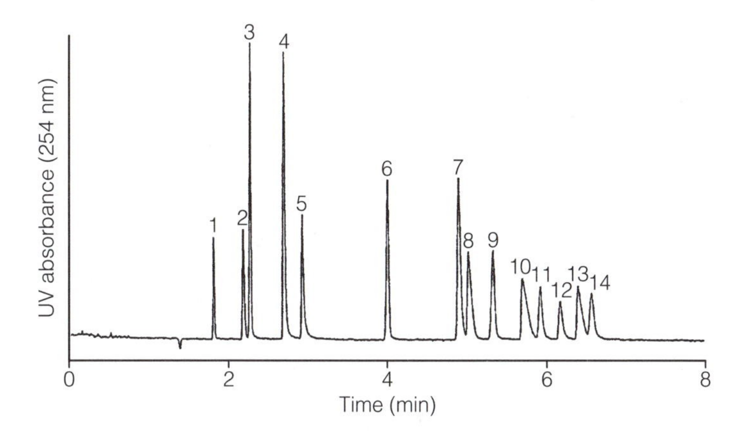

Fig. 3. Separation of 14 explosive compounds by capillary electrochromatography. Column: 30 cm × 75 µm, 17 cm packed with 1.5 µm dp non-porous ODS-II (Micra Scientific); mobile phase: 15% methanol–85% 10 mM MES; applied voltage: 12 kV; injection: 1 s at 2 kV. Peaks: 1 = octahydro-1,3,5,7-tetranitro-1,3,5,7-tetrazocine; 2 = hexahydro-1,3,5-trinitro-1,3,5-triazine; 3 = 1,3-dinitrobenzene; 4 = 1,3,5-trinitrobenzene; 5 = nitrobenzene; 6 = 2,4,6-trinitrotoluene; 7 = 2,4-dinitrotoluene; 8 = methyl-2,4,6-trinitrophenylnitramine; 9 = 2,6-dinitrotoluene; 10 = 2-amino-4,6-dinitrotoluene; 11 = 2-nitrotoluene; 12 = 4-nitrotoluene; 13 = 4-amino-2,6-dinitrotoluene; 14 = 3-nitrotoluene.

Qualitative analysis

Classical gel electrophoresis is primarily a qualitative technique, the identification of unknown solutes being accomplished by comparisons of migration distances with those of standards run simultaneously or sequentially under identical conditions. Migration distances can be expressed as R_f values analogous to those used in **TLC** (Topic D3).

Solute **migration times** in **CE** and **CEC** separations can be equated to **chromatographic retention times**, t_R, and similar methods are used to identify unknowns (Topic D2), that is:

- **comparisons of retention times** with those of known solutes under identical conditions;
- **comparisons of electropherograms** of samples **spiked** with known solutes with the electropherogram of the unspiked sample;
- **comparisons of UV-visible spectra** recorded by a diode-array detector with those of known solutes;
- **interfacing** of a **CE** or **CEC** system with a mass spectrometer. Identifications are facilitated by searching libraries of computerized spectra (Topics F3 and F4).

Quantitative analysis

Generally, only semi-quantitative information can be obtained from classical gel electrophoresis; visual comparisons can be made with standards run on the same gel. Scanning densitometry, as used in **TLC** (Topic D3) provides more reliable data, but variations in gel thickness, band spreading and reactions with reagents and staining dyes all contribute to considerable variability in the measurements.

For **CE** and **CEC**, the methods used for quantitative chromatography (Topic D2) are suitable, and alternative calibration procedures are described in Topic A5. **Peak areas** are more reliable than peak heights, as they are directly proportional to the quantity of a solute injected when working within the linear range of the detector.

E1 ELECTROMAGNETIC RADIATION AND ENERGY LEVELS

Key Notes

Electromagnetic radiation	The electromagnetic spectrum covers a very large range of wavelengths, frequencies and energies, and many analytical spectrometric techniques involve electromagnetic radiation.
Atomic energy levels	Energy levels in atoms are defined by quantum numbers, the atoms of each element possessing a characteristic set of discrete levels determined by its atomic and nuclear structure.
Molecular energy levels	Every molecule has several sets of discrete energy levels, which are associated with particular structural and behavioral properties of molecules.
Related topics	Other topics in Section E.

Electromagnetic radiation

The nature of light and other radiation was the subject of much investigation since Newton's experiments in the 17th Century. It is a form of energy and may be considered either as a continuous wave travelling through space, or as discrete photons of the same energy. For many spectrometric techniques, the wave approach is more useful. *Figure 1* shows a representation of an electromagnetic wave as an oscillating electric field of amplitude E and a magnetic field of amplitude H at right angles to each other.

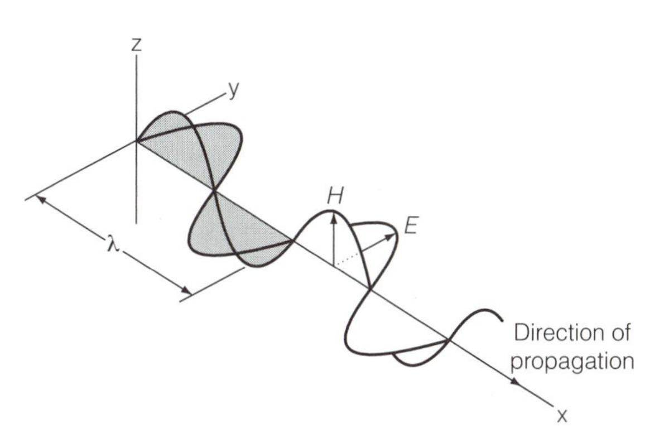

Fig. 1. Plane-polarized electromagnetic radiation traveling along the x-axis.

In a vacuum, this wave travels at a fundamental constant speed, c_o

$$c_o = 2.997\ 925 \times 10^8 \text{ m s}^{-1}$$

The wave is characterized in three ways, as shown in *Figure 1*.

The **wavelength, λ,** is the distance between equivalent points on the wave train, for example, between two consecutive positive crests, or two points where the wave increases through the zero value. The wavelength has been expressed in a variety of units, but these should now all be related to the metre, as shown in *Table 1*.

Table 1. Wavelength units

Name	Units
femtometer (fermi)	10^{-15} m = 1 fm
micrometer (micron)	10^{-6} m = 1 μm
nanometer	10^{-9} m = 1 nm

Older units such as the Ångstrom (Å) are used in earlier work.
1 Å = 10^{-10} m.

The **frequency, ν,** is the number of cycles of radiation passing a point in space per second. It is expressed as s^{-1}, or hertz (Hz).

The above definitions show that the relation between these quantities is:

$$\nu = c_o/\lambda$$

Sometimes the **wavenumber, $\bar{\nu}$,** is used where

$$\bar{\nu} = 1/\lambda$$

The wavenumber is frequently given in cm^{-1}, especially in infrared spectrometry, and it should be noted that 100 m^{-1} = 1 cm^{-1}. (Note that if there are 100 per meter, there is 1 every centimeter.)

The **energy, ε,** of the radiation is most important, since it defines the molecular or atomic processes which are involved. For a single **photon,**

$$\varepsilon = h\nu = hc_o/\lambda = hc_o\bar{\nu}$$

where h is the Planck constant, 6.62608×10^{-34} J s^{-1}.

Occasionally, the **electron-volt** is used as a unit for energy, where

$$1 \text{ eV} = 1.602 \times 10^{-19} \text{ J}$$

Thus, a wavelength of about 5.00 μm is equivalent to a frequency of 5.996×10^{13} Hz, a wavenumber of 2000 cm^{-1}, and energy 3.973×10^{-20} J. This corresponds to molecular vibrational energy. It is sometimes an advantage to consider 1 **mole** of photons. For the above example the molar energy will be:

$$N_A\,\varepsilon = 6.022 \times 10^{23} \times 3.973 \times 10^{-20} = 23.9 \text{ kJ mol}^{-1}$$

Table 2 shows the very wide range of wavelengths and energies that relate to spectrometric techniques (see Topic A3) and *Figure 2* relates this to the electromagnetic spectrum.

When electromagnetic radiation is directed at an atom or molecule, the atom or molecule can absorb photons whose energy corresponds exactly to the difference between two energy levels of the atom or molecule. This gives rise to an

Table 2. The regions of the electromagnetic spectrum[a]

Wavelength range	Frequency (Hz)	Region	Spectra
100–1 m	$3–300 \times 10^6$	Radiofrequency	Nuclear magnetic resonance
1–0.1 m	$0.3–3 \times 10^9$	Radiofrequency	Electron spin resonance
100–1 mm	$3–300 \times 10^9$	Microwave	Rotational
1–0.02 mm	$0.3–15 \times 10^{12}$	Far infrared	Vibrational
20–2 mm	$15–150 \times 10^{12}$	Infrared (IR)	Vibrational
2–0.8 mm	$150–375 \times 10^{12}$	Near infrared	Vibrational
800–400 nm	$375–750 \times 10^{12}$	Visible	Electronic
400–150 nm	$750–2000 \times 10^{12}$	Ultraviolet (UV)	Electronic
150–2 nm	$2–150 \times 10^{15}$	Vacuum UV	Electronic
2–0.1 nm	$150–3000 \times 10^{15}$	X-ray	Inner shell electronic
0.1–0.0001 nm	$3–3000 \times 10^{18}$	γ-ray	Nuclear reactions

[a]The regions overlap considerably, and the range is approximate.

absorption spectrum. It must be noted, however, that when a large change occurs (e.g. due to an alteration in the electronic structure of a molecule) the less energetic changes, such as the vibration of bonds and rotation of the molecule, will happen as well, leading to more complex spectra.

Atomic energy levels

Quantum theory shows that atoms exist only in discrete states, each of which possesses a characteristic energy, defined by **quantum numbers,** which characterize the atomic state. Transitions may occur only between these levels, and even then some transitions are unfavorable. Electrons occupy **atomic orbitals** with characteristic spatial distributions around the nucleus.

The discrete energy levels arise naturally as the allowed solutions of the wave equations for the system under consideration. Electronic energy levels in atoms may be accounted for by solving the **Schrödinger wave equation.**

Atoms have electronic energy levels and atomic orbitals that are defined by three quantum numbers that can have integer values:

n principal quantum number,
l orbital angular momentum quantum number,
m magnetic quantum number, and
m_s electronic spin quantum number which can be +1/2 or −1/2 only.

The energy of an orbital is mostly dependent on its principal quantum number n. In fact, for hydrogen, the energy depends only on n. There are only certain allowed values of the other quantum numbers. For example, l may take integer values from 0 to $(n - 1)$; m values from $+l$ to $-l$ and m_s +1/2 or −1/2.

The different orbitals are described by symbols:

s (sharp)	for $l = 0$
p (principal)	for $l = 1$
d (diffuse)	for $l = 2$
f (fundamental)	for $l = 3$

For atoms other than hydrogen, the other quantum numbers modify the energy slightly. For example, the 3*p* level where $n = 3$, $l = 1$ has a higher energy than the 3*s* with $n = 3$, $l = 0$. These are often referred to as subshells.

The atoms of the various elements are built up by adding electrons into the next empty level with the lowest energy, remembering that each level may

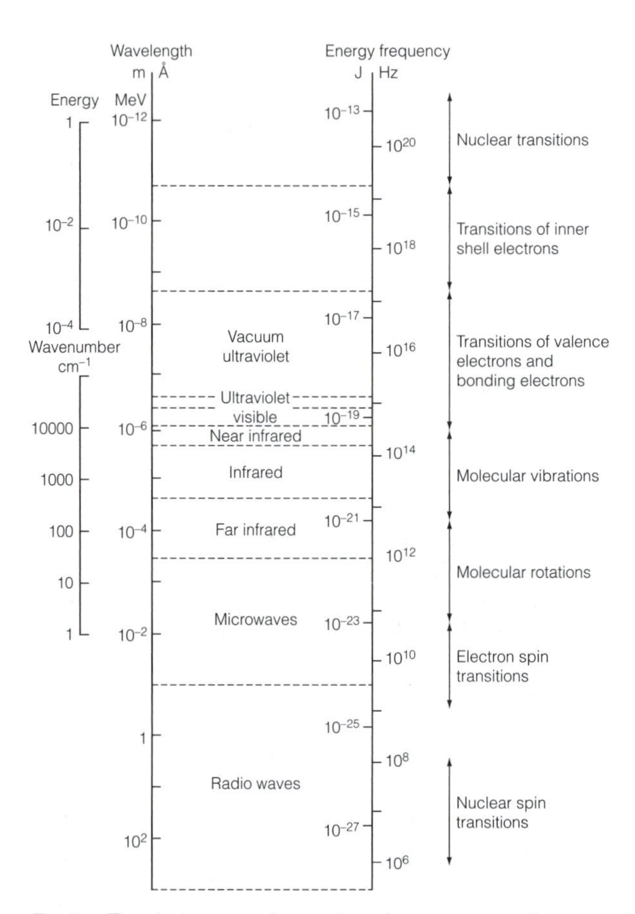

Fig. 2. The electromagnetic spectrum from γ-ray to radiowave.

contain two electrons with opposite spins ($m_s = \pm 1/2$). This is called the **Aufbau principle**. An example may be used to illustrate this. The element lithium, atomic number 3, has 3 electrons. In the unexcited or **ground** state, these must occupy the lowest energy levels, which are the 1*s* and 2*s* levels. Two electrons fill the 1*s* level and one goes into the 2*s*.

Figure 3 shows the sets of atomic energy levels with $n = 2, 3, 4, 5, 6$ and 7 and $l = 0, 1, 2$ and 3. The diagram also shows that the most favorable transitions occur when l changes by ± 1. It is worth noting that the transition shown in bold is used to measure lithium in atomic emission spectrometry (see Topic E4). In an excited state, the electron population is altered. In **transition elements** there are many low-lying energy levels and excited states with similar energies.

Molecular energy levels

Molecules also possess energy levels defined by quantum numbers. When atoms combine into molecules, their orbitals are changed and combined into **molecular orbitals**. As an example, the atomic orbitals of carbon, hydrogen and oxygen combine in the molecule of propanone, C_3H_6O, so that the three carbons are linked in a chain by single (σ) bonds, the two outer carbons are each linked by σ bonds to three hydrogens, while the central carbon is linked by a double bond to the oxygen, that is by both a σ and a π bond. Additionally, the oxygen still has unpaired or nonbonded n electrons.

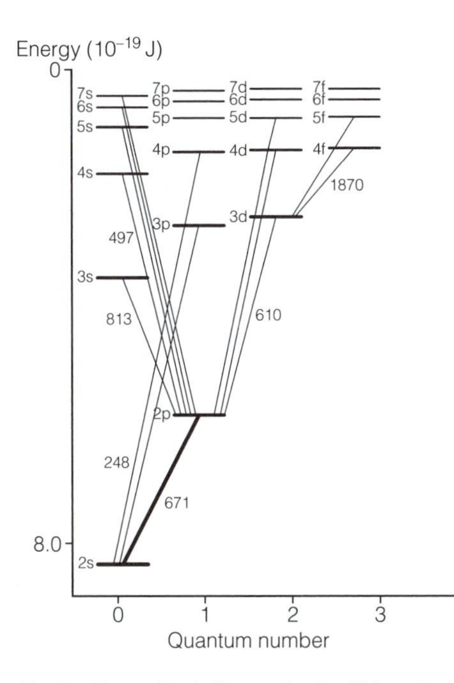

Fig. 3. Energy level diagram for the lithium atom showing the wavelength in nm for a number of transitions. Note: the transition at 671 nm is used in flame emission spectrometry.

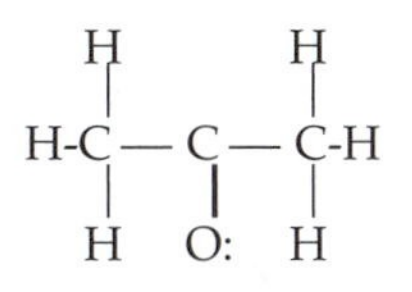

This results in a set of bonding and corresponding antibonding electronic orbitals or energy levels as shown schematically in *Figure* 4. Transitions may occur selectively between these levels, for example between the π and π^* levels.

In addition, molecules may **vibrate**, and the **vibrational energy levels** are defined by the vibrational quantum number, v. In the gaseous state, molecules may also **rotate** freely, and the **rotational energy levels** are defined by the rotational quantum number, J. In the condensed states of solid and liquid, rotation is restricted. With nuclear spin, the **nuclear spin quantum number,** I, is important.

In more complex molecules, additional rotational, vibrational, electronic and nuclear energy levels are possible, and some simplification is usually employed in interpreting their spectra, as described in the later topics in this section *(Fig. 5)*.

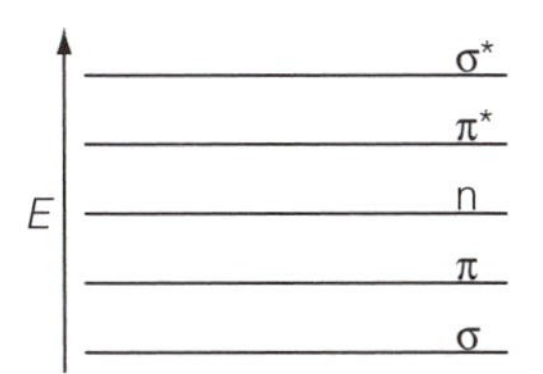

Fig. 4. Schematic of molecular orbitals or energy levels.

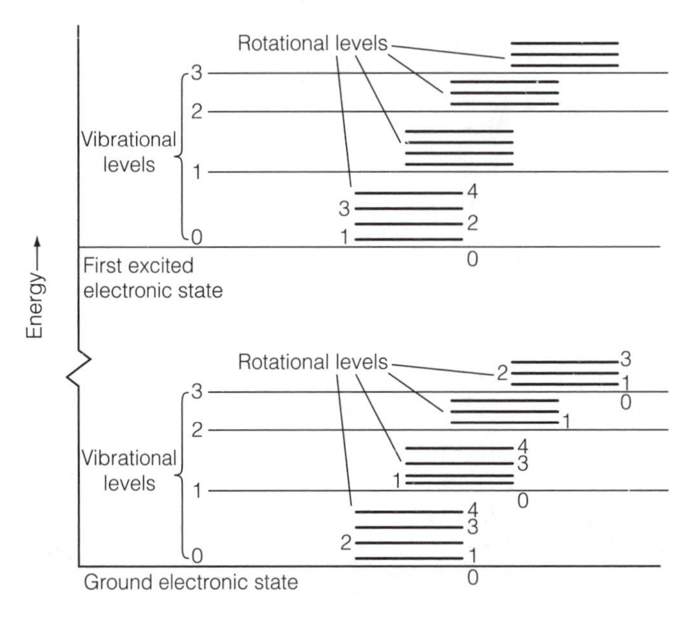

Fig. 5. Energy level diagram for a molecule.

Although the technique of **mass spectrometry** does not normally involve electromagnetic radiation, molecules or atoms are ionized and dissociated by the removal of electrons using high-energy techniques (see Topic E14). The ions are then dispersed, detected and displayed as a **mass spectrum**. The information given by measurements using mass spectrometry is helpful both in combination with atomic emission spectrometry (see Topic E5) and in molecular structure elucidation, where it is used in conjunction with other molecular spectrometric techniques (see Topic F2). For this reason mass spectrometry is often included in a discussion of spectrometry.

E2 ATOMIC AND MOLECULAR SPECTROMETRY

Key Notes

Transitions — Transitions between energy levels may occur in several ways, providing different types of spectra for analysis.

Quantitative spectrometry — The intensity of the spectral emission or the reduction of intensity by absorption is related to the concentration of the species producing the spectrum.

Beer–Lambert absorption law — An equation relating the concentration of an absorbing species and the path length through the sample to the absorbance at a particular wavelength was derived by Beer and Lambert and is used extensively in quantitative absorption spectrometry.

Selection rules — Certain transitions are favored by quantum theory rules. Others are much less favored or are forbidden. The allowed transitions are summarized by selection rules.

Related topics — Other topics in Section E.

Transitions

For a given set of energy levels, defined by their quantum numbers, there are several possible types of transition (*Figs 1 and 2*). Consider just two energy levels, the upper with energy E_U and the lower, E_L, separated by an energy difference ΔE, where

$$\Delta E = E_U - E_L$$

There is a definite statistical probability of transitions between these levels, which depends on the structure and population of the energy levels and also on ΔE. These are summarized by **selection rules** which govern allowed transitions.

If energy is supplied to the sample, for example, by passing electromagnetic radiation through it, the sample atoms or molecules may **absorb energy** and be promoted into the higher energy level. The radiation emerging from the sample

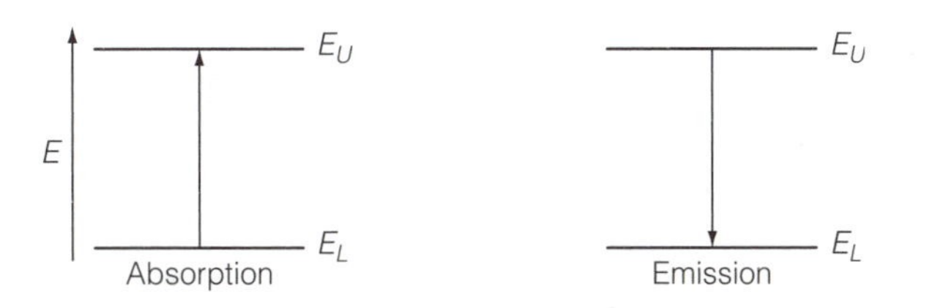

Fig. 1. Absorption and emission of radiation.

will be less intense. If the sample is activated in another way, for example thermally, then atoms or molecules are promoted to the higher energy level and **emit energy** in falling back to the lower level.

If a sample absorbs energy, and the atoms or molecules then drop back to an intermediate level, before returning to the lowest level, emitting radiation, this is called **fluorescence**. The energy of the fluorescent emission is generally lower than that initially absorbed. This is shown in *Figure 2(a)*.

Raman spectrometry is a technique in which a much higher energy radiation is applied to the sample, so molecules initially in both upper and lower energy levels are promoted, and then fall back. If they fall back to the higher level, a lower frequency radiation is emitted, called **Stokes Raman scattering**, whereas if they fall back to the higher level it is **Anti-Stokes Raman scattering**. If the exciting radiation is re-emitted unchanged, this is **Rayleigh scattering**. As shown in *Figure 2(b)*, the difference or **Raman shift** corresponds to ΔE.

Thus, if tetrachloromethane CCl_4 is excited in a Raman spectrometry experiment by a laser source of wavelength 488.0 nm, the resulting spectrum also shows lines at 499.2 and at 477.3 nm, corresponding to a wavelength difference of 21.79 μm. This means that the levels of energy E_L and E_U are separated by 459 cm^{-1}, which is due to a vibrational transition, appearing as a Raman band in the visible region. It is found to correspond to the symmetric stretching vibration of CCl_4 (Topic E10).

As noted above, the probability of a transition depends in part on the **populations** of the energy levels, which are given by the **Boltzmann distribution law**:

$$\frac{N_U}{N_L} = \frac{g_U}{g_L} \exp(-\Delta E/kT)$$

where N is the number of the species of interest in the level, g is the **degeneracy** or number of levels of that energy, k is the Boltzmann constant, 1.380×10^{-23} J and the subscripts denote the upper (U) or lower (L) level.

This law shows several features useful for analytical spectrometry:

- the greater the energy difference, the smaller the ratio of the population of the upper level with respect to the lower;
- the higher the temperature the larger the ratio;
- the higher the degeneracy ratio (g_U/g_L), the larger the ratio.

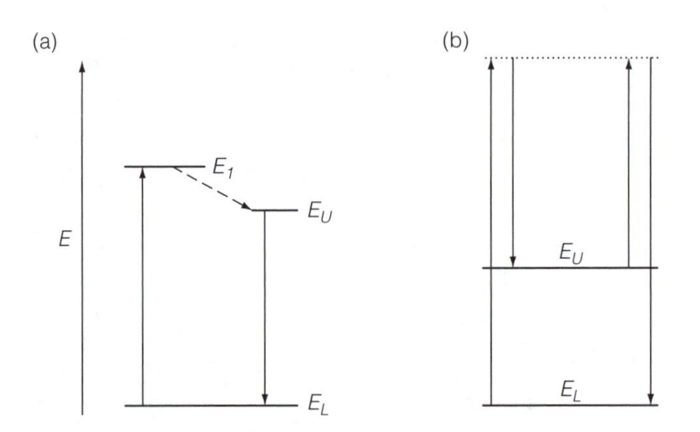

Fig. 2. Transitions for (a) fluorescence emission, and (b) Raman spectrum.

This has considerable importance in practice. For example:

- In infrared spectrometry, suppose a fundamental vibration occurs at a wave number 3000 cm^{-1} and has an **overtone** at around 6000 cm^{-1}. At room temperature, approximately 300 K, $N_U/N_L = 5.55 \times 10^{-7}$ or about 1 in 2 million molecules are in the upper level for the fundamental band, and for the first overtone, $N_U/N_L = 3.09 \times 10^{-13}$, which is very small. At 1000 K, the ratios are 0.0133 or 1 in 75 and 1.769×10^{-4}, showing that transitions involving the overtone levels will be more probable at high temperature and are referred to as '**hot bands**'.
- In proton nuclear magnetic resonance spectrometry, the levels are separated by about $\Delta E = 6.6 \times 10^{-26}$ J. Therefore, at room temperature $N_U/N_L = 0.99998$, which indicates that the levels are very nearly equally populated. An NMR spectrometer will need a very sensitive detection system, and precautions must be taken to prevent the population of the upper level becoming greater than that of the lower.

Quantitative spectrometry

In **emission spectrometry**, the intensity of the spectral line is related to the number of emitting species present in the emitting medium and to the probability of the transition. If there are N_o atoms in the ground state, then the number of excited atoms capable of emission, N_E, is given by the Boltzmann distribution law (see above).

Therefore, the emitted intensity, I, is given by an equation of the form:

$$I = A.N_o.\exp(-\Delta E/kT)$$

where A is a constant for a particular transition, incorporating the transition probability, the degeneracies and any reduction due to other unwanted transitions, such as ionization in atomic spectra.

Under constant temperature and other excitation conditions, this may be written:

$$I = k'c$$

where k' is a constant and c the concentration.

The constant, k', may vary in a complex way as c varies, and **calibration,** plus the use of an **internal standard** (see Topic B4) must be used to obtain reliable quantitative results.

Beer–Lambert absorption law

For **absorption spectrometry** the intensity of the incident (exciting) radiation is reduced when it interacts with the atoms or molecules, raising them to higher energy levels. In order to interact, the radiation must come into contact with the species. The extent to which it does this will depend on the **concentration** of the active species and on the **path length** through the sample, as shown in *Figure 3*.

As the radiation of a particular wavelength passes through the sample, the intensity decreases exponentially, and Lambert showed that this depended on the path length, l, while Beer showed that it depended on the concentration, c.

The two dependencies are combined to give the **Beer–Lambert absorption law:**

$$I_t = I_o \exp(-k' c\, l)$$

where I_o and I_t are the incident and transmitted intensities, respectively. Converting to the base 10 logarithmic equation:

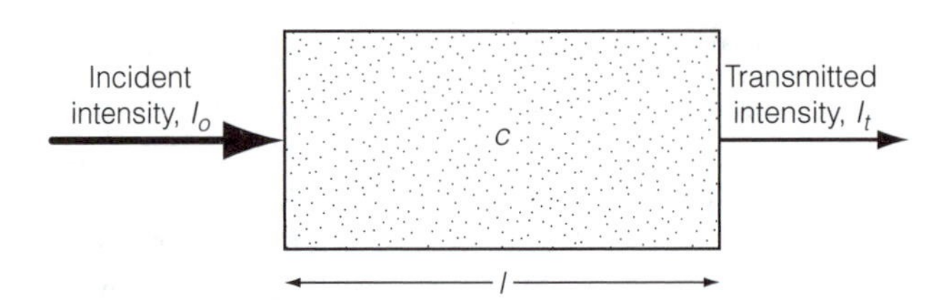

Fig. 3. The absorption of radiation by a sample.

$$\log(I_o / I_t) = A = \varepsilon c l$$

where A = the **absorbance** and **ε = the molar absorptivity**.

The value of ε (sometimes incorrectly referred to as the 'extinction coefficient'), is most usually quoted for a concentration of 1 M and a path length of 1 cm. However, if the concentration is expressed in mol m^{-3} (= 1000 × M) and l is expressed in meters (= cm/100) then the units of ε are m^2 mol^{-1}, suggesting that the absorptivity depends on the effective capture area of the species. This indicates that ε combines the transition probability and the nature of the absorbing species. (Note that ε in m^2 mol^{-1} = ε in (cm $M)^{-1}/10$.

Very high values of ε, for example over 10 000 for the UV, $\pi - \pi^*$ absorption of conjugated polyenes, indicate a favoured transition. Lower values, for example, ε is less than 100 for the $n - \pi^*$ absorption of ketones, show that the transition is less favoured or 'forbidden'.

The Beer–Lambert law applies equally to infrared absorption spectra. Spectra are plotted *either* as absorbance, A, *or* as the **transmittance,** T, against wavelength, frequency or wavenumber, where

$$T = (I_t/I_0)$$

or sometimes as **percentage transmittance** = 100 T.

If the relative molecular mass is unknown, comparison may be made using $E^{1cm}{}_{1\%}$ representing the absorbance of a 1% solution in a 1 cm cell. It is worth noting the range of values which each of these parameters may take. A can have any value from 0 to infinity. T must be between 0 and 1, and ε usually has values from about 1 to 10^6.

If several species in a solution absorb at the same wavelength without chemically interacting with each other, then the total absorbance is the sum of the individual absorbances for each species:

$$A_{\text{total}} = \varepsilon_1 c_1 l + \varepsilon_2 c_2 l + \varepsilon_3 c_3 l \ldots$$

Calibration graphs of A against c may be plotted to verify that the Beer–Lambert law applies over the range of concentrations that is to be studied.

It must be appreciated that the absorbance, and hence the transmittance and absorptivity, vary with wavelength, with the highest absorbance giving a peak in the absorption spectrum, as shown in *Figure 4*. Measurements made at the wavelength of maximum absorbance, λ_{max}, give the highest sensitivity, but care must be taken over the wavelength chosen and over the wavelength range around a selected wavelength transmitted through the optics of the spectrometer, that is the **bandpass** of the system.

There are no known exceptions to the Beer–Lambert law, but apparent deviations may arise as follows.

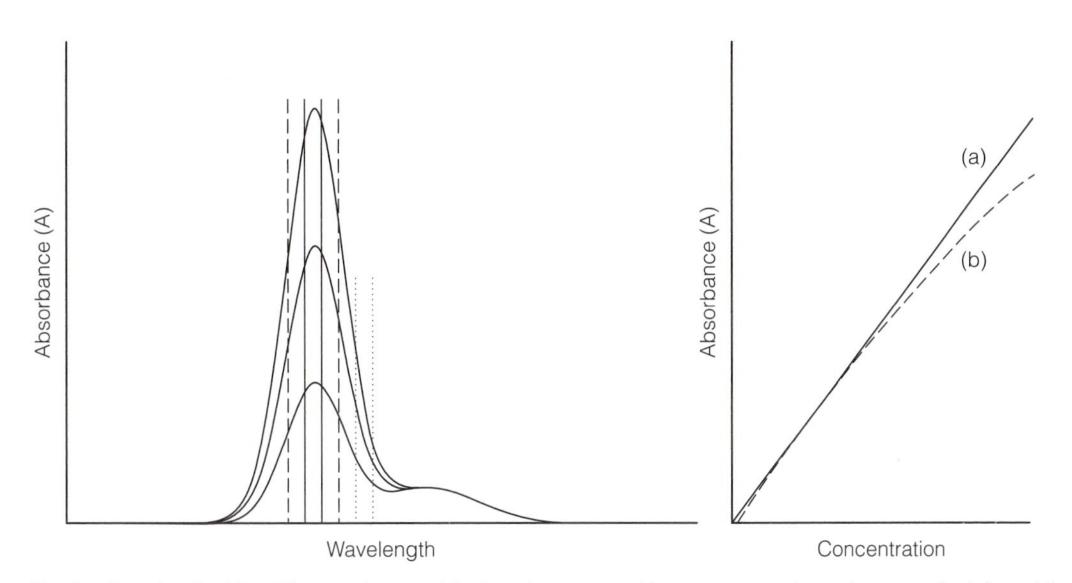

Fig. 4. Beer–Lambert law. The spectrum and the bandpasses used for measurement are shown on the left, and the relation of absorbance to concentration is shown on the right. (a) Measurements made with a narrow bandpass at an absorbance maximum (solid lines), (b) with a wide bandpass at an absorbance maximum, or with a narrow bandpass on the side of a peak (dashed lines), showing negative deviation.

(i) At high concentrations the solute may chemically affect the nature of the solvent, and hence affect the value of ε. Dilute solutions will give better linearity of results.

(ii) If chemical equilibria affect the solute species, then, since the nature of the absorbing species is changed, ε would be expected to change. For example, in the infrared spectra of hydroxyl compounds, such as alcohols, the OH stretching vibration absorbs sharply at around 3600 cm^{-1} in the gas phase spectrum. In the liquid phase, or in solution, hydrogen bonding may occur and the vibrational frequency lowered to around 3300 cm^{-1}.

(iii) If the radiation is polychromatic, or measured in a part of the spectrum other than at an absorbance maximum, the Beer's law dependence will be affected, giving a negative deviation as shown in *Figure 4b*.

The **precision** of absorbance measurements depends on the instrumentation used and on the chemical species being determined. At high absorbances, ($A>1$), very little radiation reaches the detector, so that a higher amplifier gain is needed. At very low absorbances, ($A<0.1$) the instrumental noise becomes very important. Therefore, there is a region where the relative concentration error as a percentage ($100 \times dC/C$) is at a minimum. With a photovoltaic detector, the error curve has a narrow minimum, whereas for the photomultiplier detector used in many modern instruments, the curve has a broader minimum, and therefore an extended useful working range. This is shown in *Figure 5*. In practice it is advisable to measure absorbances in the range $0.1<A<1.0$.

In **fluorescence spectrometry**, the sample must first absorb radiation before it can re-emit at a different energy. Thus, the Beer–Lambert law will be involved, together with the **quantum yield Φ_F** of the fluorescence process. The quantum yield can depend on the kinetics of the fluorescence reactions and the lifetime of the excited states. The fluorescence intensity, I_F, is given by:

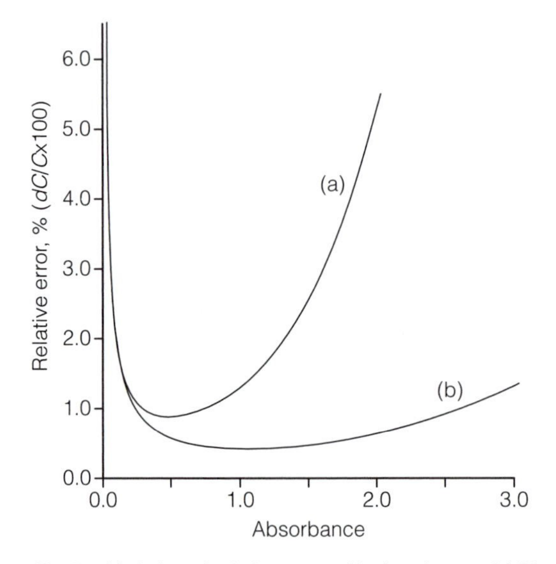

Fig. 5. Variation of relative error with absorbance. (a) Photovoltaic detector; (b) photomultiplier detector.

$$I_F = \ln(10)\,\Phi_F\, I_o\, \varepsilon\, c\, l = k''\, c$$

The proportionality to concentration applies if the absorbance is small, and hence the concentration is low. At higher concentrations, the radiation may be reabsorbed.

Selection rules

With so many atomic and molecular energy levels available, spectra are frequently very complex. Since many spectra actually show considerable regularity and simplicity, it is clear that there are rules governing the transitions that are allowed, and indicating those that are forbidden. More rigorous theoretical work showed that these rules were to be expected from the precepts of quantum mechanics. Particular rules will be discussed in the sections dealing with each spectral technique.

E3 SPECTROMETRIC INSTRUMENTATION

Key Notes

Components of spectrometers

A spectrometer consists of components to provide a source of radiation or other energy, to hold the sample, to disperse the radiation, to detect the resultant radiation intensity and process the results. Each type of spectrometry has its own specialized instrumentation.

Single-beam spectrometers

The components may be arranged so that one beam of radiation only passes along the spectrometric path. Comparisons are then made by interchanging a sample and reference.

Double-beam spectrometers

In order to compensate for parts of the sample that are not of analytical interest, to correct for changes related to the source and detector, and in order to speed up the analysis, double-beam instruments automatically pass beams through both the sample and a reference.

Fourier transform spectrometers

If radiation comprising waves of different frequency and amplitude are combined, a complex periodic function is obtained. By using the mathematical techniques of J.B. Fourier, it is possible to reverse this process and get back to the original radiation. This allows faster, more sensitive and more versatile measurements.

Related topics Other topics in Section E.

Components of spectrometers

In any spectrometer, it is necessary to have the same set of components to produce and analyze the spectrum. While each region of the spectrum and each particular technique requires its own specific modules, the basic parts of each set are the same: the **source,** the **sample,** the **dispersion element,** the **detector** and the **display** or **data processor**. Details of mass spectrometric instrumentation are given in Topic E14.

The **source** must produce radiation in the appropriate region of the spectrum, or must energize the sample in the case of emission spectrometry and mass spectrometry. Two examples will illustrate this.

A suitable source of infrared radiation is a heated rod or strip of metal, which produces an intense radiation in the mid-infrared region (see Topic E10). In X-ray spectrometry, the sample is excited by an X-ray source and the emitted X-rays are analyzed (see Topic E6).

The **sample** must be examined with as little change as possible, and sometimes measurements can be made directly with no sample preparation. Very often, a solution of the sample in a solvent suited to the spectrometric investigation is required.

The term 'spectrometry' indicates measurements made after separating the radiation using a device to **disperse** it. The first spectra were produced with visible light using **prisms** of glass, and these may still be used for some visible spectrometry. Light can also be dispersed by **diffraction gratings,** which have advantages in resolution and absorption. In nuclear magnetic resonance spectrometry, scanning a magnetic field or a range of radiofrequencies provides the dispersion. The details of these devices may be found in textbooks on instrumental analysis.

The technique of **interferometry** can also yield information about a range of frequencies, and using the mathematical technique of **Fourier transformation**, the interferogram, which is a time-domain representation, may be converted into a frequency-domain spectrum.

The **resolution** or resolving power, R, of a spectrometer is important. This may be defined for a general signal S as:

$$R = S/\Delta S$$

where ΔS is the smallest difference between two signals which may be detected and S is the average value of the two signals.

Non-dispersive analysers are occasionally useful, for example in gas analysis by infrared and ultraviolet-visible spectrometry.

After interacting with the sample, the resulting beam of radiation or ions must be **detected**. As with the source, each spectral region and technique requires its own type of detector. Photoelectric devices are useful for visible and UV radiation, and thermal detectors for infrared, whereas X-ray detectors measure the ionizing energy of the radiation. Energy dispersive X-ray instruments employ detectors, which discriminate between the energies of the photons received.

Finally, in order that the spectrometric results may be observed and recorded, some visual presentation or **display** is needed. Data processing, incorporating any corrections, and generally computerized, must give a true representation of the sample spectrum, which can then be interpreted and from which quantitative measurements can be made.

Single-beam spectrometers

The simplest type of spectrometer employs a single source to supply radiation to the sample and then to the background in turn; for example, in some UV-visible absorbance measurements or in plasma emission spectrometry.

The advantages of this system are that only a single set of components is required and that complex sampling devices may be incorporated. The main disadvantage is that correcting for the background spectrum, due to the solvent, matrix or interferences must be done separately, adding to the analysis time.

Double-beam spectrometers

In order to make rapid, accurate comparisons of a sample and a reference, double-beam instruments are frequently used. Since it is essential that the two beams are as similar as possible, a single source is used and the optics arranged to pass equal intensities of the beam through the sample area and through the reference area, and then to disperse and detect them alternately. This is shown schematically in *Figure 1* for an infrared spectrometer.

The source is reflected equally onto mirrors so that beams pass through the sample and reference areas. These beams are then selected alternately by a rotating mirror and each beam follows a common path to the diffraction grating, which disperses the radiation and directs it onto the detector. The width of the

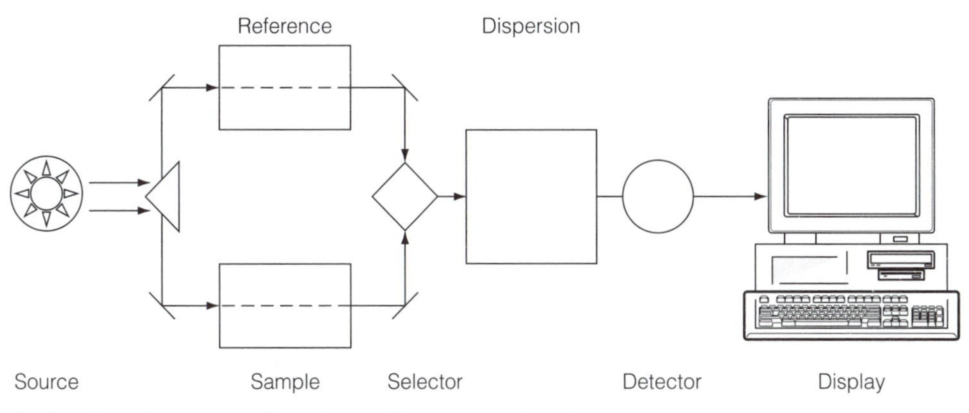

Fig. 1. Block diagram of a dispersive double beam spectrometer.

beams is controlled by slits, which determine the resolution. In a UV spectrometer the beam is dispersed before passing through the sample to avoid irradiating the sample with high energy UV radiation which could cause decomposition.

Fourier transform spectrometers

In Fourier transform spectrometry, the source supplies a wide range of frequencies. For infrared spectrometry, a polychromatic source is used, while for nuclear magnetic resonance (NMR) a powerful microsecond pulse of radiofrequency energy provides the range of frequencies required.

In infrared spectrometry, the radiation is split into two beams. One passes through a fixed optical path length, while the path length of the other beam is varied in a controlled way, as shown in *Figure 2*.

The resulting beams are combined, passed through the sample and on to the detector where they produce a signal as a function of the optical path difference, that is, an **interferogram**.

In NMR, the pulse of frequencies generates a new magnetic field when it interacts with the molecular nuclei which then gradually decays. The receiver detects this and records the **free induction decay (FID)** pattern as a function of time.

The basic mathematics for analyzing such complex patterns were devised by J.B. Fourier and are referred to as **Fourier transforms (FT)**. In brief, the initial detector signal gives an intensity signal as a function of time, either as an IR interferogram, or as an NMR FID. This is then converted, using a Fourier transform algorithm into the relation between intensity and frequency, which is the normal form in which a spectrum is viewed. A full mathematical treatment is not given here, as analytical chemists generally require only the final spectrum (*Fig. 3*).

One advantage of the Fourier transform technique is that every spectral element, M, contributes to the intensity of every data point. From this it is possible to obtain either an improvement in the signal-noise (S/N) ratio by a factor of $M^{1/2}$ or, for an equivalent S/N, the scan time may be reduced by $1/M$. This is called **Fellgett's advantage,** or the **multiplex advantage**.

Since the optical throughput that can produce a good spectrum is much wider for FT spectrometry, a tolerable spectrum can be measured for very small

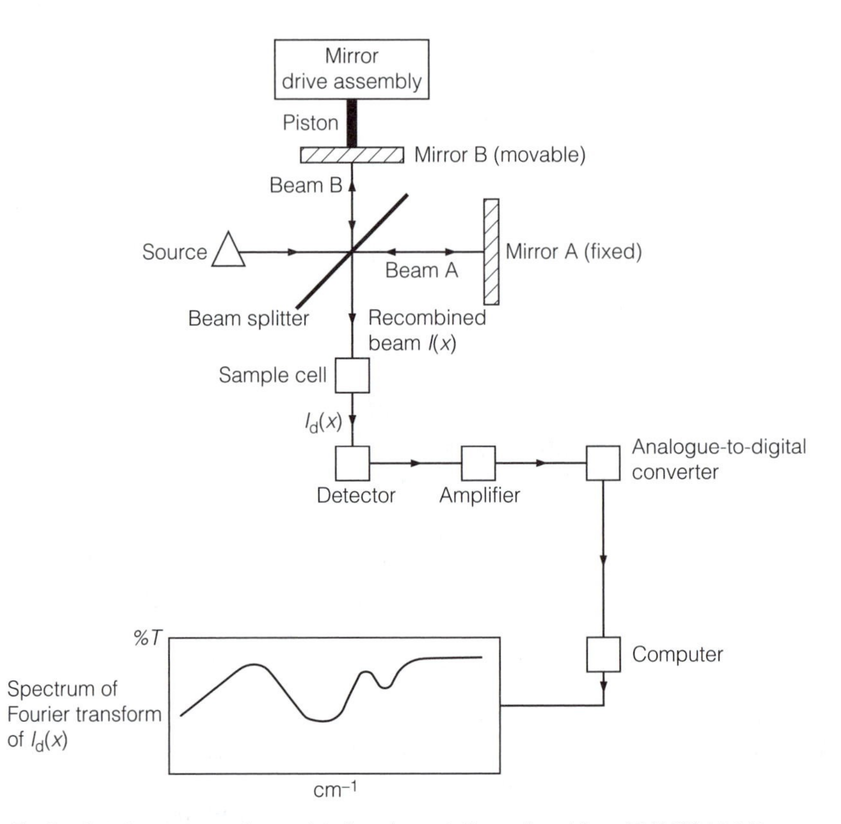

Fig. 2. Interferometer and associated equipment. Reproduced from F.W. Fifield & D. Kealey, Principles and Practice of Analytical Chemistry, *2000, 5th edn, with permission from Blackwell Science Ltd.*

signals, or low intensities, through to large signals. This is the **Jacquinot advantage**.

The Fourier transformation is carried out using fast computer processing. This produces other advantages by subtracting the background, averaging multiple scans, and also allows the treatment of the spectral data to enhance the appearance so that particular regions of the spectrum may be studied more closely.

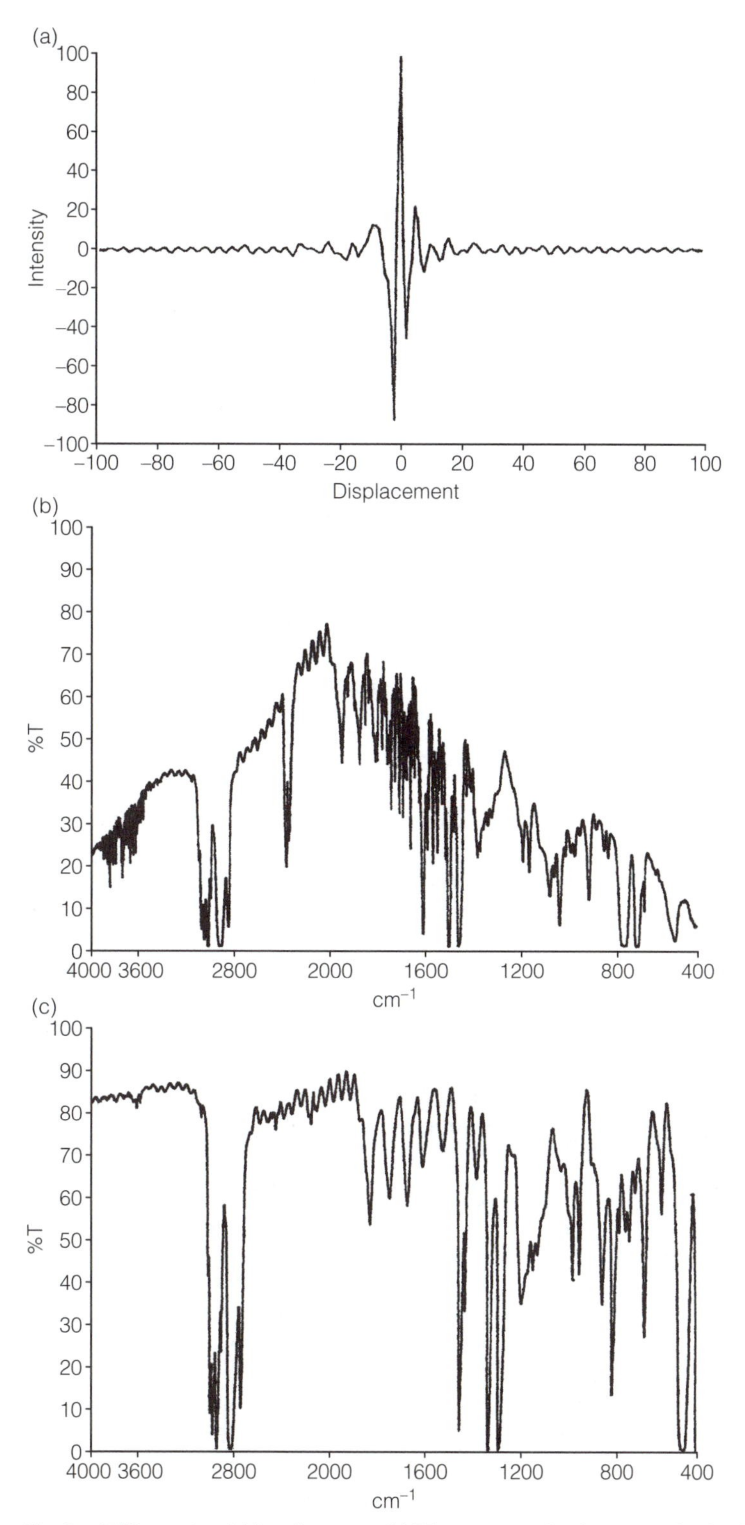

Fig. 3. FTIR spectra. (a) Interferogram; (b) FT spectrum of polystyrene plus background; (c) spectrum of polystyrene after subtraction of background.

E4 Flame atomic emission spectrometry

Key Notes

Principles

When the atoms of samples are excited to higher electronic energy levels in flames they emit radiation in the visible and UV regions of the electromagnetic spectrum. Emission intensities may be measured to analyze for metals, especially alkali and alkaline earth elements.

Instrumentation

A flame atomic emission spectrometer or flame photometer incorporates a burner, monochromator, or filters, a detector and a method of introducing the sample solution into the flame.

Applications

The technique is used primarily for the quantitative determination of alkali and alkaline earth metals in clinical, biological and environmental samples.

Related topics

Inductively-coupled plasma spectrometry (E5)

Atomic absorption and atomic fluorescence spectrometry (E7)

Principles

When the salts of some metals are introduced into a flame, they impart characteristic colors to it. For example, sodium salts give an intense yellow-orange color. This is the basis of the **'flame test'** used in qualitative analysis. The thermal energy of a gas-air flame is quite low, since the temperature is usually less than 2000 K and only those transitions of low energy are excited.

Early atomic emission instruments used electric **arc** or **spark** excitation. The higher energy of these sources produced a very great number of emission lines throughout the visible and UV regions. However, simultaneous measurement of a large number of elements was possible.

With flame excitation, electronic transitions in alkali and alkaline earth metals, as listed in *Table 1* are the most important.

Table 1. Flame excited lines of some metals

Metal	Wavelength/nm	Color
Lithium	670	Red
Sodium	589[a]	Orange–yellow
Potassium	766[a]	Red
Rubidium	780[a]	Dark red
Magnesium	285	UV
Calcium	622[b]	Orange
Strontium	461	Red
Barium	554	Green
Thallium	535	Green
Copper	513	Green

[a]These lines are doublets. [b]This line is due to emission from a calcium hydroxide band.

Table 2. Gas mixtures used in flame atomic emission spectrometry

Fuel	Oxidant	Maximum flame temperature (K)
Natural gas	Air	1800
Propane	Air	1900
Hydrogen	Air	2000
Acetylene	Air	2450
Acetylene	Nitrous oxide	2950
Acetylene	Oxygen	3100

The flame may be produced by burning various gas mixtures, some of which are listed in *Table 2*.

The structure of the flame comprises an **inner cone**, which is the primary reaction zone for combustion, and the **outer cone or mantle** where secondary reactions occur. For the best results, the optical axis is arranged to pass through the flame at the junction of the inner and outer cones. The supply of fuel and oxidant is adjusted to give an optimum **burning velocity**.

The processes that occur to transfer the sample to the flame may be summarized as follows:

(i) production of an aerosol from solution (nebulization)
(ii) removal of solvent MA(aq) → MA (solid)
(iii) vaporization of sample MA(solid) → MA(vapour)
(iv) atomization MA → M• + A•
(v) excitation M• → M*
(vi) emission M* → M•

Ionization may also occur to give the M^+ ion.

These stages each depend on the experimental parameters used in the instrument. For example:

- the viscosity of the solvent, which affects the aerosol production;
- the nature of the solvent, which may affect the vaporization;
- the rate of fuel flow, which can change the nebulization and the time the atoms spend in the flame;
- the flame temperature, which controls the evaporation, the atomization and the extent of ionization; and
- the nature of the flame.

Because of the chemical reactions taking place in the flame, various species such as OH radicals, CO, water and other combustion products are present, and may give a background emission throughout the UV-visible range. Compensation for this background must be made.

Instrumentation

Flame atomic emission spectrometers have similar optical systems to those of UV-visible spectrometers, but the source of radiation is provided by the sample itself. A **flame photometer** is a simpler instrument employing narrow bandpass optical filters in place of a monochromator (*Fig. 1*). The sample is prepared as a solution, which is drawn into a **nebulizer** by the effect of the flowing oxidant and fuel gases. The fine droplets produced pass into the flame where sample atoms are progressively excited. The emitted radiation passes through the monochromator or filter and is detected by a photocell or photomultiplier tube.

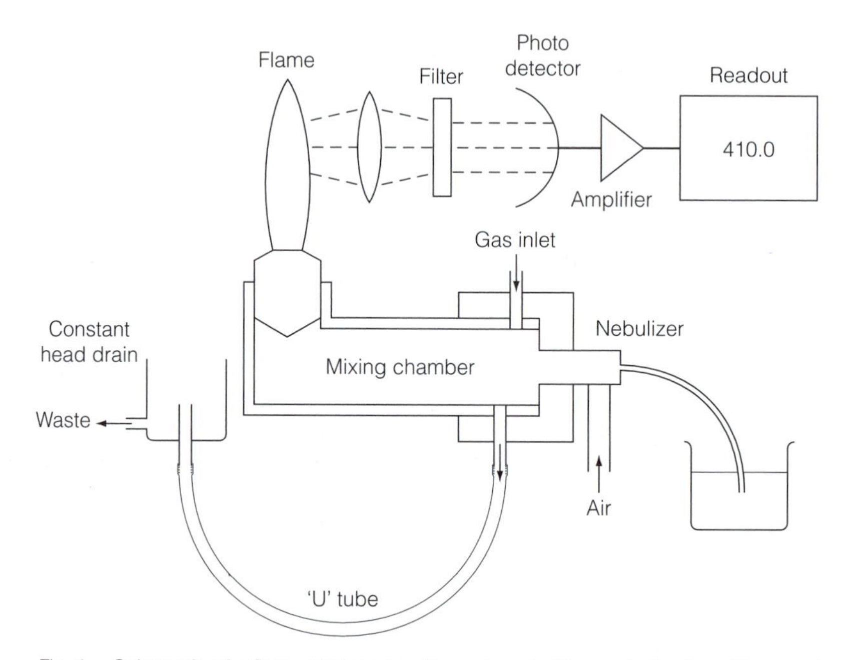

Fig. 1. Schematic of a flame photometer. Reproduced with permission from Sherwood Scientific Ltd.

Applications

Flame atomic emission spectrometry (FAES) and **flame photometry** are used widely for the determination of alkali and alkaline earth metals. The rapid determination of Na, K and Ca in biological and clinical samples is one of the most important applications; for example, calcium in beer, milk or biological fluids. The usual solvent is water, but organic solvents may be used to enhance the intensity, since they produce smaller droplets, and have a smaller cooling effect on the flame.

The instrument is calibrated with standard solutions of the elements to be determined and the intensity of the emission recorded at each characteristic wavelength. A calibration graph is constructed after correcting for background (blank) emission.

Interferences may affect the linearity of the calibration and are chiefly due to the emission lines produced by other species close to those of the analyte. They may be minimized by selecting a different spectral line for the analysis, or by altering the spectral resolution or filter. The presence of anions that form very stable compounds with the metal ions, such as sulfate and phosphate may interfere with some determinations.

At high analyte concentrations, the concentration of atoms in the flame may be high enough to cause **self-absorption**. That is, the emission is reabsorbed by the ground state atoms in the cooler outer layers of the flame. This sometimes causes a loss in sensitivity at higher concentrations.

The **advantages** of FAES and flame photometry are that the instrumentation is relatively simple and measurements can be made quickly. A **disadvantage** is the sensitivity of the emission intensities to changes in flame temperature due to variations in gas flow, or cooling by the solvent.

E5 Inductively coupled plasma spectrometry

Key Notes

Principles

A gas plasma provides a very high temperature excitation source for atomic spectrometry. Quantitative analysis for a large number of elements may be achieved rapidly. By combination with a mass spectrometer, individual isotopes may be identified and quantified.

Inductively coupled plasmas

A high-voltage discharge into an argon flow creates a plasma, which is sustained by induction heating due to the field of a radiofrequency coil. The sample solution is nebulized into the plasma. The emitted radiation is analyzed using a monochromator and photomultiplier detector.

Inductively coupled plasma-mass spectrometry

If part of the sample stream from the plasma is directed into a mass spectrometer, the resulting mass spectrum is used to analyze for elements and to determine isotopic ratios.

Applications

Over 70 elements may be determined using these techniques, many down to ultra-trace levels.

Related topics

Flame atomic emission spectrometry (E4)
Atomic absorption and atomic fluorescence spectrometry (E7)
Mass spectrometry (E14)

Principles

When heated to temperatures above 6000 K, gases such as argon form a plasma – that is a gas containing a high proportion of electrons and ions. The plasma may be produced by a DC arc discharge or by inductive heating in an **inductively coupled plasma (ICP)** torch.

Discharge of a high voltage from a Tesla coil through flowing argon will provide free electrons which will 'ignite' the gas to a plasma. If the conducting plasma is enclosed in a high frequency electromagnetic field, then it will accelerate the ions and electrons and cause collisions with the support gas, argon, and the analyte. The temperature rises to around 10 000 K. At such temperatures, energy transfer is efficient and the plasma becomes self-sustaining. It is held in place by the magnetic field in the form of a **fireball**. The sample aerosol enters the fireball at high speed and is pushed through it, becoming heated and emerging as a **plume**, which contains the sample elements as atoms or ions, free of molecular association. As they cool to around 6000–7000 K, they relax to their ground state and emit their characteristic spectral lines. This technique is known as **ICP-atomic emission spectrometry (ICP-AES)** or sometimes as **ICP-optical emission spectrometry (ICP-OES)**.

If part of the plume is diverted into a mass spectrometer, the isotopic masses of individual elements present may be identified. This is the technique of **ICP-mass spectrometry (ICP-MS)**.

Quantitative measurements are possible with both ICP-AES and ICP-MS.

Inductively coupled plasmas

Argon gas is supplied at 10–15 l min^{-1} through the three concentric quartz tubes of the torch, shown in *Figure 1(a)*. The tangential flow of gas in the outer tube contains the plasma, while the central tube carries the nebulized sample droplets suspended in argon.

The plasma is established by high-voltage ignition and sustained by the magnetic field of the radiofrequency generator providing 2 kW of power at about 27 MHz. The sample is pumped into the nebulizer and the finest droplets carried forward by the gas, while other, larger drops flow to waste from the spray chamber. Viscous solvent systems should be avoided. High-solids nebulizers, where particulate matter and slurries are introduced into the ICP, have been developed. Laser ablation, where the sample is vaporized by a laser

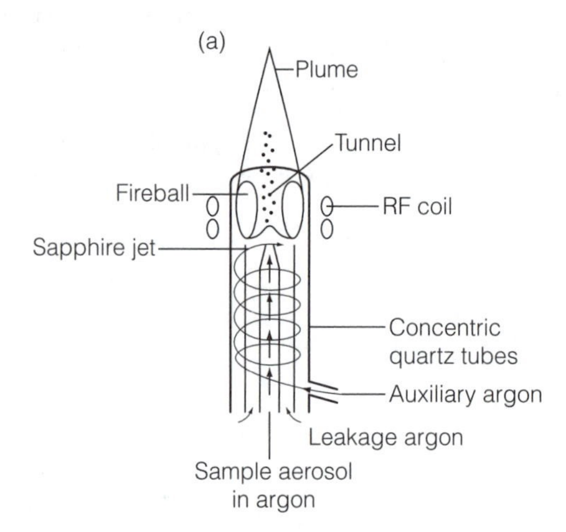

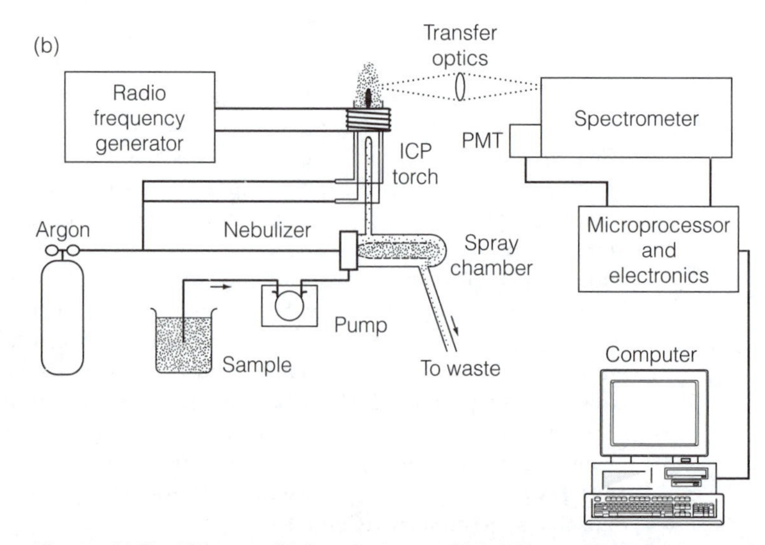

Fig. 1. (a) The ICP torch. (b) Schematic of an ICP-AES spectrometer.

focused on it, and hydride generation (Topic E7) are also used. Electrothermal vaporization may also be employed for solids. As detailed in Topic E4, the sample undergoes a sequence of processes to generate excited atoms.

The optics are aligned with the base of the plume where atomic relaxation is most prevalent. The emitted radiation from the ICP torch is focused into the monochromator and detected by a **photomultiplier tube (PMT)** or 'polychromator' detector. The output is then processed and displayed under computer control as the inductively coupled plasma-atomic emission spectrum (ICP-AES).

ICP-AES can detect a greater number of elements at low concentrations than other atomic emission or atomic absorption techniques. For example, at 1–10 ppb ICP-AES can measure over 30 elements, while AES and AAS are restricted to around ten.

Inductively coupled plasma-mass spectrometry (ICP-MS)

By extracting the atoms from the cooling plasma, the high sensitivity and selectivity of the mass spectrometer (see Topic E14) may be exploited. *Figure 2* shows a schematic of an ICP-MS system.

A horizontal ICP torch is placed next to a water-cooled aperture placed in the sampling cone. The sample, initially at atmospheric pressure, is skimmed down through water-cooled nickel cones through small orifices into progressively lower pressure regions until the sample ions enter the mass spectrometer (see Topic E14).

Usually a quadrupole mass spectrometer is used, but double focusing instruments are also possible. Two modes of operation are employed. *Either* the mass spectrometer may be set to select a single m/z ratio and monitor a single ion, *or* the mass spectrum may be scanned to provide a complete overview of all m/z ratios and ions.

Since the ICP torch can produce ions as well as atoms from the sample, it provides a ready source for the mass spectrometer. Problems may arise due to interferences.

- **Isobaric interference** occurs where different elements produce ions of the same m/z ratio, for example at $m/z = 40$, Ca and Ar both produce abundant ions, as does ^{40}K. At $m/z = 58$, ^{58}Ni and ^{58}Fe mutually interfere.
- **Polyatomic interference** occurs when molecular species, or doubly charged

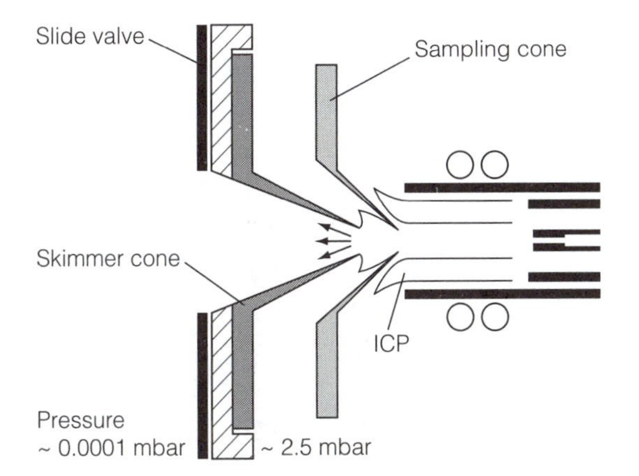

Fig. 2. Schematic of an ICP-MS system. Reproduced from J. Baker, Mass Spectroscopy, *1999, 2nd edn, with permission from Her Majesty's Stationery Office.*

ions occur at the same *m/z* ratio as the analyte ion. For example, $^{32}S^{16}O^+$ and $^{31}P^{16}O^{1}H^+$ both interfere with ^{48}Ti, and $^{40}Ar^{16}O^+$ interferes with $^{56}Fe^+$.

- **Matrix effects** may occur due to excess salts or involatile solids.

Some of these interferences may be removed by the use of **reaction cell** technology where a gas such as helium is added so that, by ion-molecule reactions, interfering ions may be converted into noninterfering species or removed by a multipole filter device.

The sensitivity is generally very high for a large number of elements, typically ten times more sensitive than ICP-AES. Since ICP-MS can scan over a wide mass range, every element is detected simultaneously. Additionally, the isotopes are separated so that changes in isotope ratios produced from radioactive or other sources, or required for geological dating, may be measured accurately. If interferences occur, an alternative isotope may be available for quantitative analysis.

Applications

With ICP-AES there is little interference from ionization, since there is an excess of electrons present. The high temperature ensures that there is less interference from molecular species or from the matrix. Since a large number of elemental emission lines are excited, line overlap, though rare, may occur. *Figure 3* shows the simultaneous emission of a number of elemental lines from a sample. Up to 70 elements, both metals and nonmetals can be determined.

Table 1 gives details of the comparative detection limits of the various atomic spectrometric techniques. The ICP-AES technique provides a wide linear range of detection. For example, for lead, the linear range extends from below 0.01 ppm to 10 ppm.

Mercury in waste water may be determined by ICP-MS, using the most abundant mercury isotope, ^{202}Hg. Since lead from different sources may have different isotopic compositions, ICP-MS can be used to identify sources of environmental contamination. Tracer studies and measurements of isotopes after chromatographic separation of species have also proved the value of ICP-MS.

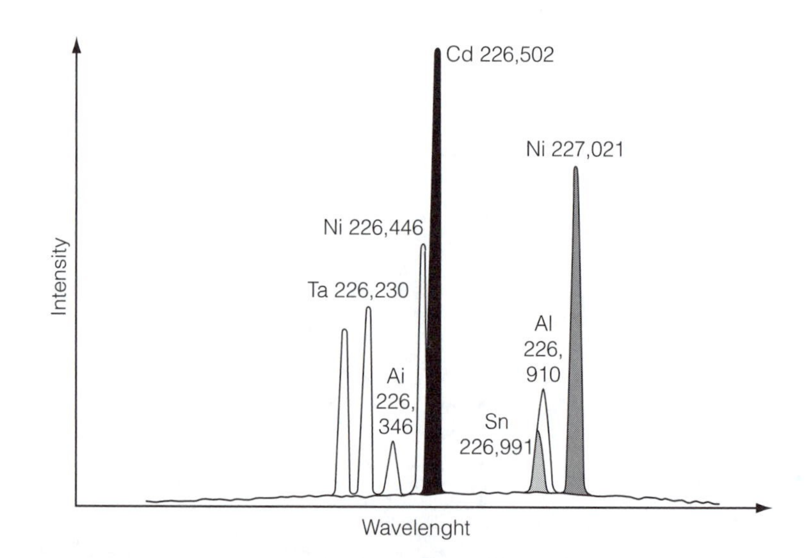

Fig. 3. Simultaneous determination of 7 elements by ICP-AES.

Table 1. Detection limits for atomic emission spectrometry/ppb (µg l^{-1})

Element	Flame	ICP-AES	ICP-MS
Al	10	4	0.1
Ba	1	0.1	0.02
Be	–	0.06	0.1
Cu	10	0.9	0.03
Mn	10	0.4	0.04
P	–	30	20
Pb	100	20	0.02
Zn	10	1	0.08

Flame, flame atomic emission (see Topic E4); ICP-AES, inductively coupled plasma-atomic emission spectrometry; ICP-MS, inductively coupled plasma-mass spectrometry.

E6 X-RAY EMISSION SPECTROMETRY

Key Notes

Principles — Excitation of the inner electrons of atoms promotes some to higher energies. In falling back to lower levels, they emit radiation in the X-ray region, characteristic of the element concerned.

Instrumentation — Excitation by high-energy electrons, radioactive particles or X-rays may be used. Analysis of the emitted X-rays using crystal analyzers is followed by detection using gas ionization detectors or scintillation counters. Nondispersive semiconductor detectors and multichannel pulse height analyzers are often used in conjunction with scanning electron microscopes.

Applications — Elemental analysis of metal and mineral samples as well as surface studies and the determination of heavy metals in petroleum are typical uses.

Related topics — Other topics in Section E.

Principles

Each element has electrons occupying specific energy levels, characterized by quantum numbers. A simple description refers to the lowest energy level as the **K shell,** the next as the **L shell**, **M shell** and so on. Although with elements of low atomic number these electrons may be involved with bonding, for high atomic number elements, such as nickel and copper, they are inner electrons, and largely unaffected by valency changes and bonding.

A simple example is shown in *Figure 1*. A target atom is bombarded with high-energy radiation, for example accelerated electrons or radioactive particles. This causes the excitation of an inner (K shell) electron completely out of the atom, leaving a vacancy in the K shell (*Fig. 1(a)*). An electron from a higher energy shell (L) can relax into the lower level, emitting primary X-rays whose wavelength corresponds to the difference between the energies of the L and K shells (*Fig. 1(b)*). Similar behavior will occur if an electron is excited out of the L shell or higher levels. Since the L shell has two slightly different energy levels, corresponding to 2s and 2p orbitals, the emission is actually a doublet, the **Kα_1 and Kα_2 lines**. If the electron relaxes from the M shell, **Kβ** lines are produced, and if from M to L, **Lα** and so on. High-energy X-ray photons may then interact with the sample and are absorbed, causing the ejection of inner electrons as shown in *Figure 1(c)*. This produces an **X-ray fluorescence** emission spectrum. A competitive process involves the **Auger effect** where the photon is internally converted and an electron emitted.

Moseley's law states that the reciprocal of the wavelength of each characteristic series of X-rays (for example, the Kα_1 series) is related to the atomic number Z of the element by the formula:

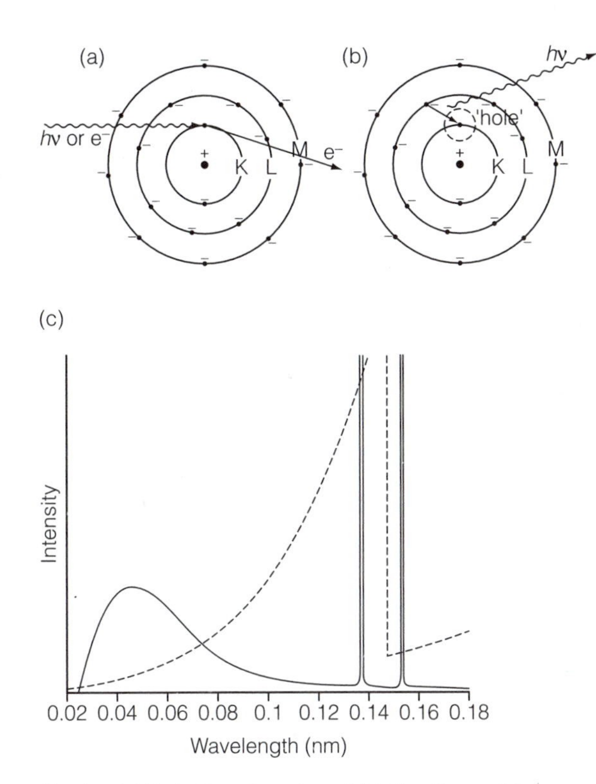

Fig. 1. (a) Excitation of an atom. (b) Relaxation and fluorescent emission of X-rays. (c) X-ray emission spectrum of copper (full line) plus the X-ray absorption for nickel (dashed line).

$$1/\lambda = a(Z - b)^2$$

For example, the copper-Kα_1 line has a wavelength of 0.154 nm, and Z = 29, whereas for nickel the values are 0.166 nm and 28.

The **absorption** of X-rays must be considered, since certain elements may act as **filters** for removing components of the X-ray spectrum, and also the sample itself will absorb. The **absorption** of X-rays depends upon the thickness of penetration into the target and a **mass absorption coefficient** determined by the atomic number of the element and the wavelength of the X-rays. However, the absorption does not follow a smooth curve, but shows a series of **absorption edges** which appear when the ionization energy for a K, L or M electron is reached. This happens because more energy is absorbed in exciting the electrons in the target. For nickel, the edge occurs at 0.148 nm, which means that a nickel filter will absorb the copper Kβ lines, around 0.139 nm, strongly, but absorb the copper Kα lines at 0.154 nm very little. This is shown in *Figure 1(c)*. In a complex matrix, all the elements will contribute to the absorption.

Instrumentation

Two major types of instrumentation are used for X-ray emission spectrometry. These are illustrated in *Figures 2(a)* and *(b)*.

In a dispersive instrument, the specimen is the target for bombardment by high energy X-rays from the source, generally an X-ray tube containing a target such as tungsten, onto which electrons are accelerated by a 50 kV potential difference. These primary X-rays excite the specimen to produce X-rays

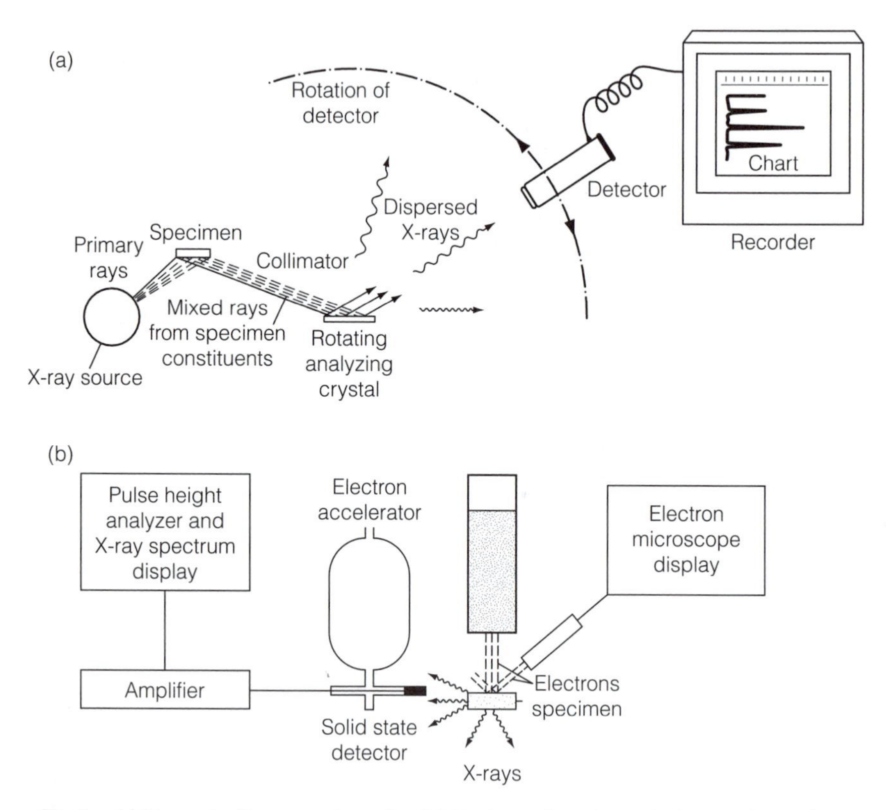

Fig. 2. (a) Dispersive X-ray spectrometer. (b) Electron microscope-electron probe analyzer. Reproduced from F.W. Fifield & D. Kealey, Principles and Practice of Analytical Chemistry, *2000, 5th edn, with permission from Blackwell Science Ltd.*

characteristic of the elements in it. These are diffracted by a crystal analyzer, for example lithium fluoride, or quartz, which separates the X-rays according to the **Bragg equation**:

$$n\lambda = 2d\sin\theta$$

where n is the order of the diffraction, λ is the wavelength of the X-rays produced by the specimen, d is the lattice spacing of the crystal and 2θ is the angle through which the beam is diffracted.

By rotating the crystal, each successive X-ray may be directed onto the detector, which is either a Geiger-Müller tube, a scintillation or a proportional counter. Since the signal generated by a proportional counter depends on the energy of the incident X-ray photons, this may be used to enhance the resolution and reduce the background. This is illustrated in *Figure 3*, where the intensity of the characteristic X-rays is plotted against the 2θ.

In **electron probe microanalysis,** the specimen itself is used as the target for an electron beam, as shown in *Figure 2(b)*, where the beam is that of a **scanning electron microscope (SEM)**. It is especially valuable to have the ability to produce a visual image of the surface of the specimen and to analyze the elemental content of that surface. Here, a **solid-state detector** is used, which detects all the emitted radiations simultaneously, and the multichannel analyzer then produces a mixture of voltage pulses corresponding to the X-ray spectrum. This is also known as **energy dispersive analysis of X-rays (EDAX)**.

Applications

The detection and measurement of the elements present in a sample is valuable and an example in *Figure 3* shows the analysis of a metallurgical sample, containing silver, copper, nickel and chromium.

The ability of X-ray fluorescence to analyze complex solid samples without recourse to time-consuming wet chemical methods has considerable advantages for metallurgy, mineral and cement analysis, as well as petrochemical products.

Electron probe analysis and EDAX allow accurate analysis of tiny areas, of the order of 1 μm diameter. For example, the distribution of additives in polymers, and the presence of high concentrations of elements of interest in biological samples are readily studied.

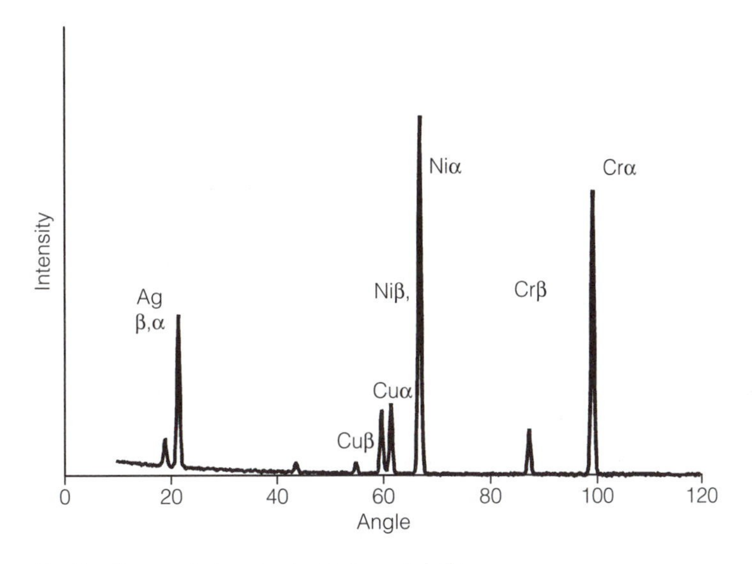

Fig. 3. X-ray emission spectrum of a metal alloy.

E7 Atomic absorption and atomic fluorescence spectrometry

Key notes

Principles

The absorption of electromagnetic radiation by atoms allows both qualitative and quantitative determination of a wide range of elements. Alternatively, fluorescence radiation may be emitted and measured.

Instrumentation

Narrow band sources of radiation specific to particular elements irradiate the atomic vapor produced by flame, furnace or other methods and the absorption is measured. For fluorescence, both continuum and element specific sources are used.

Applications

Atomic absorption spectrometry is used widely for the quantitative determination of metals at trace levels. Atomic fluorescence spectrometry is usually limited to mercury and other volatile species.

Related topics

Flame atomic emission spectrometry (E4)

Inductively coupled plasma spectrometry (E5)

Principles

The energy levels of atoms are specific and determined by the quantum numbers of the element. If ground state atoms are excited, as described in Topic E1, some will be promoted to higher energy levels, the transitions being characteristic of the element involved. Atoms may be excited by incident UV or visible electromagnetic radiation, and if the wavelength (or frequency) corresponds to that of the transition, it will be absorbed. The degree of absorbance will depend on concentration, in the same way as with other spectrometric techniques. This technique is known as **atomic absorption spectrometry (AAS)**.

The sample is generally volatilized by a flame or furnace. The temperature is not usually sufficient to produce ionization, so that the vapor contains largely atoms. These atoms absorb the characteristic incident radiation resulting in the promotion of their electrons to an excited state. They may then undergo transitions to other energy levels and re-emit radiation of another, but still characteristic, wavelength as **fluorescence**. This allows determination by **atomic fluorescence spectrometry (AFS)**.

Ideally, the lines of an atomic spectrum should be very narrow, of the order of 10^{-5} nm. However, thermal movement causes **Doppler shifts** and **pressure broadening** due to collisions among the atoms, and also electrical and magnetic fields in the sample vapor all tend to broaden the lines slightly to about 10^{-2} nm.

If a source emitting a broad band of wavelengths were used, the fraction absorbed by a narrow line would be small. It is therefore important to use a source producing a sharp emission line characteristic of the element to be analyzed, for example, a zinc source lamp to determine zinc. These are called **resonance line** sources, and may be a **hollow cathode lamp (HCL)** or an **electrodeless discharge tube**, both described below. Continuum source instruments require very high-resolution monochromators.

In a similar way to the flame emission spectrometers described in Topic E4, a flame may be used to volatilize the sample. Nonflame atomizers, particularly the **graphite furnace**, are very useful especially when only small volumes of sample are available.

Instrumentation

The **source** is often a **hollow cathode lamp (HCL)** as shown in *Figure 1(b)*. This has a glass envelope with a quartz window and contains a gas such as argon, which is excited by an electric discharge. The excited argon atoms bombard the cathode, which is made of the element to be determined and the atoms of that element are then excited in the discharge too. The excited atoms decay back to their ground state, emitting the characteristic radiation. A turret with several lamps allows multi-element determinations.

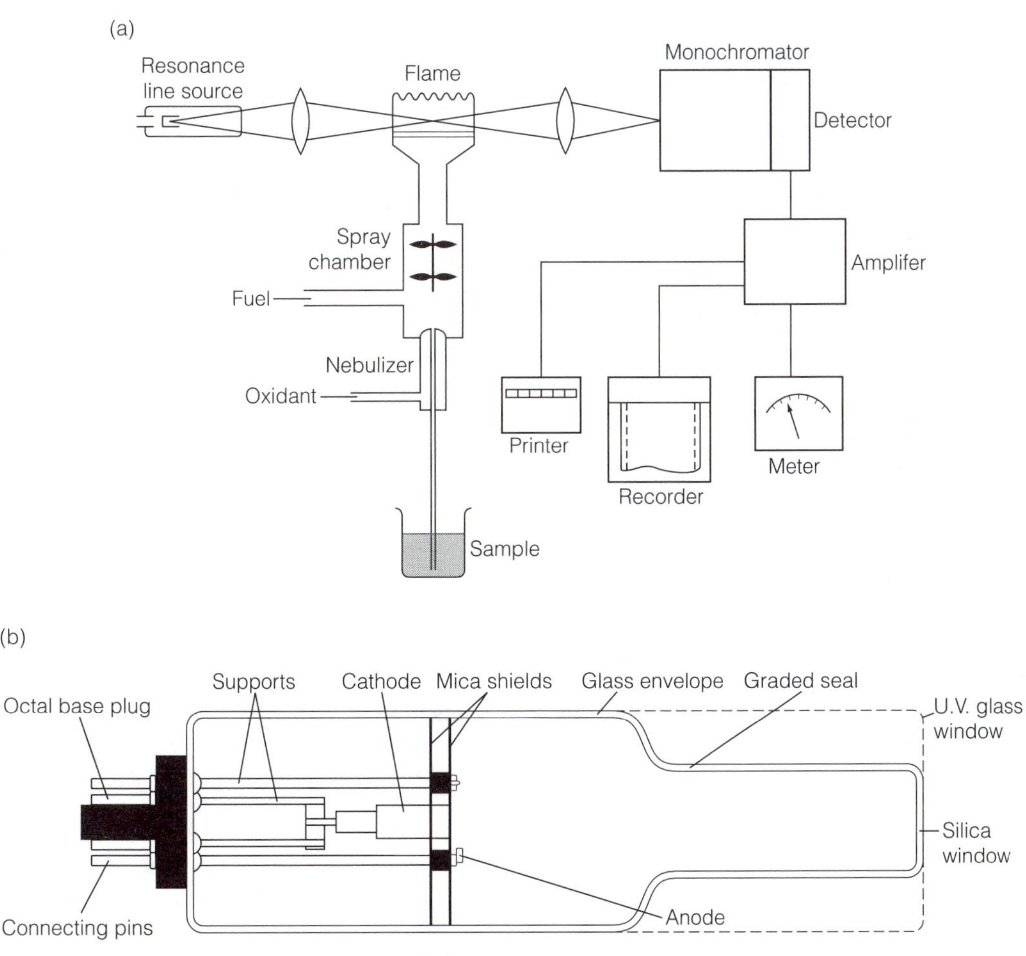

Fig. 1. (a) Schematic of a flame atomic absorption spectrometer. (b) A typical hollow cathode lamp. Reproduced from F.W. Fifield & D. Kealey, Principles and Practice of Analytical Chemistry, *5th edn, 2000, with permission from Blackwell Science Ltd.*

Electrodeless discharge lamps have a small, sealed cavity containing a minute amount of the element to be measured, or its salt. The cavity is excited by a radiofrequency signal passed through a coil, and tuned to resonate at a chosen frequency. This excites the atoms to produce the desired emission spectrum.

For flame vaporization, the **sample** is usually prepared as a solution that is sprayed into the burner. A flow spoiler removes large droplets, and the sample undergoes a similar sequence of events to that described in Topic E4, converting it into gaseous atoms. The signal reaches a constant value proportional to the concentration of the analyte element in the sample.

The **flame** is generally produced using one of the gas mixtures given in Topic E4. Air-acetylene gives a flame temperature of about 2400 K, while air-propane is cooler (~1900 K), and nitrous oxide-acetylene hotter (~ 2900 K).

The structure and temperature of the flame is most important, as is the alignment of the optical path with the region of the flame in which an optimum concentration of the atoms of analyte is present. Variations in flame temperature, including those caused by cooling due to the sample, will affect the sensitivity of the technique. Flame sources have advantages for analysis where large volumes of analyte are available.

An alternative vaporization method is to use a **graphite furnace,** which is an open-ended cylinder of graphite placed in an electrically heated enclosure containing argon to prevent oxidation. Temperatures in the region of 2500 K are achieved, and the heating program is designed to heat the sample, deposited on a smaller tube or **L'Vov platform,** by radiation. The graphite furnace produces a peak signal whose area is proportional to the total amount of vaporized element in the sample.

The use of graphite furnace atomic absorption **(GFAA)** has a number of advantages:

- it avoids interactions between the sample components and the flame since atomization takes place in an inert gas stream;
- it gives increased sensitivity because of the longer residence time of the sample in the beam from the source;
- the sensitivity is further increased because a higher proportion of atoms are produced;
- it has the ability to handle small volumes of samples, down to 0.5–10 μl, such as clinical specimens;
- the results are more reproducible than flame AAS.

One disadvantage is that it is rather slower that flame AAS.

Another approach is **hydride generation,** as certain elements, such as arsenic, tin and selenium, have volatile hydrides. By removing all organic matter by oxidation, and then reducing the sample with sodium borohydride, $NaBH_4$, the volatile hydride is produced and can be swept out into the radiation path using argon.

Table 1 gives comparitive sensitivities for the detection of a selection of elements. It should be noted that more than 50 elements may be determined by AAS.

The monochromator and detector are similar to those used in other forms of spectrometry. AAS is essentially a single-beam technique. Since the flame and matrix may produce background radiation in the region of interest, correction

Table 1. Sensitivity of atomic absorption spectrometry analysis in μg cm^{-3} (ppm)

Element	Wavelength/nm	Flame AA	Graphite furnace AA
Al	309.3	0.03	0.00001
Bi	223.1	0.02	0.0001
Cu	324.8	0.001	2×10^{-5}
Zn	213.9	8×10^{-4}	1×10^{-6}

for this is important. This can be accomplished by using a deuterium lamp as a source of continuous radiation and by modulating the signal to observe the radiation from the HCL and from the deuterium lamp alternately. Signal processing then corrects for the background. Double beam instruments are also used (Topic E2).

An alternative approach is to utilize the **Zeeman effect,** which is observed when a magnetic field is applied to a source of radiation. With no magnetic field, a single line might be observed (e.g., at 285.2 nm for magnesium). When a magnetic field is applied, this is split into several components. If the magnetic quantum number of the element is 1, then the original line is split into three, one component (π) at the original wavelength, polarized in one plane, and two components ($\pm\sigma$), shifted away from the original, by about $\pm$ 0.004 nm and polarized perpendicular to the π line. By changing the polarization of the light, the intensity of the background absorption alone and of background plus sample signal may be measured.

Interferences may be troublesome in AAS. Background absorption by smoke particles or solvent droplets may be removed as detailed above.

Matrix interference, such as any reaction that prevents the sample getting into the flame, may reduce the sensitivity. It is always preferable to run the standards in the same matrix, or to use standard addition procedures (Topic B4).

Chemical interference due to the production of thermally stable compounds, such as involatile phosphates of calcium, may sometimes be dealt with by adding a releasing agent such as EDTA, or by using a hotter flame or a reagent that preferentially forms stable, volatile compounds.

Ionization interference due to the production of ions is most troublesome with alkali metals because of their low ionization potentials. Occasionally, ionization suppressors such as lithium or lanthanum salts, which are easily ionized, are added.

Spectral interference is rare because of the sharpness of the atomic elemental lines, but is difficult to overcome. For example, the zinc line at 213.856 nm is too close to the iron line at 213.859 nm, but the iron line at 271.903 nm could be used to determine iron instead.

In **atomic fluorescence spectrometers** the detector is placed so that no radiation from the lamp reaches it, often at right angles to the incident radiation path. Both HCL and high intensity continuum sources, such as mercury lamps, may be used since the fluorescence intensity depends on the intensity of the primary radiation.

Applications

The spectrometer should be **calibrated** by using standard solutions of the element to be determined, prepared in the same way as the sample solution. Usually, acidic aqueous solutions are used; for example, many elements are commercially available as 1000 ppm certified standards in dilute perchloric acid. Organic solvents, such as white spirit (~C10 alkanes) or methyl isobutyl ketone, may be used if the sample is insoluble in water (e.g., a lubricating oil). Standards can then be made of metal complexes soluble in the organic solvent. The standard solutions should be chosen to 'bracket' the concentration to be determined and, ideally, the calibration curve should be linear in that region.

Multi-element standard stock solutions are useful and permit more rapid determination of several elements of interest.

The techniques are used for many industrial and research purposes, especially:

- **agricultural** samples, particularly the analysis of soils – metal pollutants in soil and water samples are often determined by AAS;
- **clinical and biochemical** determinations, e.g., the measurement of sodium, potassium, lithium and calcium in plasma and serum, and of iron and lead in whole blood;
- **metallurgical samples** may be assayed to measure impurities;
- **oils and petrochemical samples** can be analyzed for metals in feedstocks and to detect metals in used oils due to corrosion and wear.
- **water** samples are extremely important, since pollution may be a health hazard. Nickel, zinc, mercury and lead are among the metals determined.

Atomic fluorescence has the advantage, compared to AAS, that with a continuous source, several elements may be determined simultaneously. There are, however, problems due to scattered radiation and quenching, but detection limits are lower than for AAS.

It is particularly suitable for elements that are volatile, such as mercury, or form volatile hydrides, such as arsenic or bismuth, or for elements forming stable volatile organic derivatives.

E8 Ultraviolet and visible molecular spectrometry: principles and instrumentation

Key Notes

Absorption spectrometry

Absorption in the ultraviolet and visible regions of the electromagnetic spectrum corresponds to transitions between electronic energy levels and provides useful analytical information for both inorganic and organic samples.

Photoluminescence

Molecules may possess several excited states. After excitation by absorption of radiation, rapid transitions can occur to lower energy excited states, which then revert to the ground state, emitting electromagnetic radiation at a lower energy and by slower processes referred to as photoluminescence.

Instrumentation

The components of ultraviolet and visible spectrometers include a source of radiation, a means of dispersion and a detector specific to this spectral region.

Related topics

Other topics in Section E.

Absorption spectrometry

The ultraviolet (UV) and visible region of the electromagnetic spectrum covers the wavelength range from about 100 nm to about 800 nm. The **vacuum ultraviolet** region, which has the shortest wavelengths and highest energies (100–200 nm), is difficult to make measurements in and is little used in analytical procedures. Most analytical measurements in the UV region are made between 200 and 400 nm. The **visible** region occurs between 400 and 800 nm.

The energy levels involved in transitions in the UV-visible region are the electronic levels of atoms and molecules. For example, although light atoms have widely spaced energy levels, some heavy atoms have their outer orbitals close enough together to gives transitions in the visible region. This accounts for the colors of iodides. Transition metals, having partly occupied *d* or *f* orbitals, often show absorption bands in the visible region and these are affected by the bonding of ligands (see Topic C2). For example, iron(III) reacts with the thiocyanate ion to produce an intense red color due to the iron(III) thiocyanate complex, which may be used to determine iron(III) in the presence of iron(II).

Organic molecules contain carbon–carbon bonds, and bonds between carbon and other elements such as hydrogen, oxygen, nitrogen, sulfur, phosphorus and the halogens. Single bonds correspond to the bonding σ orbital, which has an associated antibonding σ^* orbital. Multiple bonds may also be formed and correspond to the π bonding and π^* antibonding orbitals. Bonding orbitals have lower energy, while antibonding orbitals have higher energy. Lone pair electrons on atoms such as oxygen are little changed in energy. Thus, a molecule such as propanone (acetone) has the structure:

```
  H        H
  |        |
H-C — C — C-H
  |   |    |
  H   O:   H
```

The single C—H and C—C bonds relate to σ orbitals, the carbonyl double bond to π orbtitals and the unpaired electrons on the oxygen to the nonbonding n-levels. The energy levels may be grouped approximately as shown in *Figure 1*. Transitions between σ and σ^* levels, and between π and π^* are favored, and those of the n electrons to the higher levels also occur.

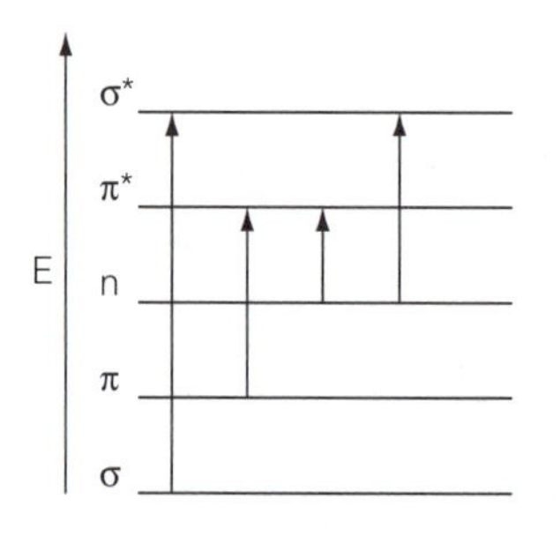

Fig. 1. Typical transitions for organic molecules.

Figure 1 shows that the σ–σ^* transitions require the largest energy change and occur at the lowest wavelengths, usually less than 190 nm, which is below the wavelengths measurable with most laboratory instrumentation. The π–π^* transitions are very important, as they occur in all molecules with multiple bonds and with conjugated structures, such as aromatic compounds. The transitions occur around 200 nm, but the greater the extent of the conjugation, the closer the energy levels and the higher the observed absorption wavelength. Transitions involving the lone pairs on heteroatoms such as oxygen or nitrogen may be n–σ^*, which occur around 200 nm, or n–π^*, which occur near 300 nm. These values are considerably altered by the specific structure and the presence of substituents (**auxochromes**) in the molecules, as discussed in Topic E9.

Since electronic transitions may occur between states with many vibrational and rotational sublevels, and since these may also be affected by sample–solvent interactions, UV and visible spectra of solutions do not generally give sharp lines, but broad bands, as shown in *Figure 3*. Generally, the peak wavelength, λ_{max}, is specified for analytical purposes. The absorbances obey the Beer–Lambert law, which is described in Topic E2.

Photoluminescence

Fluorescence and **phosphorescence** are both forms of **photoluminescence**. Most molecules in their ground state have electrons that are **paired** – that is, they occupy the same orbital, but have opposed electron spins. This is referred to as a **singlet state,** S_0. Absorption of energy may promote one electron to a higher energy level (*Fig. 2*). Usually the spins are still opposed, so this is an **excited singlet state, S_1**. Deactivation may then occur in a number of ways.

- The energy may be re-emitted at the same wavelength as that absorbed. This is a rapid process of **resonance fluorescence**.
- The excited state may undergo **internal conversion** with the loss of vibrational energy to reach a lower energy singlet state S_2. This may then emit energy at a longer wavelength to return to the ground state S_0. This is fluorescence, and is also a rapid process with a lifetime of 10^{-6} s or less.
- Vibrational relaxation may be followed by **inter-system crossing** into the **excited triplet state, T_1,** where the electron spins are **unpaired**. This may radiate at a longer wavelength in the much slower process of phosphorescence to return to the ground state.

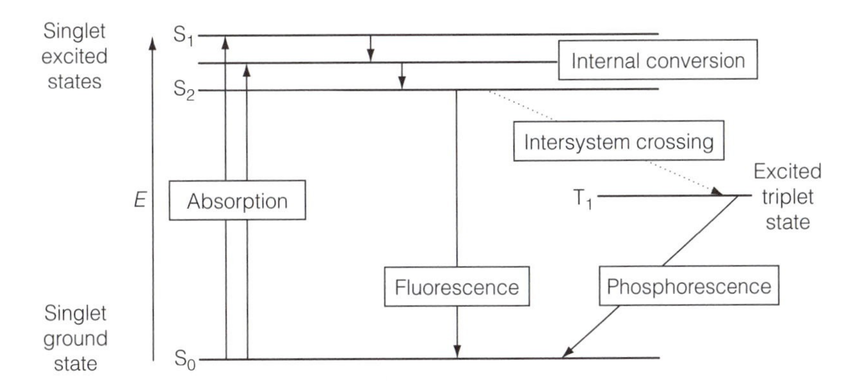

Fig. 2. Molecular energy levels associated with fluorescence and phosphorescence.

Instrumentation

The source of visible light (400–800 nm) is generally a tungsten filament lamp or a tungsten halogen bulb. For the UV (200–400 nm) region, the source most often used is a deuterium lamp and arrangements are made to switch between these sources at an appropriate wavelength, often around 380 nm. Xenon arc lamps may also be used.

The **sample** is generally a dilute solution of the analyte in a solvent with a low absorbance in the region of interest. The nature of the solvent may alter the position of the spectral peaks, as discussed in the next section. The sample solution is contained in a thin-walled silica glass cell, usually with a path length of 1 cm, while a matched reference cell contains the solvent alone or a blank of solvent and reagents. If measurements are restricted to the visible region, ordinary glass or plastic cells may be used, but these should not be used for UV work. Occasionally, **reflectance** measurements are used on opaque surfaces, such as thin layer chromatography plates or materials with surface coatings.

The sample should be positioned in the spectrometer after dispersion of the radiation to avoid UV photochemical decomposition. Suitable solvents for UV and visible spectrometry are listed in *Table 1*.

Dispersion of the spectrum is achieved using silica glass prisms or diffraction gratings. Occasionally, filters are used to select a narrow band of radiation for

Table 1. Solvents for UV-visible spectrometry

Solvent	Minimum usable wavelength (nm)
Acetonitrile	190
Water	191
Cyclohexane	195
Hexane	201
Methanol	203
Ethanol	204
Diethylether	215
Dichloromethane	220

quantitative analysis. An example of a filter system is a modern fiber-optic photometer. Light from a tungsten lamp passes down the quartz fiber and into the sample solution. A mirror set a fixed short distance away reflects the light back up the fiber through a suitable interference filter set to the wavelength of interest and onto a photodiode detector. This type of system may be used for continuous on-line analysis of sample (see Topic H1).

For **detection** in the UV-visible region, photomultipliers or other photoelectric devices are used. Some instruments may use a multi-channel diode array detector. An array of typically 300 silicon photodiodes detects all the wavelengths simultaneously with a resolution of about ±1 nm. This provides a great saving in time and an improved signal/noise ratio.

The recorded spectrum is generally displayed by plotting **absorbance** against wavelength, as shown in *Figure* 3. This allows direct quantitative comparisons of samples to be made.

The instrumentation for detecting and measuring fluorescence is similar to that for absorption spectrometry, except that two dispersion monochromators are needed, one for the excitation wavelength and the other for analyzing the resulting fluorescence. Note that the emitted radiation is detected at 90° to the excitation radiation, as shown in *Figure 4*.

For practical purposes, it is important that the sample solution is dilute and that it contains no particulate matter. More concentrated samples may reabsorb the emitted radiation, either due to the sample itself or due to the presence of some other **quenching** agent. Particles will cause the radiation to be scattered.

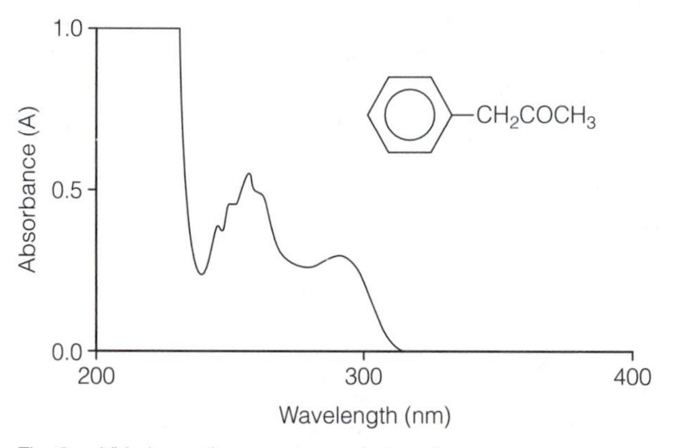

Fig. 3. UV absorption spectrum of phenyl propanone.

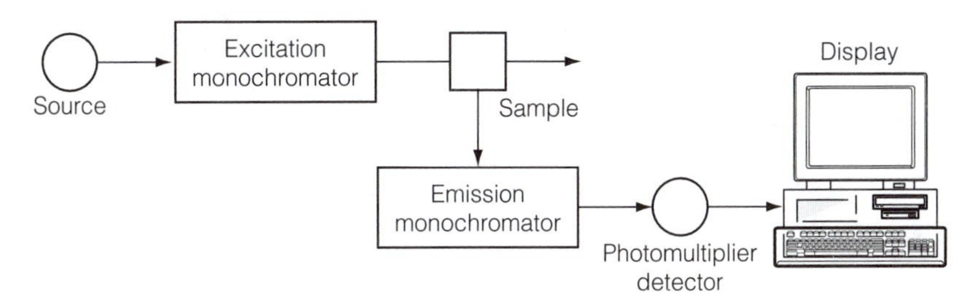

Fig. 4. Schematic of a fluorescence spectrometer.

As noted in Topic E2, the fluorescent intensity, I_F, is directly proportional to the concentration, c, provided that the absorptivity is small and that quenching and scattering are minimal.

$$I_F = k'c$$

A typical fluorescence spectrum is shown in *Figure 5*.

Phosphorescence methods have fewer applications than fluorescence. Since sample molecules may show both fluorescence and phosphorescence, it is necessary to measure the slower phosphorescence by introducing a finite delay between excitation and measurement. This is done using a shutter system.

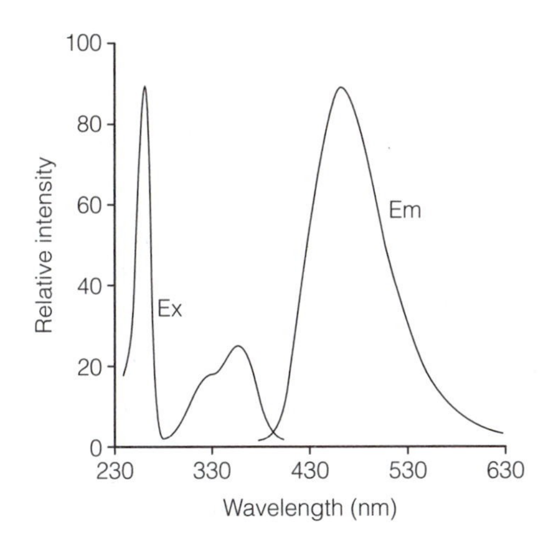

Fig. 5. Fluorescence emission of quinine sulfate. Excitation (Ex) at 250 nm yields a fluorescence emission (Em) peak at 450 nm.

E9 ULTRAVIOLET AND VISIBLE MOLECULAR SPECTROMETRY: APPLICATIONS

Key Notes

Structure effects — The structure of a molecule determines the nature of its UV or visible spectrum and facilitates qualitative analysis of a sample.

Quantitative applications — Measurement of the relation between concentration and absorbance allows quantitative analysis using the Beer–Lambert Law.

Related topics — Ultraviolet and visible molecular spectrometry: principles and instrumentation (E8); Other topics in Section E

Structure effects

The structures of organic molecules may be classified in terms of the functional groups, which they contain. Where these absorb UV or visible radiation in a particular region they are called **chromophores**. Some of the chromophores important for analytical purposes are listed in *Table 1*. This shows that the absorption by compounds containing only σ bonds such as hexane, or with lone pairs, such as ethanol, will occur below 200 nm. These compounds are therefore useful solvents.

The effects of conjugation, that is, the interaction of the molecular orbitals between alternate single and double bonds in a chain or ring, greatly affect the absorption characteristics of the chromophore. Since each isolated double bond has a π-bonding energy level and a higher energy π^*-antibonding level, when two double bonds are separated by one single bond, the molecular orbitals overlap, and there are two π levels and two π^* levels. The separation in energy

Table 1. Absorption of simple unconjugated chromophores

Chromophore	Transition	Approximate wavelength (nm)
σ-bonded		
⩾C—C⩽/⩾C—H	σ–σ^*	~150
Lone pair		
O:	n–σ^*	~185
⩾N:	n–σ^*	~195
>C=O:	n–σ^*	~195
	n–π^*	~300
—N=N—	n–π^*	~340
π-bonded		
>C=C< (isolated)	π–π^*	~180

between the highest filled π level and the lowest empty π^* level is smaller, and hence the wavelength of that absorption is greater.

Thus, ethene $CH_2{=}CH_2$ absorbs at about 180 nm ($\varepsilon = 1500$ m^2 mol^{-1}), whereas butadiene, $CH_2{=}CH{-}CH{=}CH_2$ absorbs at 210 nm (ε = 2100). Long-chain conjugated polyenes, such as the carotenes absorb in the visible region with a very high ε. There is a general rule that states 'the longer the chromophore, the longer the wavelength at which is absorbs'. It is also generally true that molecules possessing extended conjugation have more intense absorption bands and more complex spectra.

Similar arguments apply to conjugation between carbonyl double bonds and carbon–carbon double bonds.

Solvent effects are important, both in considering the position of the absorption maximum and also the nature of the spectral transition involved. For π–π^* transitions, the excited state is more polar than the ground state, so it will tend to form dipole–dipole bonds with a polar solvent, such as water or ethanol. This will lower the transition energy and raise the absorption peak wavelength. This is called a **red shift** (or a **bathochromic shift**). Tables of solvent corrections are available in specialist texts.

For n–π^* transitions, the ground state is often more polar and may form hydrogen or dipole bonds with polar solvents. This increases the transition energy and lowers the peak wavelength, causing a **blue shift** (or **hypsochromic shift**).

pH will affect the structures of compounds with acidic or basic groups, and may cause considerable wavelength shifts. This is most evident in acid–base indicators such as described in Topic C4.

Substituents that alter the wavelength or absorptivity of a chromophore significantly are called **auxochromes**, and tables of the effect of substituents plus rules for their application in particular structures are to be found in specialist texts. For example, an unsubstituted, unsaturated ketone would have a peak maximum at about 215 nm. Substitution of a hydroxyl group on the carbon next to the carbonyl (α) raises the peak to 250 nm, and two alkyl groups on the next (β) carbons would raise it to 274 nm.

Table 2 lists a few of the substituent effects for aromatic compounds. It should be noted that the phenoxide ion ($-O^-$), which is present in alkaline solutions of phenols, absorbs at a considerably longer wavelength than the parent phenol (-OH). Generally electron donating and lone-pair substituents cause a red shift and more intense absorption. More complex shifts arise when there is more than

Table 2. Absorption maxima for some monosubstituted benzenes Ph-R (in methanol or water)

R	Maxima/nm	
-H	204	254
$-CH_3$	207	261
-Cl	210	264
-OH	211	270
$-OCH_3$	217	269
$-CO_2^-$	224	271
-COOH	230	280
$-NH_2$	230	280
$-O^-$	235	287

one substituent present, and tables are given in standard spectrometry texts listing these.

Identification of unknown organic samples can be considerably aided by considering the UV-visible absorption spectra. The following general rules may be used as a guide:

Observation	Possible conclusion
No UV absorption present. Isolated double bond	σ bonds or lone pairs only
Strong absorption between 200 and 250 nm ($\varepsilon \sim 1000$)	Aromatic ring
Weak absorption near 300 nm ($\varepsilon \sim 1$)	Carbonyl compound.

Example

An organic compound, $C_7H_{14}O$, gave a UV spectrum with a peak at 296 nm and $\varepsilon = 3.7\ m^2\ mol^{-1}$. Is it more likely to be a ketone or an alkene?

The formula allows the possibility of only one double bond. It must therefore be an alkene with an isolated double bond, absorbing below 200 nm, or a ketone with a weak n–π* transition near 300 nm. The value of both the absorption maximum and of the absorptivity suggest a **ketone**.

Fluorescence is also related to structure. The most intense fluorescence is found in compounds with conjugated structures, especially polycyclic aromatic compounds. Fluorescence is particularly favored for rigid structures, or where the sample is adsorbed onto a solid surface, or in a highly structured metal ion complex. This is useful for the quantitative determination of metals.

Metal complexes involving organic or inorganic ligands are important in analytical determinations.

The transitions which are responsible for visible and UV absorption by complexes may be classified as follows.

(i) *d–d* transitions due to a transition metal ion. These give rise to the color of many compounds of transition metals and are modified by changing the ligands that are bound to the central atom. The **spectrochemical series** relates the increasing strength of the ligand field to the shift of the absorption band towards the UV. However, these transitions often have low absorptivity constants, which makes them suitable only for determining high concentrations of metals.

(ii) The absorption bands of an organic ligand may be modified when it is complexed with a metal. The complexes formed with dyes such as **eriochrome black T**, used in complexometric titrations as an indicator, and metals such as magnesium, have high absorptivities. Complexing agents such as dithizone ($C_6H_5.NH.NH.CSN{=}N.C_6H_5$) form highly colored complexes with several metals.

(iii) **Charge-transfer bands** arise because of transitions between the levels of an electron donor, often the σ or π orbitals of the ligand, and electron acceptor levels, such as the empty orbitals of a transition metal. The intense red color of the iron(III) thiocyanate complex and the purple color of the permanganate ion both arise as a result of such transfers.

Examples of the quantitative determination of metals by the direct absorptiometry of complexes formed by the addition of chromogenic reagents and also through the fluorescence of complexed species using fluorogenic reagents are given below in *Table 3*.

Table 3. Examples of UV-visible and fluorimetric analysis

Analyte	Reagent	Wavelength(s)/nm[a]	Application
Absorptiometry			
Fe	*o*-Phenanthroline		Water, petroleum
Mn	Oxidation to MnO_4^-	520	Steel
Aspirin	–		Analgesics
Sulfonamides	Diazo derivatives		Drug preparations
Fluorimetry			
Al	Alizarin garnet R	470/580	Water, soil
Borate	Benzoin	370/450	Water, soil
Quinine	–	250/450	Water, drugs
Codeine	–	200/345	Drugs, body fluids
Vitamin A	–	250/500	Foods
Polyaromatic hydrocarbons	–	200/320–550	Environmental

Quantitative analysis

Many organic compounds and inorganic complexes may be determined by direct absorptiometry using the Beer–Lambert Law (Topic E2). It is important to recognize that for the most accurate work, or determination of trace amounts, three criteria must be observed.

(i) The absorptivity of the species to be determined must be reasonably large. While it is possible to determine metals such as copper or cobalt in water as the aquo complex, this will give accurate results only down to about 1% since $\varepsilon \sim 10$ m^2 mol^{-1}. However, for anthracene, $C_{14}H_{10}$, which has three fused aromatic rings, $\varepsilon = 18\,000$ m^2 mol^{-1} and, thus, even a solution of about 0.5 ppm will give an absorbance of approximately 0.1 in a 1 cm cell.

(ii) The species must be stable in solution. It must not oxidize or precipitate or change during the analysis (unless the analysis intends to study that change).

(iii) Calibration must be carried out over the range of concentrations to be determined. Agreement with the Beer–Lambert law must be established.

(iv) In complex matrices, it is not possible to analyze for all the species present using a few spectra. It is necessary to separate the components using one of the techniques described in Section D, or to use combined methods such as those in Section F.

It should be noted that it is possible to determine two (or more) species in an analytical sample by measuring the absorbance at several wavelengths. Calibration and measurements at two wavelengths enables two components to be determined simultaneously, but if more wavelengths are measured, a better 'fit' of the experimental data is achieved.

Example

Two organic components X and Y have absorption maxima at 255 and 330 nm, respectively.

For a pure solution of X, $\varepsilon(255) = 4.60$; $\varepsilon(330) = 0.46$
For a pure solution of Y, $\varepsilon(255) = 3.88$; $\varepsilon(330) = 30.00$
For a mixture of X and Y in a 0.01 m cell, $A(255) = 0.274$ and $A(330) = 0.111$

Calculate the concentrations of X and Y in the mixture.

Using the Beer–Lambert law at each wavelength:

$$A = \varepsilon_X c_X l + \varepsilon_Y c_Y l$$

At 255 nm: $0.274/0.01 = 4.60c_X + 3.88c_Y$
At 330 nm: $0.111/0.01 = 0.46c_X + 30.0c_Y$

Solving these simultaneous equations gives

$$c_X = 5.71 \text{ mol m}^{-3} = 5.71 \times 10^{-3} \text{ M}$$

$$c_Y = 0.288 \text{ mol m}^{-3} = 2.88 \times 10^{-4} \text{ M}$$

Table 3 gives some examples of the use of UV-visible spectrometry and of fluorimetry for quantitative analysis.

E10 INFRARED AND RAMAN SPECTROMETRY: PRINCIPLES AND INSTRUMENTATION

Key Notes

Principles — Vibrational transitions in molecules cause absorption in the infrared region of the electromagnetic spectrum. They may also be studied using the technique of Raman spectrometry, where they scatter exciting radiation with an accompanying shift in its wavelength.

Group frequencies — Vibrational spectra give information about the functional groups in molecules, and the observed group frequencies are affected by molecular interactions such as hydrogen bonding.

Instrumentation — Infrared and Raman instruments include a radiation source, a means of analyzing the radiation and a detection and data processing system. Additionally, sampling methods to deal with gases, liquids, solids, microsamples and mixtures are available.

Related topics
Infrared and Raman spectrometry: applications (E11)
Gas chromatography-infrared spectrometry (F4)

Principles

The vibrational levels of molecules are separated by energies in the infrared (IR) region of the electromagnetic spectrum. That is, in the wavenumber range from 13 000 to 10 cm^{-1}, or between 0.8 and 1000 μm on the wavelength scale. For convenience, this large region is divided into **near IR**, or **NIR** (13 000–4000 cm^{-1}), **mid IR** (4000–400 cm^{-1}) and **far IR** (400–10 cm^{-1}).

Molecules contain bonds of specific spatial orientation and energy. These bonds are seldom completely rigid, and when energy is supplied, they may bend, distort or stretch. A very approximate model compares the vibration to that of a **harmonic oscillator**, such as an ideal spring. If the spring has a **force constant, *k*,** and masses m_A and m_B at the ends, then the theoretical vibration frequency ν is given by:

$$\nu = (1/2\pi)\sqrt{(k/\mu)}$$

where $\mu = m_A . m_B/(m_A + m_B)$ is called the **reduced mass**.

Each type of molecular vibration is characterized by a vibrational quantum number, v. For a simple stretching vibration, there is a series of levels whose energy is given approximately by

$$E = h\nu_0 .(v + ½)$$

This means there is a set of levels spaced in energy by $h\nu_0$ or in wavenumber by $\bar{\nu}_0$. The selection rule for an ideal harmonic oscillator allows transitions where $\Delta v = \pm 1$, giving a single, fundamental vibrational absorption peak.

However, when bonds are stretched they weaken, so a better model takes this into account, and the molecules are treated as **anharmonic oscillators**. Thus, where high energies are involved, larger energy transitions may occur, where $\Delta v = +2, +3$, etc., giving the first overtone at a wavenumber approximately double that of the fundamental, and so on.

The electrical field associated with the electromagnetic radiation will interact with the molecule to change its electrical properties. Some molecules (for example, HCl) have a **dipole moment** due to charge separation and will interact with the field. Others may acquire a dipole when they vibrate. For example, methane, CH_4, has no dipole, but when one of the CH bonds stretches, the molecule will develop a temporary dipole.

Even if the molecule does not have a dipole, the electric field, E, may distort the electron distribution and **polarize** the molecule:

$$\mu_{ind} = \alpha E$$

where μ_{ind} is the dipole induced by the field, E, and α is the **polarizability** of the molecule.

The rules governing transitions in the infrared region of the spectrum require that, in order to absorb, the dipole moment of the molecule must change during the vibration. Such vibrations are said to be **IR active**. For transitions to be active in the Raman region, it is required that the polarizability must change during the vibration. The transitions are then **Raman active**, or **R active** (*Fig. 1*).

Consider two simple diatomic molecules, nitrogen and carbon monoxide. These molecules have only one fundamental vibration frequency, ν_o. For nitrogen it is 2360 cm^{-1}, and for carbon monoxide 2168 cm^{-1}.

Since carbon monoxide has a permanent dipole, which will increase and decrease as the molecule stretches and compresses, the vibration will interact with IR radiation, and an absorption peak will be observed close to 2168 cm^{-1}. Nitrogen has no dipole, and vibration does not produce one. Therefore, it will not absorb IR radiation. This is of great importance, since IR spectra may be recorded in air without interference.

However, when the nitrogen molecule vibrates, the bonding electrons are distorted and the polarizability is changed. Therefore, it will give a spectrum using the Raman technique.

In order to excite Raman transitions, energy comparable to the difference between electronic energy levels must be supplied. This may be visible laser light

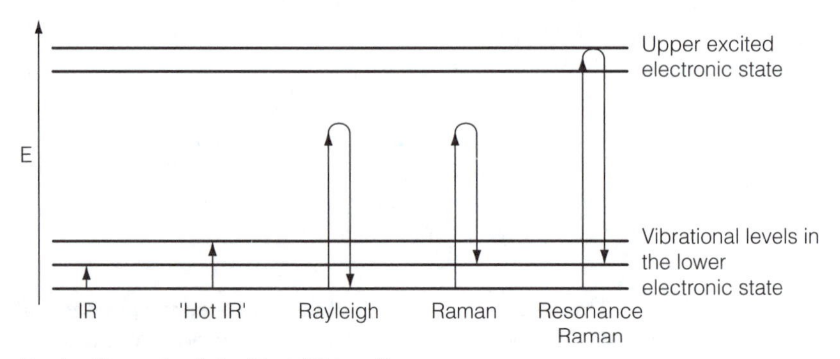

Fig. 1. Energy levels for IR and R transitions.

or NIR radiation. If the exciting wavelength matches the energy difference between the electronic levels of the sample, the Raman signal is greatly enhanced by the resonance Raman effect. Rayleigh scattering re-emits the exciting line. The more intense emission due to fluorescence effects may mask the weak Raman signal, but with NIR radiation fluorescent interference is much less.

As molecules become more complex, the number of possible vibrations increases. For example, carbon monoxide, CO_2, has three atoms arranged in a line: O=C=O. This molecule does not have a dipole and may vibrate in three ways.

(i) The symmetric stretch, denoted by v_1, is where both oxygens are equidistant from the central carbon, but the C–O bonds lengthen and contract together. The dipole does not change, but the polarizability does, so this vibration is IR inactive, but R active.

(ii) The antisymmetric stretch, v_3, has one C–O bond stretching, while the other contracts. The carbon atom moves as well so that the center of mass of the molecule remains stationary. The dipole changes, but the polarizability does not, so this is IR active but R inactive.

(iii) The bending vibrations, v_2, may be resolved into two identical and mutually perpendicular components corresponding to two transitions of the same energy (**degenerate**). It is necessary to think in three dimensions: considering the page as a plane, then if the two oxygens go equally 'down the page', while the carbon goes 'up the page' to balance, this is **in-plane bending**. If the oxygens go 'into the page' and the carbon 'out of the page', this is **out-of-plane bending**. These changes will be reversed as the vibration progresses. This vibration is IR active and R inactive.

The triangular molecule of water, H_2O, also has three different vibrations, corresponding to the same vibrational types. However, each involves a change in dipole so all three are IR active. The Raman spectrum shows only one line due to the symmetric stretch. These vibrations are shown schematically in *Figure* 2.

It is possible to extend these arguments to more complex molecules, but this is only of value for studies of structural parameters such as the length and

Schematic	Dipole	Polarizability	Activity	Wavenumber (cm^{-1})
<O = C = O>	No change	Change	R active	1320
O <:>C = O>	Change	No change	IR active	2350
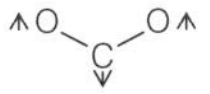 ∧O C O∧ (C ∨)	Change	No change	IR active	668
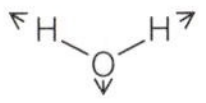 H O H	Change	Change	IR & R active	3650
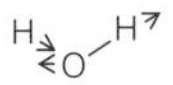 H <O H	Change	No change	IR active	3756
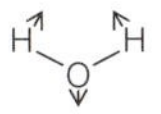 H O H	Change	No change	IR active	1600

Fig. 2. Schematic of the vibrations of the CO_2 and H_2O molecules.

strength of bonds and their interaction. For analytical purposes, it is sufficient to note that there are $3N - 6$ fundamental vibrations for a molecule with N atoms (or $3N - 5$ if the molecule is linear). If the molecule has a **center of symmetry** (e.g. CO_2, ethyne, CH≡CH or benzene, C_6H_6) then those bands that are IR active are *not* R active, and vice versa.

Besides the fundamental vibrations, it is important to note that some IR absorptions may correspond to **combinations of vibrations** and also to **overtones**.

For example, for HCl, which has a fundamental stretching band around 2800 cm^{-1}, there is a first overtone at just below 5600 cm^{-1} in the NIR. Carbon dioxide, whose fundamentals are listed above, has overtones of ν_1 at 6980 and 11 500 cm^{-1} in the NIR and several combination vibrations such as $(\nu_2 + \nu_3)$ at 2076 cm^{-1}.

In complex molecules, the structure is dominated by **functional groups**. For example, a large number of compounds contain the **carbonyl group, C=O**. These include aldehydes, ketones, acids, esters, amides and quinines. Almost every organic compound has **C–H bonds**, although they may differ considerably in their behavior. Alcohols, acids and phenols contain the **–OH group**. These groups and also aromatic rings and other larger structural units may be considered as giving rise to **characteristic group frequencies**.

Group frequencies

There are three main types of functional group that give rise to absorptions in IR (and R) spectra which are highly characteristic.

(i) The stretching of bonds between a heavier atom and hydrogen, H–X–. Relating these bonds to the harmonic oscillator model, the reduced mass depends chiefly on the mass of the H atom, because this has such a low mass compared to all the other atoms, and the force constant is high because the bonds connecting the hydrogen and other atoms are strong. This accounts for the H–X– stretching vibrations, which almost all occur in the region between 2000 and 4000 cm^{-1}. The X-atoms and the structure to which they are attached will determine the exact frequency, as shown in *Table 1*. It should be noted that **hydrogen bonding**, which occurs with electronegative atoms such as oxygen or nitrogen, but *not* with carbon, will make the H–X– bond both weaker and more variable. Consequently, whereas the free OH stretch of gaseous alcohols has a sharp absorption at 3600 cm^{-1}, liquid alcohols show a very broad band nearer 3200 cm^{-1}.

(ii) **Double** and **triple bonds**; **aromatic systems**. Because these are bonds with high bond energies, they are less affected than single bonds by the structures to which they are attached. It is worth noting that the higher the bond order, the higher the IR absorption frequency or wavenumber, as shown for C–C and C–N bonds as shown in *Table 2*.

(iii) The **bending** vibrations of organic molecules also give characteristic group frequencies. As a general 'rule of thumb', it is usually found that the bending vibrations occur at the lowest frequency or wavenumber, the symmetric stretch next and the antisymmetric stretch at the highest value. *Figure* 3 and *Table 3* give a selection of the most useful group frequencies in the mid-IR and NIR regions respectively.

Raman spectra have similar group frequency correlations, but two features are of special interest. The R spectrum of water is much less intense than the IR

Table 1. H–X group frequencies

X	Wavenumber (cm^{-1})	Comment
H–C (aliphatic)	2960–2900	Strong
H–C (aromatic)	3050–3000	Strong
H–C (alkyne)	3300	
H–O	3600	Free OH
H–O	3500–2500	H-bonded
H–N	3500–3300	Broad
H–S	2600–2500	Weak

Table 2. Multiple bonds

Bond		Wavenumber (cm^{-1})	Comment
Single	C–C	Approx. 1200	Variable
Aromatic	C:::C	1600, 1500	
Double	C=C	1650	
Triple	C≡C	2200	Weak
Single	–C–N	1100	Variable
Double	–C=N–	1670	
Triple	–C≡N	2250	Strong

spectrum, and therefore aqueous solutions may readily be studied. Overtones and combination vibrations occur less often. Vibrations of symmetrical structures, such as R–C≡C–R, which are weak in the IR, appear as strong bands in the R.

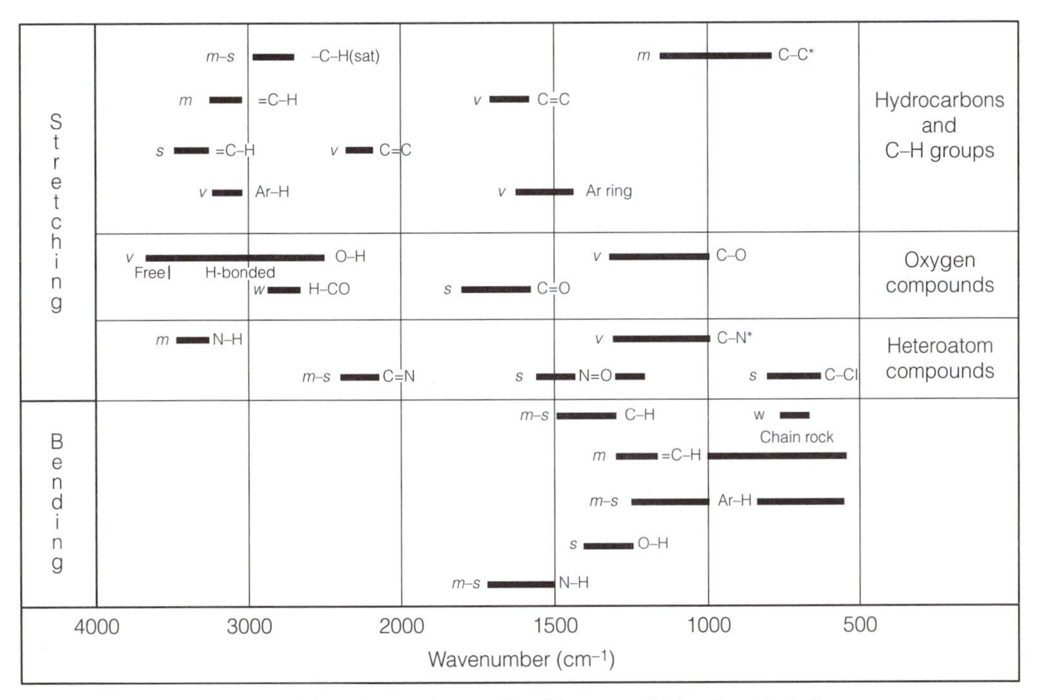

*Fig. 3. Chart of characteristic infrared group frequencies. *Not so useful for characterization.*

Table 3. Selected group frequencies in the NIR region

Frequency	Group
A Combination bands 5000 to 4000 cm^{-1}	
5300, 4500	Water
4700–4600	RNH_2
4650–4150	CH
4700–4600	CC
B First overtone region 7500–5000 cm^{-1}	
7300	Water, ROH
5500–5100	RCOOR, RCOOH
5400–4800	CONHR, $CONH_2$
7500–5700	CH, CH_2, CH_3
C Higher overtone region 14 000 to 8000 cm^{-1}	
13 600, 10 700	ROH, ArOH
14 000	ArCH
14 000–11 000	CH, CH_2, CH_3

Vibrations that appear in the **near infrared (NIR)** region are the overtones and combination vibrations of those in the mid-IR. They are generally less intense, and are useful in quantitative measurements. For example, water has fundamental vibrations at 3700, 3600 and 1600 cm^{-1}. In the NIR of water, combination bands occur at 5300, and overtones around 7000, 11 000 and 13 500 cm^{-1}.

For compounds with an –OH group, overtones at 7000, 10 500 and 13 600 occur, while for aliphatic hydrocarbons, there are absorptions at 4200–4700, 5700–6300, 7000–7500 and so on.

Instrumentation

For mid-IR, NIR and Raman spectrometry, the instrumentation is different, but the main components of spectrometers are all required.

The sources used for mid-IR are heated rods, such as a nichrome ribbon or a 'globar', which is a rod of silicon carbide. The Opperman source is a rhodium heater in an alumina tube packed with alumina and zirconium silicate. When heated to above 1000°C, these sources emit energy over a wide range, resembling a black-body radiator with a maximum intensity at about 1000 cm^{-1}. For NIR, tungsten or tungsten halogen lamps are used.

In Raman spectrometry, a high intensity source is required, since Raman scattering yields low intensity lines. Laser sources such as the Ar^+ laser give strong, sharp lines at 488.0 nm and 514.5 nm. One disadvantage is that these wavelengths may cause fluorescence. This is avoided by using NIR laser sources. A schematic of an FT-Raman spectrometer is shown in *Figure 4.*

Sampling for vibrational spectroscopy is an extensive subject. Only the basic methods will be discussed here. Since glass and polymers absorb strongly in the IR region, it is necessary to use ionic materials to contain samples. Typical examples are listed in *Table 4.*

The types of samples that may be analyzed by IR are very varied and will be considered in turn.

For **gases,** because they are present at much lower concentrations than pure liquids or solids (e.g., 0.04 M for nitrogen in air, 17.4 M for liquid ethanol), longer path lengths are required. The gas-phase spectrum of HCl at 0.2 atm may be studied in a 10 cm glass cell with NaCl windows. Low concentrations of

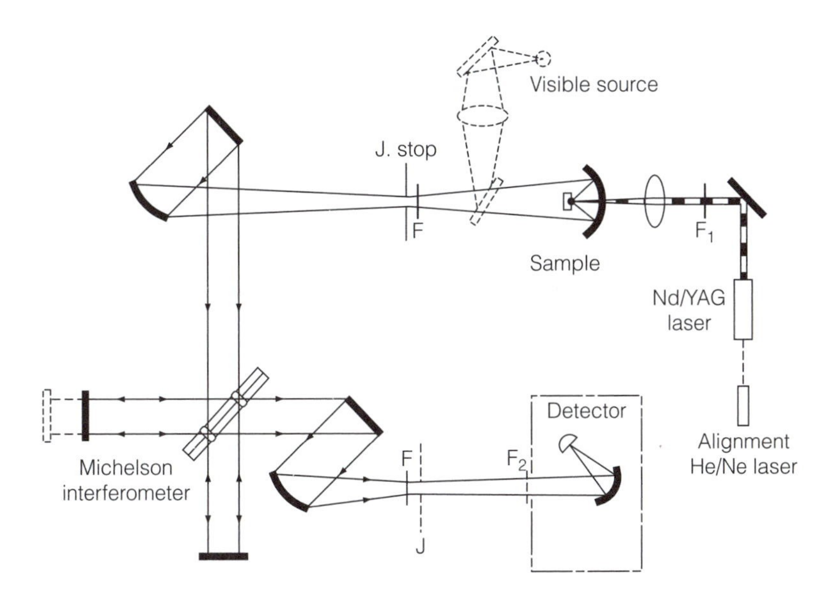

Fig. 4. Schematic of an FT-laser Raman spectrometer.

Table 4. Window materials for infrared spectrometry

Material	Range (cm^{-1})	Comment
NaCl	40 000–625	Soluble in water
KBr	40 000–385	Soluble in water
TlBr/I (KRS-5)	16 600–250	Slightly soluble
ZnSe (Irtran-4)	10 000–515	Insoluble

exhaust gases may need a 10 m cell, which reflects the IR beam to achieve the long pathlength.

Two points should be stressed if gas phase spectra are to be studied.

(i) The fine structure of the IR stretching band of HCl shows many lines. These are due to rotational transitions superimposed onto the vibrational one. The vibration-rotation spectrum of water vapor shows a complex pattern of lines across much of the IR region. With heavier molecules, the moments of inertia of the molecule are larger, and therefore the rotational lines are closer together. For gaseous CO_2, one vibration-rotation band occurs at about 2300 cm^{-1} and, at low resolution, shows two lobes only. It is therefore important to exclude water vapor and carbon dioxide from the spectrometer as far as possible, and this is usually done by purging with dry nitrogen.

(ii) The second cautionary point is that, with gas-phase spectra, hydrogen bonding is much less important. For the –OH stretch in ethanol, therefore, instead of the broad band between 3500 and 3000 cm^{-1} seen in the spectrum of a liquid, a single sharp peak occurs at 3600 cm^{-1} when the gaseous spectrum is recorded.

Liquids or melts are more concentrated, and may be studied directly as a thin film between NaCl plates. For more quantitative work, accurately prepared solutions in solvents which do not absorb in the region of analytical interest, such as CCl_4 or CS_2 in NaCl cells, with a known path length provided by a spacer, may be used. Most of these are also applicable to NIR, and short path length silica cells may also be used there.

Since Raman spectra are recorded in the visible and NIR regions, glass or quartz cells may also be used. The routine setting-up and calibration of a Raman spectrometer can be carried out using liquid CCl_4 in a thin glass tube, which gives a strong peak at a Raman shift of 458 cm^{-1} and weaker peaks at 218, 314 and 760–790 cm^{-1}. Since water is a weak Raman scatterer, aqueous solutions are readily studied.

If a **solid** organic powder sample is placed in an IR beam, the particles scatter the light, and little is transmitted. Therefore, for routine analysis, the sample is usually ground to a fine powder and mixed with paraffin oil ('Nujol') to form a paste or **mull**. This reduces the scattering at the powder surface and gives a good spectrum, with the disadvantage that the bands due to the oil (at approximately 2900, 1450, 1380 and 750 cm^{-1}) are superimposed on the spectrum. Alternatively, the fine powder may be mixed with about 10–100 times its mass of dry, powdered KBr and the mix pressed in a hydraulic press between smooth stainless steel dies to give a clear KBr disk.

Solutions of solids may also be used, and tetrachloromethane, CCl_4 is often used as solvent, since it has few IR-active bands, mostly at the low wavenumber end of the spectrum. These must be ignored when the spectrum is interpreted. Thin films of solids such as polymers may be supported directly in the IR beam. Polystyrene is a useful calibration sample to check the performance of an IR spectrometer (see *Fig. 2* in Topic E11).

Raman spectra of solids may be obtained by placing the sample directly in the beam so that the radiation is scattered correctly into the dispersion system.

Reflectance spectra can be measured in three ways. A powder is placed in the incident beam and allowed to interact by **diffuse reflectance**. The reflections are collected by a mirror, as shown in *Figure* 5*(a)*, or for NIR by an integrating sphere surrounding the sample.

If the beam is reflected off a flat sample surface, **specular reflectance** results, and this may give good spectra.

If the sample is placed in good contact with the surface of an optical device of high refractive index (such as a prism of KRS-5) and illuminated through the prism by IR, the beam passes into the layers in contact and is **attenuated** before being **totally internally reflected** by the system, as shown in *Figure 5(b)*. This is called **attenuated total reflectance** or **ATR**. If the beam interacts several times, then we have **multiple internal reflectance (MIR)** and if the surface is horizontal, which is an advantage in setting up the sample, then it is **horizontal attenuated total reflectance (HATR)**.

It should be noted that the detail of spectra obtained by reflectance methods might be different from that obtained in solution or with KBr disk techniques. Modern instruments possess software to convert reflectance spectra to resemble the more usual transmission spectra.

Analysts must often deal with samples of very small size, or analyze a small area of a large sample. One technique is to reduce the size of the IR beam using a **beam-condensing** accessory. This has lenses of CsI, which focus the beam down to about 1 mm diameter and permit the study of micro-KBr disks.

A more versatile modern development is the **IR microscope,** or Raman microscope. This is an adaptation of a conventional stereo microscope which

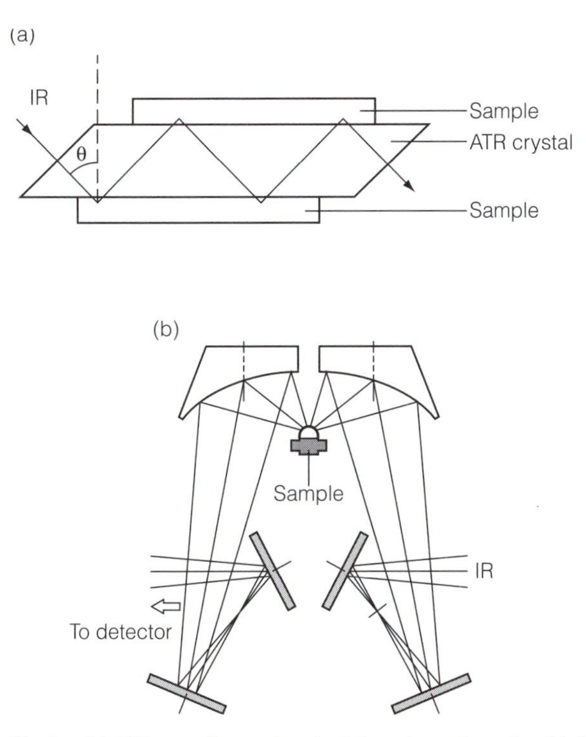

Fig. 5. (a) ATR sampling system for infrared spectrometry. (b) Diffuse reflectance sampling for infrared spectrometry.

permits location of the area of interest using visible illumination, and which may then be altered so that the IR (or R) spectrum of that area is obtained by reflectance or transmission.

Dispersion, or separation, of the IR radiation is generally done by IR diffraction gratings, or by Fourier transformation. Resolution for routine IR and NIR spectra of 2–4 cm^{-1} is adequate, but to resolve vibration-rotation bands, a resolution of better than 1 cm^{-1} is needed. For Raman spectra, the introduction of Fourier transform techniques has been most valuable (Topic E3).

Detectors for IR radiation relate to the source, heat sensors being suitable. Older instruments used thermocouple detectors. Pyroelectric detectors, which change electrical properties when exposed to IR radiation, especially doped triglycine sulfate, are often used.

Semiconductors, such as mercury cadmium telluride (MCT) have a more sensitive response in the mid IR.

E11 Infrared and Raman spectrometry: applications

Key Notes

Structural identification	The group frequencies observed in an infrared or Raman spectrum of organic compounds are useful indicators of molecular structure. Inorganic materials may also give specific infrared bands characteristic of their structure and bonding.
Quantitative measurements	The intensity of infrared absorbances obeys the Beer–Lambert law and may be used for quantitative analysis, especially for time-resolved measurements and kinetic studies. Near infrared spectrometry is used extensively for quantitative work.
Related topic	Infrared and Raman spectrometry: principles and instrumentation (E10)

Structural identification

Comparison of the infrared (or Raman) spectra of related compounds illustrates the importance of the group frequency concept. The following examples will illustrate this.

Example 1. The spectrum of compounds containing aliphatic chains exhibit bands due to CH_3–, –CH_2– and =CH– groups, both in their stretching and bending modes. The IR spectrum of a **liquid paraffin**, or Nujol, used as a mulling agent is shown in *Figure 1*.

The –CH stretching vibrations all absorb just below 3000 cm^{-1}, and provide at least four peaks. The bending vibrations of –CH_3– and –CH_2– absorb around 1450 cm^{-1} whereas the characteristic band at 1380 cm^{-1} is due to the methyl

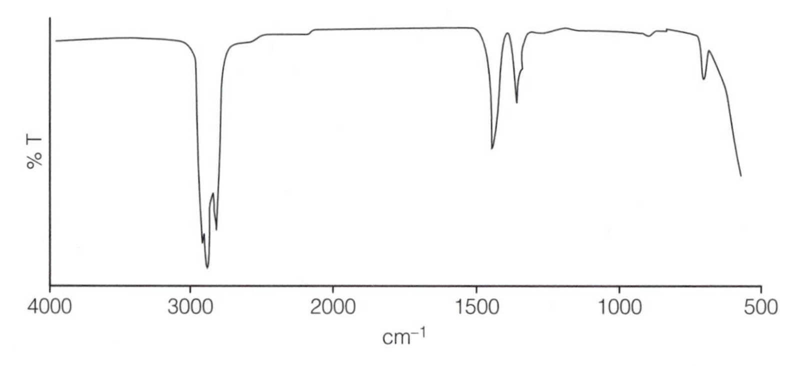

Fig. 1. Liquid paraffin: (Nujol).

groups. With long chains, a weak band around 750 cm^{-1} is due to rocking of the $(-CH_2-)_n$ groups. A similar spectrum is observed with polyethylene film.

If the compound has no methyl groups, for example cyclohexane, the spectrum is simplified, as only the methylene absorptions are observed. If hydrogens are substituted by halogens, peaks corresponding to the halogen occur at about 700 cm^{-1} for –C–Cl, 600 cm^{-1} for –C–Br or 500 cm^{-1} for –C–I.

Example 2. Aromatic compounds, especially those derived from benzene, show bands characteristic of their structure and substitution. A thin film of **polystyrene**, $(C_6H_5-CH-CH_2)_n$ shown in *Figure* 2 is often used for calibrating the wavenumber scale of an IR spectrometer.

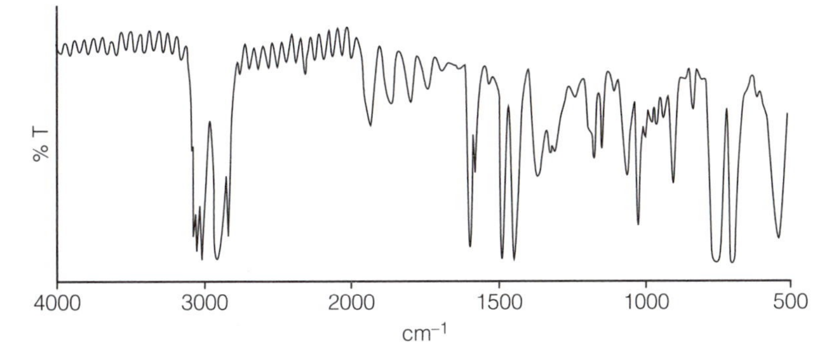

Fig. 2. Polystyrene film.

The aromatic –CH– stretch occurs just above 3000 cm^{-1} and the aliphatic just below.

The weak absorptions due to overtones in the 2000–1800 cm^{-1} region are typical of a **monosubstituted** benzene, as are the bands in the 800–600 cm^{-1} region due to out-of-plane (OOP) bending of the C–H bonds. Ring vibrations give bands between 1600 and 1500 cm^{-1} whereas the rest of the spectrum can be regarded as a "**fingerprint**" of the entire structure.

A particularly important peak in an IR spectrum is the **carbonyl, >C=O** absorption around 1700 cm^{-1}. It is nearly always a very strong peak. *Figure 3*

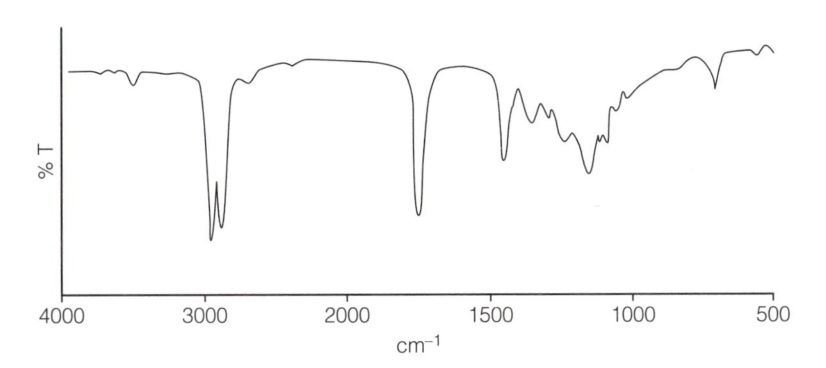

Fig. 3. Tribehenin: (triglyceride ester of C_{22} carboxylic acid).

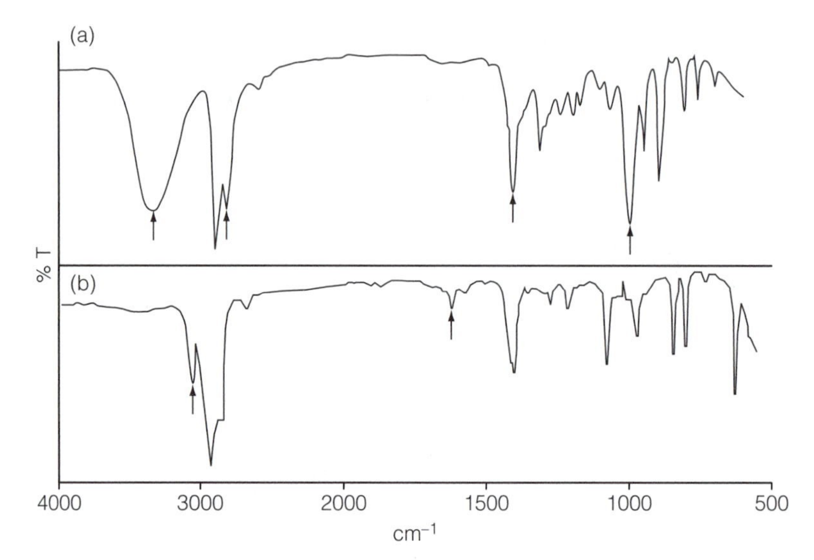

Fig. 4. (a) Cyclohexanol (liquid film); (b) cyclohexene (liquid film).

shows the IR spectrum of a triglyceride fat, with the ester group R–O–CO–R′. The long-chain of the carboxylic acid gives –CH stretching vibrations just below 3000 cm^{-1} and chain rocking at 720 cm^{-1}. The prominent peak near 1750 cm^{-1} clearly shows the presence of the carbonyl group.

The changes that may occur during a reaction are readily studied. For example, the dehydration of an alcohol to an alkene is observed by the disappearance of the strong, broad –OH bands at 3300 and 1350 cm^{-1} and the –C–O– stretch near 1100 cm^{-1}, arrowed in the spectrum of **cyclohexanol** in *Figure 4a* and the appearance of the –C=C– bands, especially around 1600 cm^{-1} as shown in the spectrum of **cyclohexene** shown in *Figure 4b*.

Acids are rather exceptional, as in the solid or liquid state the IR bands are broadened and weakened by hydrogen bonding. As seen in the spectrum of **acetyl salicylic acid (aspirin)** of *Figure 5*, the –OH peak is now so broad that it overlaps with the –CH peaks at 3000 cm^{-1}. The acid carbonyl peak is also broadened.

Raman spectra are equally useful for the interpretation of group frequencies. *Figure* 6 shows the Raman spectrum of **methyl benzoate**. Although there may

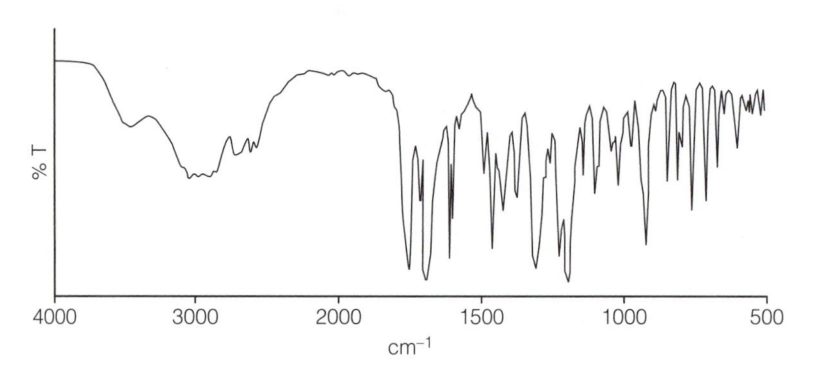

Fig. 5. Acetylsalicylic acid (KBr disk).

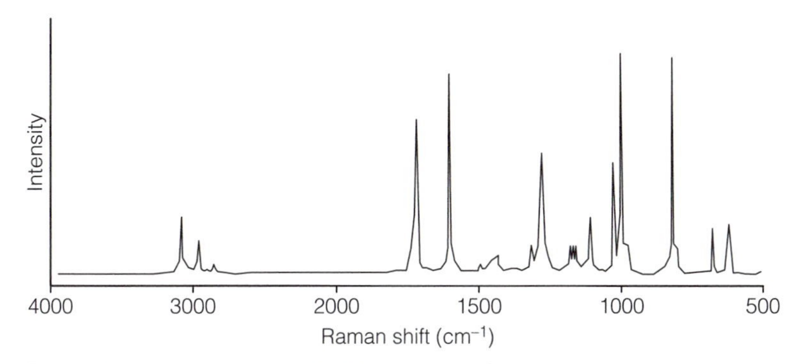

Fig. 6. Raman spectrum of methyl benzoate (liquid).

be some differences, the major peaks are in similar positions, and some groups, such as substituted alkynes R–CC–R′ and –S–S–, give stronger peaks in the Raman than in the IR spectrum.

Further examples are given in Topic F2.

One technique which may be used for the identification of materials such as drugs or polymers is a **"decision tree"**. For example, in considering the spectrum of a polymer shown in *Figure* 7, the first decision might be based upon the presence of a carbonyl band near 1730 cm^{-1}. In this spectrum, such a band is observed, and the decisions are simplified by using the chart shown in *Figure 8*.

From the above chart, the sample is readily identified as polymethylmethacrylate as shown by the heavier lines in *Figure 8*. Similar charts are available for the polymers not possessing a carbonyl band and for other classes of compounds.

Computerized analysis has enhanced the possibilities for identifying samples by IR spectrometry. The spectrum is analysed to identify the peaks, which may then be listed. This is much easier and more accurate than visual examination of the hard copy spectrum.

Two procedures are then followed, in much the same way that the scheme in *Figure 8* uses. Peaks, or groups of peaks associated with a particular structure are assigned. Thus, a peak at 1750 cm^{-1} would strongly indicate a carbonyl compound, and peaks at 2950–2800 cm^{-1} would probably belong to –CH stretching of an aliphatic or alicyclic compound. This approach yields a list of **probable structural units (PSU)**.

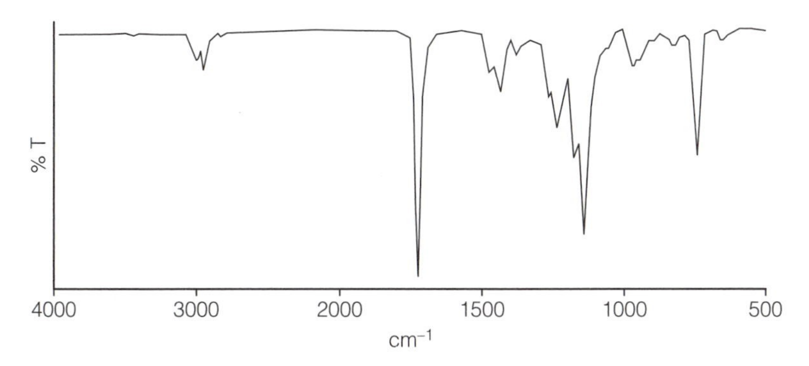

Fig. 7. Polymethylmethacrylate.

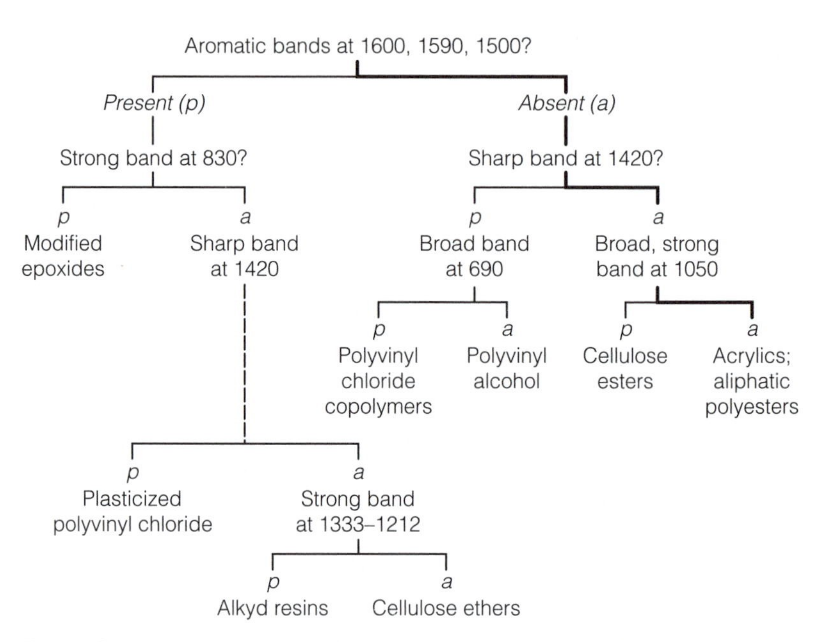

Fig. 8. Decision tree for the identification of a polymer when a carbonyl absorption is present.

The second stage is to conduct a computer search of an appropriate database or library, for example of drugs, to test for an exact match or to compile an inverted list (Topic H4).

Quantitative measurements

The use of IR and NIR spectrometry for qualitative measurements is extensive and wide ranging, and for this purpose **transmission spectra** are conventionally recorded as a function of **wavenumber**. In order to make **quantitative** measurements, it is necessary to convert the transmittance readings to **absorbance**, A, the relation between the two being:

$$A = \log(100/T\%)$$

This also allows any absorbance by solvents or other components of the sample to be subtracted from the analyte peak.

$$A(total) = A(sample) + A(background)$$

This also allows the **proper** subtraction of solvents or other components.

For example, if the spectrum of a machine oil without additives is measured in a 0.1 mm NaCl cell, and then the same procedure is followed for a sample with small amounts of additives, subtraction of the absorbance spectra will give the spectrum of the additives in absorbance form.

Gas analysis by IR spectrometry using long path length cells has been used to measure concentrations of anaesthetic gases. For example, nitrous oxide, N_2O, shows a strong absorbance at 2200 cm^{-1} at which the wavenumber of neither water vapour nor carbon dioxide interfere. Measurement of the concentrations between 2 and 50 ppm is possible with a 15 m path length gas cell. Trichloromethane (chloroform) gives a strong, sharp peak at 770 cm^{-1} and may be measured down to 0.1 ppm.

In addition, IR spectrometry has been used to measure the mineral contents of rocks, asbestos and to study residual solvents in pharmaceuticals. Mixtures can be analyzed directly, although chromatographic methods are normally preferred (see Topics F3–5).

NIR spectrometry has been used widely for the analysis of agricultural, food and pharmaceutical products. It is a rapid technique and may be adapted to the quality control of process streams as, using fiber optics, remote sampling in industrial environments is possible. One of the most useful NIR methods uses **diffuse reflectance** to analyze solid materials. The sample, usually as a powder, is placed in an **integrating sphere** and illuminated from an NIR source that directs radiation onto it.

A typical use of this type of instrument is shown in *Figure* 9, for measuring the starch, protein, water and oil components of wheat. A similar technique has been used to quantify separated analytes in samples separated on TLC plates directly from their NIR spectra and for analyzing the fat, moisture and protein content of meat. The technique of **multivariate analysis** (see Topic B5) has proved most valuable for this type of work.

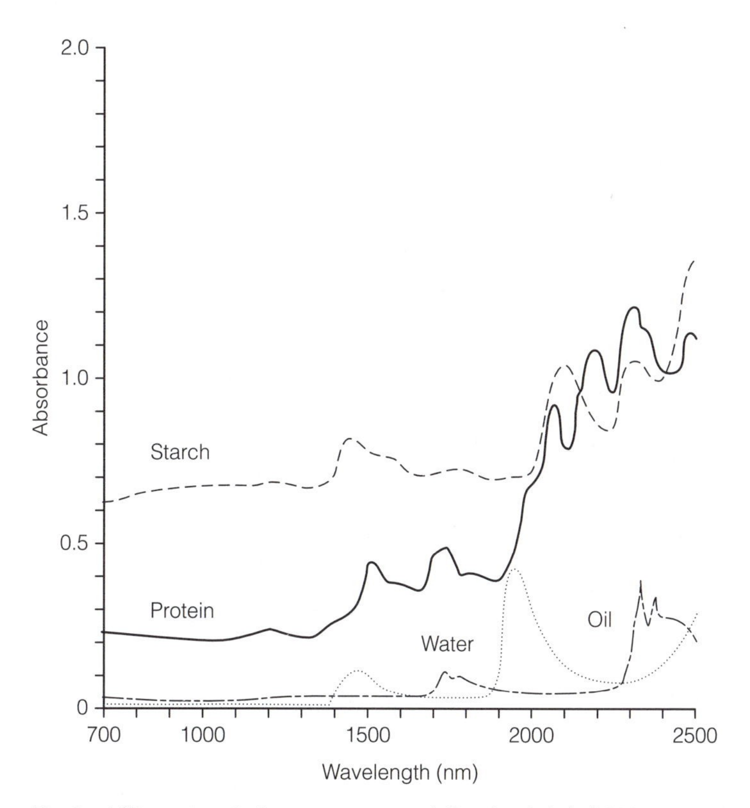

Fig. 9. NIR spectra of wheat components, (offset for clarity); (2000 nm = 5000 cm^{-1}).

E12 Nuclear magnetic resonance spectrometry: principles and instrumentation

Key Notes

Principles

Nuclear magnetic resonance (NMR) spectrometry is based on the net absorption of energy in the radiofrequency region of the electromagnetic spectrum by the nuclei of those elements that have spin angular momentum and a magnetic moment. For the nuclei of a particular element, characteristic absorption, or resonance frequencies, and other spectral features provide useful information on identity and molecular structure.

Nuclear and electron spin

Nuclei of elements that possess spin angular momentum and generate a magnetic moment are assigned a half-integral or integral spin quantum number. This determines the number of orientations in space that can be adopted by the spinning nuclei when subjected to an external magnetic field. Electrons also possess spin angular momentum, which generates a magnetic moment that affects the magnitude of the external field experienced by nuclei.

Chemical shift

Nuclei of a particular element that are in different chemical environments within the same molecule generally experience slightly different applied magnetic field strengths due to the shielding and deshielding effects of nearby electrons. As a result, their resonance frequencies differ, and each is defined by a characteristic chemical shift value.

Spin–spin coupling

The spin states of one group of nuclei can affect the magnetic field experienced by neighboring groups through intervening bonds in the molecule in such a way that the absorption peaks of each group are split into a number of components. This effect can provide useful information for spectral interpretation.

NMR spectrometers

Spectrometers comprise a superconducting solenoid or electromagnet to provide a powerful, stable and homogeneous magnetic field, a transmitter to generate the appropriate radiofrequencies, and a receiver coil and circuitry to monitor the detector signal. A dedicated microcomputer controls the recording of spectra and processing of the data.

Related topics Electromagnetic radiation and energy levels (E1) Atomic and molecular spectrometry (E2)

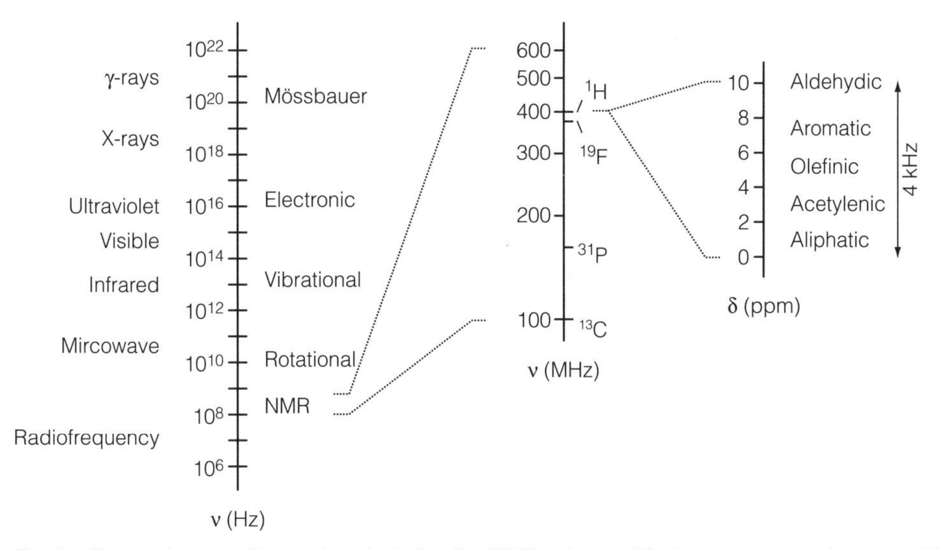

Fig. 1. The electromagnetic spectrum including the NMR region and the frequency range for proton NMR.

Principles

Nuclear magnetic resonance (NMR) transitions can be observed in the radiofrequency region of the electromagnetic spectrum (*Fig. 1*). For elements whose nuclei have **spin angular momentum** and a magnetic moment, or dipole, the application of an **external magnetic field** creates two or more quantized energy levels (*Fig. 2*). (Note: in the absence of the external field, the energy levels are **degenerate**, and no spectroscopic transitions can be observed.)

The energy difference, ΔE, between these levels is extremely small, and corresponds to radiofrequency energy (*Fig. 1*), the relation being expressed by the Planck equation

$$\Delta E = h\nu \quad (1)$$

where h is the Planck constant and ν is the corresponding radiofrequency.

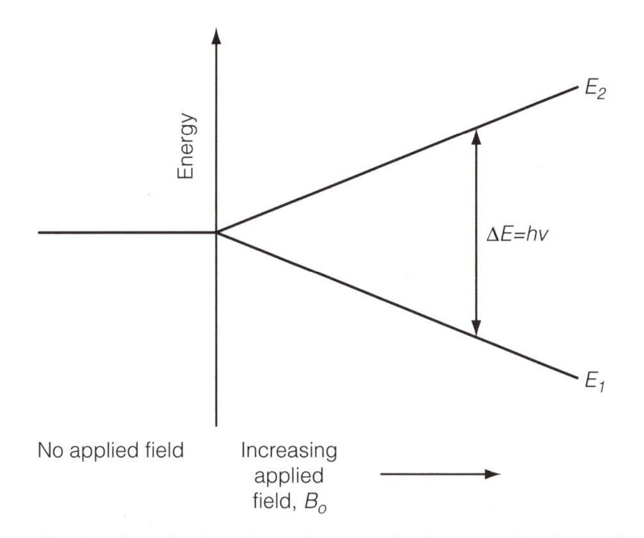

Fig. 2. Quantized nuclear spin energy levels created by the application of an external magnetic field to nuclei with a half-integral spin quantum number.

A consequence of the small difference between the energy levels is that their populations are almost equal at room temperature. When the nuclei are irradiated with radiation of frequency ν, both upward (**absorption**) and downward (**emission**) transitions occur, and the system is said to be in **resonance**. However, initially there is a small excess of a few nuclei per million in the lower energy level, which results in a **net absorption** of energy at the resonance frequency. As only this small excess of nuclei is detectable, NMR spectrometry is basically a much less sensitive technique than ultraviolet/visible (electronic) and infrared (vibrational) spectrometry (Topics E8 to E11). Following the net absorption of energy during resonance, the equilibrium ground state populations are re-established by the excited nuclei **relaxing** to the lower energy level, the process normally taking only a few seconds in liquids and solutions. Computer control and data processing enables sensitivity to be enhanced considerably by using pulsed techniques and accumulating scans.

For the nuclei of each element, the magnetic moment, μ, is directly proportional to the spin angular momentum, I, the proportionality constant, γ, being known as the **magnetogyric** or **gyromagnetic ratio**, i.e.

$$\mu = \gamma \boldsymbol{I} \quad \text{or} \quad \gamma = \mu / \boldsymbol{I}$$

The magnitude of the resonance frequency, ν, and hence of ΔE, is directly proportional to the strength of the applied magnetic field, B_o, being related by the equation

$$\nu = (\gamma / 2\pi) \cdot B_o \qquad (2)$$

Figure 2 illustrates the proportionality and *Table 1* lists some values of γ, ν and B_o for a number of nuclei along with their natural isotopic abundances.

Table 1. Natural abundance, magnetogyric ratio and resonance frequencies for protons, carbon-13, fluorine-19 and phosphorus-31

Nucleus	Natural abundance %	Magnetogyric ratio, γ 10^7 T^{-1} s^{-1}	Resonance frequency, ν (MHz) at applied field, B_o (Tesla)		
			2.3	7.1	11.7
^{1}H	99.985	26.75	100	300	500
^{13}C	1.108	6.73	25.1	75.4	125.7
^{19}F	100	25.18	94.1	282.3	470.5
^{31}P	100	10.84	40.5	121.4	202.4

The resonance frequency of the nuclei of one element is sufficiently different from those of others to enable their spectra to be recorded independently, as shown in *Figure 1* for hydrogen (protons), carbon-13, phosphorus-31 and fluorine-19, each of which provides useful information on the identity and structure of organic compounds. Small variations of the resonance frequency for the nuclei of a particular element due to different chemical environments within a molecule enable structural groups and features to be identified; for example, for proton spectra it is possible to distinguish between acidic, saturated and unsaturated molecular groups, etc., as indicated in *Figure 1*. Note that the proton has the largest nuclear magnetogyric ratio and a very high natural abundance making it the nucleus that can be detected with the highest sensitivity. Both proton and carbon-13 spectra have been the most widely studied, and it is in the analysis of organic compounds that NMR spectrometry has had the greatest impact.

Nuclear and electron spin

All nuclei are assigned a spin quantum number, I, on the basis of the number of protons and neutrons in the nucleus, the values being zero, half-integral or integral. The number of permitted orientations in space and hence quantized energy levels, or **spin states**, that can be adopted by a nucleus subjected to an externally applied magnetic field is given by $(2I + 1)$. Thus, for $I = \frac{1}{2}$, two orientations, or energy levels, are possible. The following should be noted.

- A spinning nucleus, being charged, generates a **magnetic moment vector**, or **dipole**, along its axis of rotation. This is analogous to the magnetic field associated with a current flowing in a loop of wire (*Fig. 3a*).
- Nuclei with even numbers of both protons and neutrons, e.g. ^{12}C, ^{16}O and ^{32}S, have a spin quantum number of zero, no spin angular momentum or magnetic moment, and can not give an NMR spectrum.
- Nuclei with a spin quantum number of ½, and therefore two spin states, include the proton, carbon-13, phosphorus-31 and fluorine-19 (*Table 1*).
- An electron also has spin, but being of opposite charge to a nucleus, it produces a **magnetic moment vector** in the opposite direction (*Fig. 3a*). Electrons also circulate under the influence of an applied field, generating additional fields that modify those experienced by the nuclei (*Fig. 3b*).

Chemical shift

The applied magnetic field experienced by a nucleus is affected by the fields generated by surrounding electrons, which may augment or oppose the external field. The **effective field**, B_{eff}, at the nucleus may be defined as

$$B_{eff} = B_o(1 - \sigma) \tag{3}$$

where σ is a **shielding constant** which may be positive or negative. The resonance frequency, defined by equation (2), is therefore given by

$$v = (\gamma/2\pi) \cdot B_o(1 - \sigma) \quad \text{or} \quad v = (\gamma/2\pi) \cdot B_{eff} \tag{4}$$

Consequently, nuclei in different chemical environments will have slightly different resonance frequencies due to **shielding** or **deshielding** by nearby

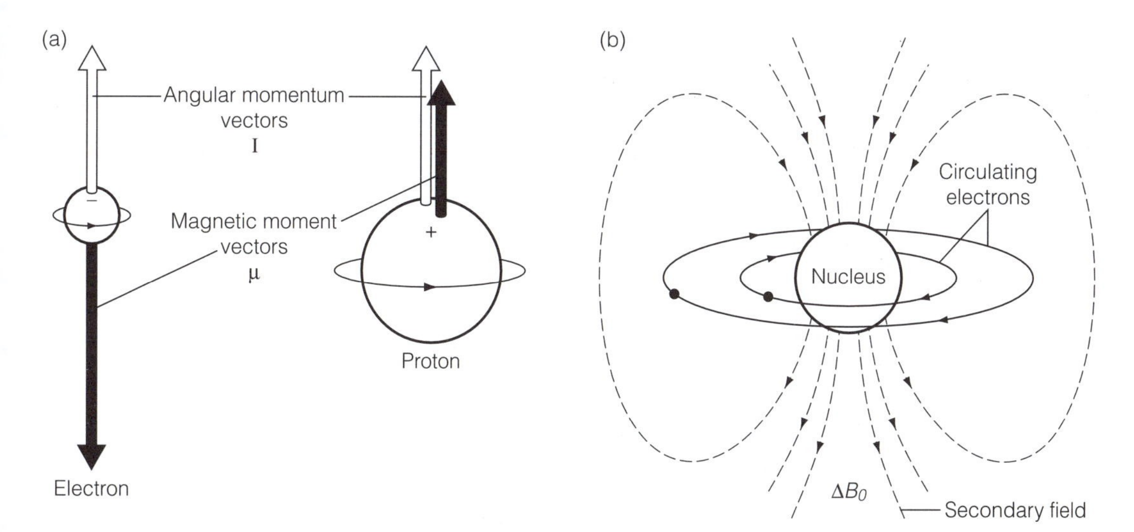

Fig. 3. (a) Spin angular momentum and magnetic moment vectors for a nucleus and an electron; (b) magnetic field induced by circulating electrons.

electrons, and this determines the value of σ, and hence ν. The exact position of each resonance frequency is referred to as its **chemical shift**, which is characteristic of the chemical nature of the particular nucleus.

Chemical shift is conventionally measured relative to the frequency of a reference compound, using a dimensionless parameter, δ, defined as the ratio of their chemical shift difference divided by the operating frequency of the spectrometer and multiplied by 10^6 to give more convenient numerical values, i.e.

$$\delta = \frac{(\nu - \nu_{ref})}{\nu_{spectrometer}} \cdot 10^6 \text{ ppm} \tag{5}$$

where ν_{ref} and $\nu_{spectrometer}$ are the resonance and operating frequencies of the reference compound and the spectrometer respectively.

It should be noted that:

- δ values are assigned units of ppm (parts per million) because the ratio is multiplied by 10^6;
- by definition, the chemical shift of the reference compound is assigned a value of zero and, conventionally, δ values are presented as a scale that increases from right to left (*Fig. 4*);
- the greater the shielding of the nucleus (larger σ), the smaller the value of δ and the further to the right, or **upfield**, the resonance signal appears;
- the less the shielding of the nucleus (smaller σ), the larger the value of δ and the further to the left, or **downfield**, the resonance signal appears;
- because field and frequency are directly proportional, it follows that upfield, or **high field** resonance signals correspond to **lower frequencies** than downfield, or **low field** signals, and *vice versa;*
- δ values are independent of the operating frequency of the spectrometer, enabling chemical shifts in ppm from spectra recorded on instruments with different operating frequencies to be compared (*Fig. 4*);
- for recording proton and carbon-13 spectra in nonaqueous solvents, the reference compound is normally **tetramethylsilane** (**TMS**, $(CH_3)_4Si$) which gives a single, high field (low frequency) resonance signal for the twelve

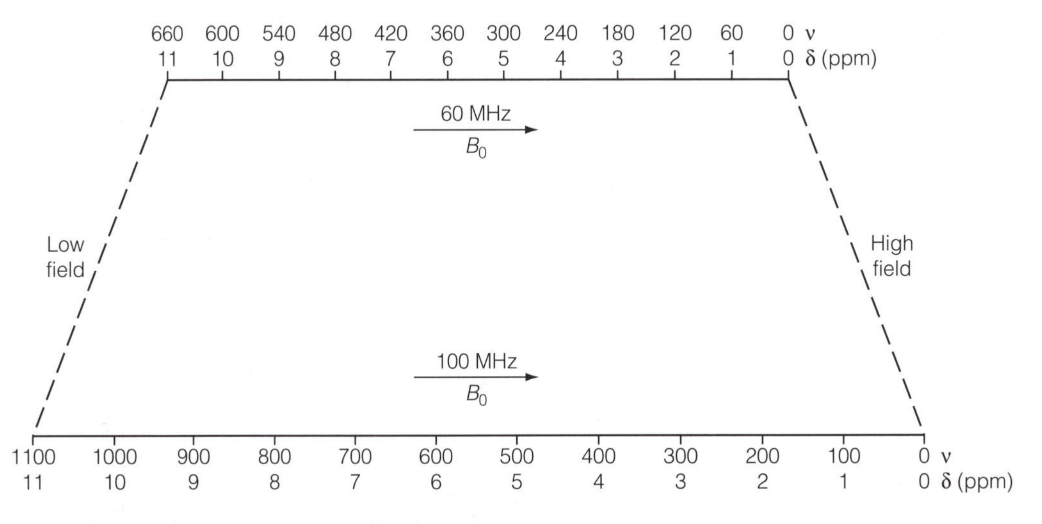

Fig. 4. Proton chemical shift scales in δ/ppm at spectrometer frequencies of 60 MHz and 100 MHz.

identical protons or the four identical carbon-13 nuclei. The protons and the carbons are all highly shielded due to the relatively electropositive silicon atom, consequently the TMS signal is observed at a higher field (lower frequency) than those of most other organic compounds;

- two or three drops of TMS, which is an inert, low-boiling liquid miscible with most organic solvents, are added to each sample. For aqueous samples, deuterated **sodium 3-trimethylsilylpropanoate (TSP,** $(CH_3)_3SiCD_2CD_2COONa$) is used.

Four principal factors affect the chemical shifts of nuclei in organic compounds.

- **Diamagnetic shielding**, which opposes B_o. The degree of shielding (positive σ) is related to the electron density around the nucleus and is therefore determined by the electronegativity of neighboring elements in the structure. This is illustrated in *Table 2*, which shows how the chemical shift for the protons of a methyl group in a series of compounds moves progressively downfield (increasing δ) as the electronegativity of the adjacent group increases and the protons become more deshielded. This **inductive effect** is reinforced when there is more than one electronegative group adjacent to a proton group (*Table 3*), but falls off rapidly with distance in saturated structures, i.e. with the number of intervening bonds (*Table 4*). In unsaturated structures, the effects travel further and are less easy to predict.
- **Diamagnetic anisotropy**, which may oppose or augment B_o. An anisotropic property varies with direction in three-dimensional space, the effect in NMR spectrometry being associated with the circulation of π-electrons in unsaturated structures. This circulation, induced by the applied magnetic field, B_o, creates conical-shaped shielding (+) and deshielding (–) zones in the space surrounding the group or structure. *Figure 5* shows the zones associated with an alkene double bond, a carbonyl group, an aromatic ring and an alkyne

Table 2. Variation of the chemical shift of methyl protons and carbons with the electronegativity of the adjacent group

CH_3X	Electronegativity of X	δ_H	δ_C
$CH_3Si(CH_3)_3$	1.90	0	0
CH_3H	2.20	0.23	–2.3
CH_3COCH_3	2.60	2.08	29.2
CH_3I	2.65	2.16	–20.7
CH_3Br	2.95	2.68	10.0
CH_3NH_2	3.05	2.50	28.3
CH_3Cl	3.15	3.05	25.1
CH_3OH	3.50	3.40	49.3
CH_3F	3.90	4.26	75.4

Table 3. Variation of the chemical shift of methyl, methylene and methine protons with the number of adjacent electronegative groups

CH_nX	X=CH_3	Cl	C_6H_5	OCH_3
CH_3X	0.9	3.05	2.35	3.25
CH_2X_2	–	5.30	4.00	4.50
CHX_3	–	7.25	5.55	5.00

triple bond. The magnitude of the effect varies with the orientation of the molecule relative to the applied field, and the illustrations show the maximum effect in each case, where the molecular axis is parallel or perpendicular to the field. In reality, the molecules in a liquid or solution are constantly tumbling due to thermal motion, so that only a net effect is observed, being the average over all possible orientations.

Table 4. Variation of the chemical shift of methylene protons with distance (number of bonds) from an electronegative group

$-(CH_2)_nBr$	δ_H
$-\mathbf{CH_2}Br$	3.30
$-\mathbf{CH_2}-CH_2Br$	1.69
$-\mathbf{CH_2}-CH_2-CH_2Br$	1.25

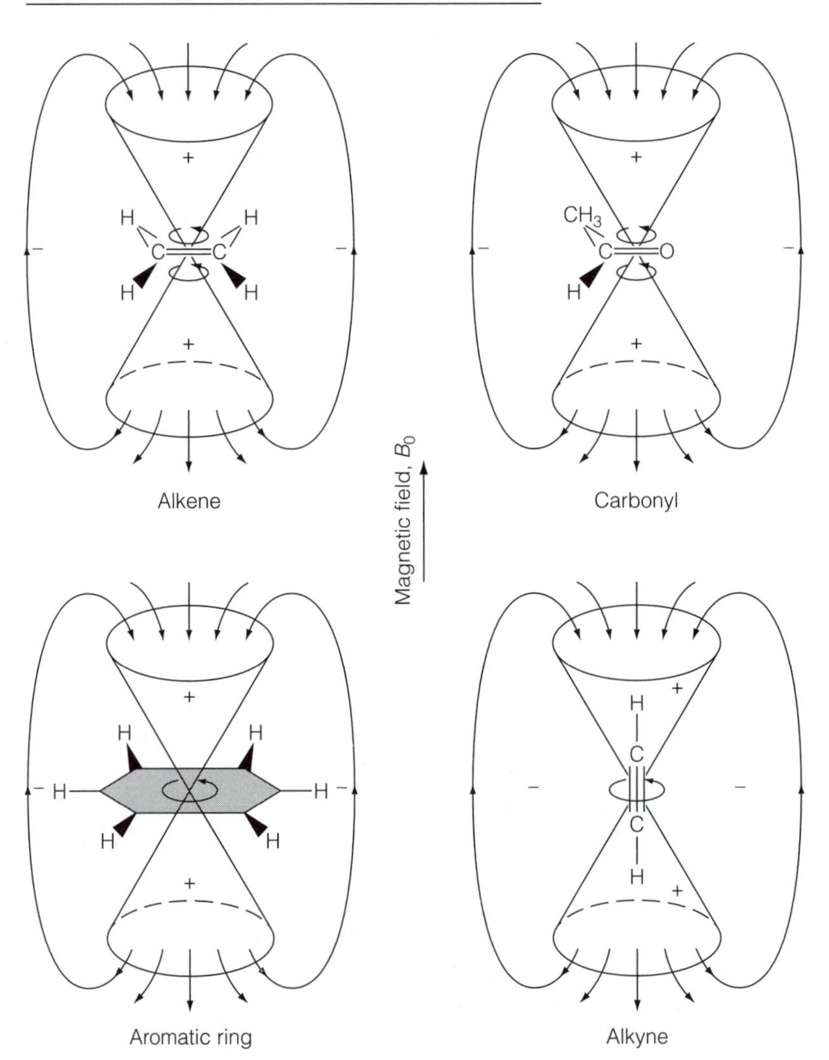

Fig. 5. Shielding and deshielding effects in unsaturated groups and structures due to diamagnetic anisotropy.

- The illustrations show that alkene (~5.3 δ/ppm) and aromatic protons (~7.3 δ/ppm), and those close to a carbonyl double bond (~9.5 δ/ppm) all lie in deshielding zones (–) where the fields associated with the circulating electrons augment the applied field, B_o. These proton resonances therefore appear further downfield than would be expected. The effect is particularly pronounced for the aromatic ring due to the generation of a strong **ring current** by the circulating π-electrons. An alkyne triple bond shows the greatest effect when it is aligned parallel to the applied field, the proton lying in a shielded zone and therefore appearing further upfield (1.5 δ/ppm) than would be expected.
- **Hydrogen-bonding**, which leads to the deshielding of protons. This reduces the electron density around the proton involved, thereby decreasing the degree of diamagnetic shielding. The effect is observed in the proton spectra of alcohols, phenols, carboxylic acids and amines where both **inter-** and **intramolecular** H-bonding (X–H$\cdots$Y where X and Y = O, N or S) can occur. The OH, NH, NH_2 or SH proton resonances show variable downfield shifts due to additional deshielding by the Y atom, the effect being concentration-dependent for intermolecular bonds but not for intramolecular bonds. Choice of solvent also has an influence if it can bond to the protons. Carboxylic acid dimers and intramolecularly bonded structures, such as the enolic form of β-diketones and 1,2-substituted aromatic rings with appropriate groups, form particularly strong H-bonds, resonances often appearing between 12 and 16 δ/ppm. *Figure 6* illustrates four spectra of ethanol recorded at different concentrations, where the OH proton resonance moves progressively upfield as the degree of hydrogen bonding diminishes with dilution in tetrachloromethane.
- **Paramagnetic deshielding**, which augments B_o. The degree of deshielding (negative σ) is variable and more complex in origin. It is mainly of importance in structures with unpaired electrons, such as transition metal complexes.

Charts and tables of chemical shifts are used to aid the interpretation of proton and carbon-13 spectra, and examples of these are given in Topic E13.

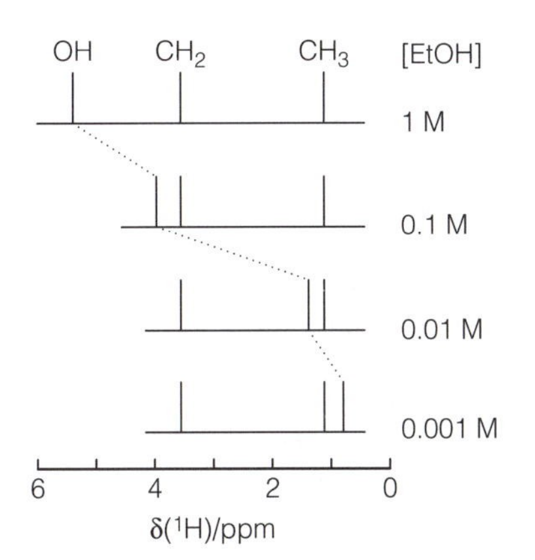

Fig. 6. Hydrogen-bonded OH resonance shifts in ethanol as a result of dilution in tetrachloromethane.

Spin–spin coupling

The spins of neighboring groups of nuclei in a molecule are said to be **coupled** if their spin states **mutually** interact. The interactions, which involve electrons in the intervening bonds, result in small variations in the effective magnetic fields experienced by one group of nuclei due to the different orientations of the spin angular momenta and magnetic moments of those in the neighboring group or groups, and *vice versa*. These lead to the splitting of the resonance signal into two or more components that are shifted slightly upfield and downfield respectively from the position in the absence of coupling, the probabilities of each being roughly the same because the permitted nuclear spin energy levels are almost equally populated. Thus, the resonance signals for two single adjacent nuclei with substantially different chemical shifts are each split into two component peaks of equal intensity. *Figure 7* shows the splitting of the two proton resonances in dichloroethanal, $CHCl_2$–CHO, into a doublet, each with the same separation between the component peaks, known as the coupling constant, *J*, which is measured in Hz. The chemical shift of each doublet is taken to be the mean value of those of its component peaks.

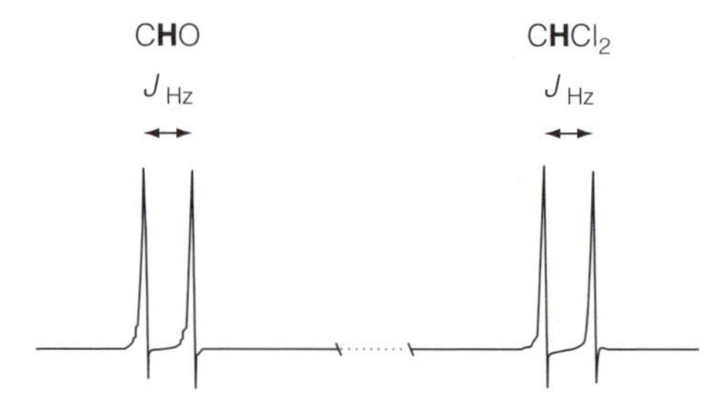

Fig. 7. First order spin-spin coupling of the two adjacent protons in dichloroethanal.

Where there is more than one nucleus in a group, all possible combinations of spin orientations must be considered, and this leads to further small upfield and downfield shifts with increased multiplicity of the observed resonances (*Fig. 8*). Statistical considerations also lead to variations in the relative intensities of the components of each multiplet. The following general rules are applicable to spin–spin coupling between nuclei with the same spin quantum number.

- The number of components in a multiplet signal is given by $2nI+1$, where n is the number of identical neighbouring nuclei in an adjacent coupled group, and I is the spin quantum number of the nuclei involved. For proton and carbon-13 nuclei, whose spin quantum number is ½, the number of components is $n+1$, and this is known as the ***n*+*1* rule**.
- The relative intensities of the components in a multiplet signal are given by **Pascal's triangle**, which is based on the coefficients of the expansion of $(a+b)^n$. For proton spectra, the $n+1$ rule and Pascal's triangle lead to the multiplicities and relative intensities for an observed resonance signal when there are ***n* adjacent identical nuclei** as shown in *Table 5*.
- In saturated structures, the effect is generally transmitted through only three bonds, but in unsaturated structures it is transmitted further, e.g. around an aromatic ring.

Commonly encountered spin–spin splitting patterns for two coupled groups in the proton spectra of saturated molecules are illustrated in *Figure 8*. These are

Table 5. Multiplicity and relative intensities of resonance signals from coupled groups of nuclei in saturated structures (I = ½)

Number of adjacent nuclei	Multiplicity of observed resonance	Relative intensities of components of multiplets
0	Singlet	1
1	Doublet	1 1
2	Triplet	1 2 1
3	Quartet	1 3 3 1
4	Quintet	1 4 6 4 1
5	Sextet	1 5 10 10 5 1
6	Septet	1 6 15 20 15 6 1

described as **first order** when the numbers of components and their relative intensities are as predicted by the *n+1* rule and Pascal's triangle. This generally requires that the ratio of the chemical shift difference between the resonance signals of the two groups of protons to their coupling constant, *J*, is at least seven. As this ratio diminishes, the relative intensities within each multiplet become distorted, and additional splitting is observed making interpretation more difficult. Such patterns are then described as **second order**. Capital letters at opposite ends of the alphabet, with subscripted numbers to indicate the numbers of protons in each set of nuclei, are used to designate first order patterns, and these are included in *Figure 8*.

It should be noted that:

(i) protons within the same group are normally **chemically** and **magnetically** equivalent, and do not show any splitting, although they are coupled;
(ii) coupling constants, *J*, are independent of the magnitude of the applied magnetic field, whereas chemical shift differences (measured in Hz) increase in proportion to it. Hence, NMR spectra can be simplified by using more powerful spectrometers, as complex spectra become closer to first order. Proton coupling constants range from 0 to over 20 Hz.

Spin–spin splitting patterns for three coupled groups are observed in unsaturated structures because the coupling extends through more than three bonds. Each of the groups interacts with the other two, giving three coupling constants and, depending on their relative magnitudes, the spectra may be complex due to second order effects.

Examples of spectra are given in Topic E13.

NMR spectrometers

Spectrometers were originally designed to scan and record an NMR spectrum by progressively changing (**sweeping**) the applied magnetic field at a fixed radiofrequency (RF), or sweeping the frequency at a fixed field. Sample resonances were recorded as a series of sharp absorption peaks along the frequency/field axis, which is calibrated in ppm. These **continuous wave (CW)** instruments have been largely superseded by pulsed **Fourier transform (FT)** spectrometers. Samples are subjected to a series of rapid, high-energy RF pulses of wide frequency range, between which a decaying emission signal from nuclei excited by the pulse and then relaxing to the ground state is monitored by the receiver circuit. The detector signal, or **free induction decay (FID)**, contains all of the spectral information from the sample, but in the form of a time-dependent **interferogram**. This can be digitized and converted into a conventional

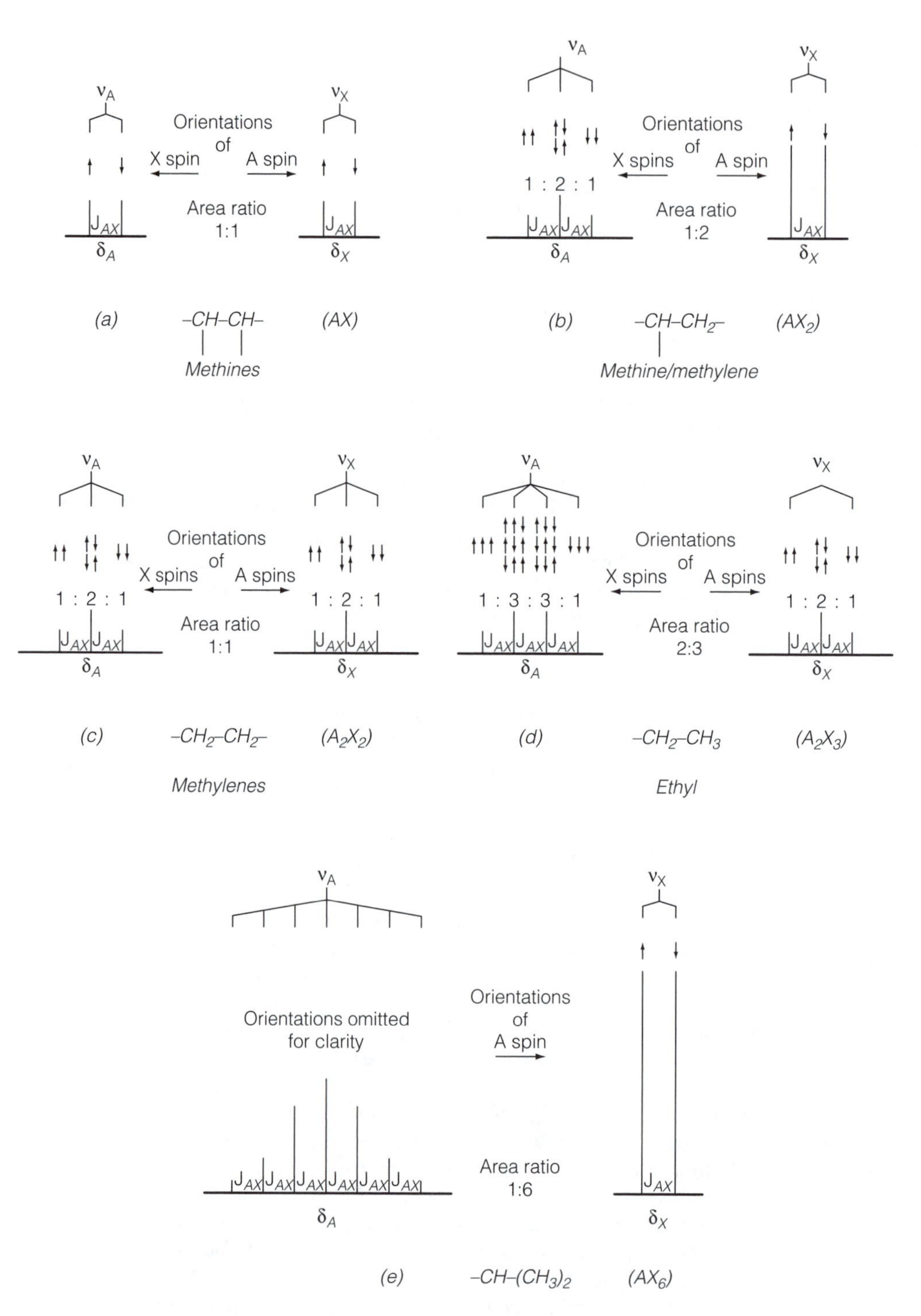

Fig. 8. Diagrammatic representations of first order proton-splitting patterns for two coupled groups.

spectrum mathematically in less than a second by a computer using a **fast Fourier transform** (**FFT**) algorithm. Multiple interferograms can be rapidly accumulated and averaged to increase sensitivity by as much as three orders of magnitude.

A block diagram of a typical NMR spectrometer is shown in *Figure 9* and comprises five main components:

- a superconducting solenoid or electromagnet providing a powerful magnetic field of up to about 17 Tesla;
- a highly stable RF generator and transmitter coil operating at up to about 750 MHz;
- a receiver coil with amplifying and recording circuitry to detect and record sample resonances;
- a sample probe positioned between the poles of the magnet;
- a dedicated microcomputer for instrument control, data processing (FFT of interferograms) and data storage.

The homogeneity and stability of the magnetic field should be at least 1 in 10^9 to ensure narrow absorption bands and good resolution. Sample tubes are made to spin in the sample probe at about 50 Hz by an air turbine so as to increase the apparent field homogeneity further. The direction of the magnetic field and the orientations of the transmitter and receiver coils must be mutually perpendicular to detect sample resonances and eliminate spurious signals in the detector cicuit as shown in *Figure 9*.

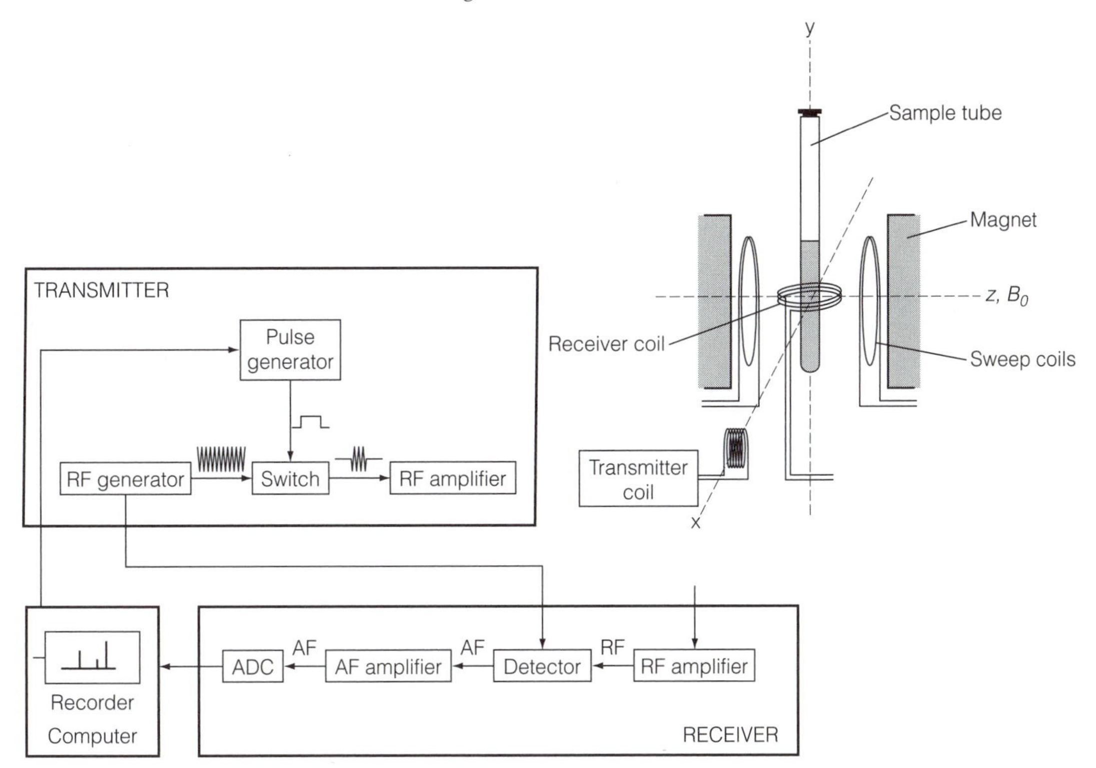

Fig. 9. Diagrammatic representation of an NMR spectrometer. Reproduced from R. Kellner et al., Analytical Chemistry, *1998, with permission from Wiley-VCH, and from* Nuclear Magnetic Resonance *by P.J. Hore (1995) by permission of Oxford University Press © P.J. Hore, 1995.*

Stability of operation is improved considerably by **locking** the field and frequency together to correct for drift. This is achieved by constantly monitoring the resonance frequency of a reference nucleus, usually deuterium in a deuterated solvent. For carbon-13 studies in particular this is essential, as accumulating large numbers of scans can take several hours.

E13 Nuclear magnetic resonance spectrometry: interpretation of proton and carbon-13 spectra

Key Notes

Chemical shift data The interpretation of proton and carbon-13 spectra is facilitated by reference to published chemical shift data in the form of charts and tables of numerical values for a range of nuclear environments, principally in organic structures.

Peak areas The relative areas of resonance peaks in a proton spectrum are directly proportional to the numbers of nuclei responsible for each signal, which enables the presence of specific groups of nuclei to be confirmed.

Proton spectra These are the most widely studied NMR spectra, the proton being the nucleus with the highest sensitivity. Information from chemical shifts, spin–spin coupling and peak areas enables the structural features of organic compounds to be recognized and their identities established.

Carbon-13 spectra Due largely to the very low natural isotopic abundance of carbon-13, spectra can be recorded only by a pulsed Fourier transform spectrometer. Chemical shifts are much greater than in proton spectra and are of particular value in establishing the skeletal structures of organic molecules.

Related topic Nuclear magnetic resonance spectrometry: principles and instrumentation (E12)

Chemical shift data

Examples of chemical shift charts for protons and carbon-13 nuclei are given in *Figure 1*. They give a general indication of the ranges in δ/ppm within which the resonances for different types of structure or functional groups will occur. More detailed tabulated numerical data can be used to establish the presence or absence of specific molecular structural features or groups. The chemical shift range for protons that can be involved in hydrogen bonding is particularly wide, as the variation in degree of shielding of these protons may be highly concentration-dependent. Factors affecting chemical shifts are described in Topic E12.

Peak areas

Peak areas are measured by **electronic integration** of the resonance signals in a spectrum. For proton spectra, the total area, or **integral**, of a multiplet is directly proportional to the number of protons in the group. Integrals are recorded as a

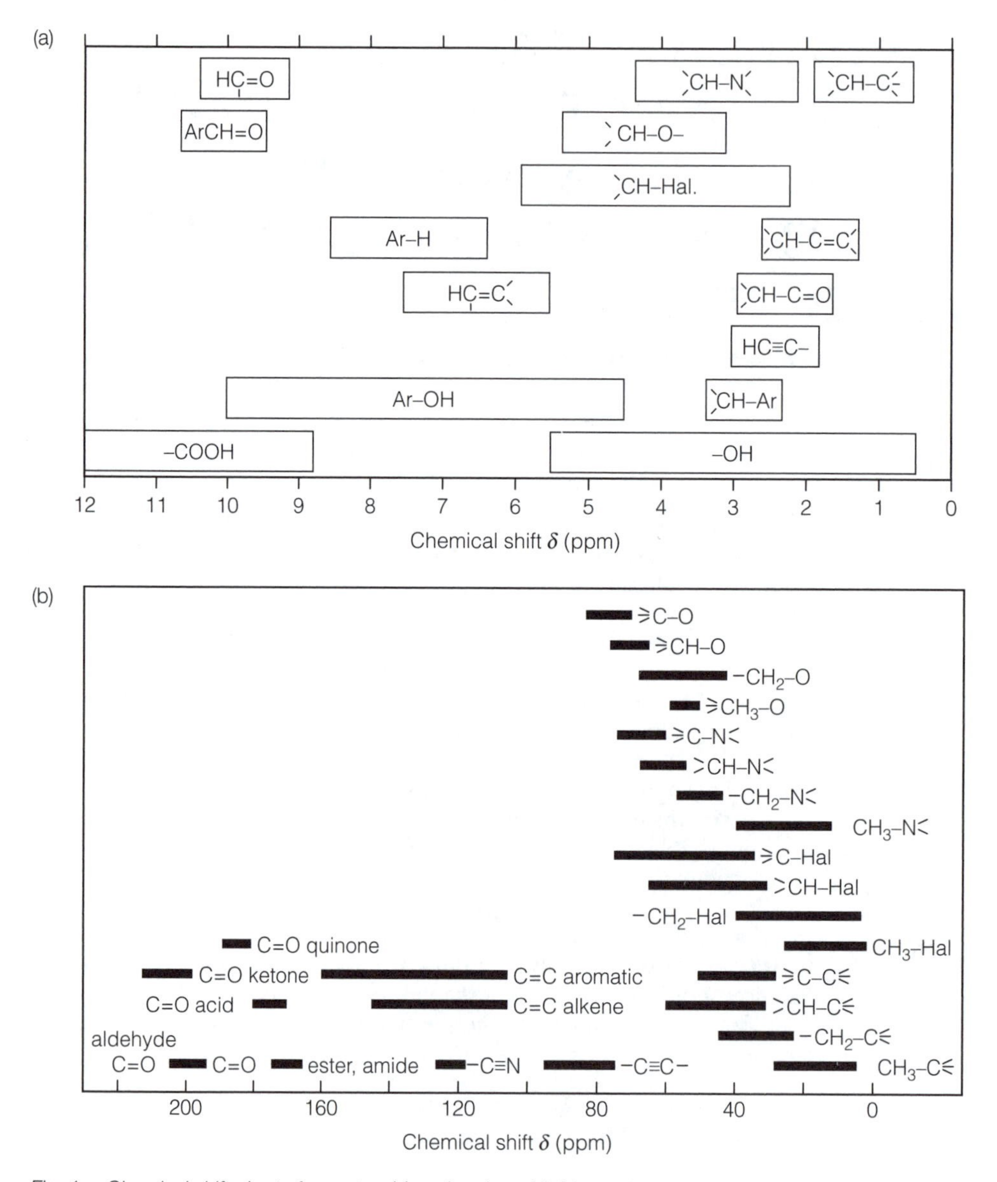

Fig. 1. Chemical shift charts for proton (a) and carbon-13 (b) spectra.

series of steps, generally displayed above each resonance signal. The vertical height of each step, in arbitrary units, gives the relative number of protons associated with the signal. Integration of carbon-13 resonances is not sufficiently reliable to be of value.

Proton spectra

The interpretation of proton spectra depends on three features: **chemical shifts**, **multiplicities of resonances** and **integrated peak areas**. These are exemplified in the spectrum of ethanol, C_2H_5OH, shown in *Figure 2*.

The following is a general approach to spectral interpretation, which should be augmented by reference to chemical shift data, coupling constants and the spectra of known compounds.

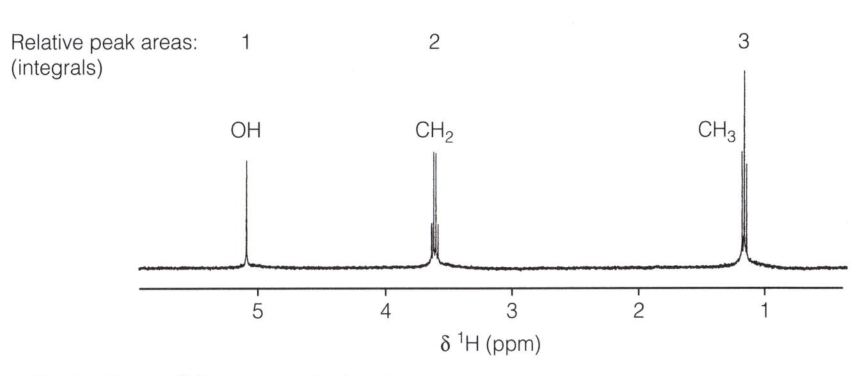

Fig. 2. Proton (^1H) spectrum of ethanol.

- Note the presence or absence of **saturated structures**, most of which give resonances between 0 and 5 δ/ppm (trichloromethane at 7.25 δ/ppm is a notable exception).
- Note the presence or absence of **unsaturated structures** in the region between about 5 and 9 δ/ppm (**alkene protons** between 5 and 7 δ/ppm and **aromatic protons** between 7 and 9 δ/ppm). (N.B. **Alkyne protons** are an exception, appearing at about 1.5 δ/ppm.)
- Note any very low field resonances (9 to 16 δ/ppm), which are associated with **aldehydic** and **acidic protons**, especially those involved in strong hydrogen bonding.
- Measure the **integrals**, if recorded, and calculate the numbers of protons in each resonance signal.
- Check for spin–spin splitting patterns given by adjacent alkyl groups according to the *n+1* rule and Pascal's triangle. (N.B. The position of the lower field multiplet of the two is very sensitive to the proximity of electronegative elements and groups such as O, CO, COO, OH, Cl, Br, NH_2, etc.)
- Examine the splitting pattern given by **aromatic protons**, which couple around the ring and are often complex due to second order effects.
- 1,4- and 1,2-disubstituted rings give complex but symmetrical looking patterns of peaks, whereas mono-, 1,3- and tri-substituted rings give more complex asymmetrical patterns.
- Note any broad single resonances, which are evidence of **labile** protons from **alcohols, phenols, acids** and **amines** that can undergo **slow exchange** with other labile protons. Comparison of the spectrum with another after shaking the sample with a few drops of D_2O will confirm the presence of an exchangeable proton by the disappearance of its resonance signal and the appearance of another at 4.7 δ/ppm due to HOD.

Some examples of proton spectra with their resonances assigned are shown in *Figures 3–8*.

The two aromatic protons, **A** and **X**, in **cytosine** (*Fig. 3*) are coupled to give an **AX** pattern of two doublets. The **A** proton is deshielded more than the **X** proton due to its closer proximity to nitrogens and the oxygen atom. The intensities of the doublets are slightly distorted by second order effects. The **OH** and $\mathbf{NH_2}$ protons have been exchanged with D_2O, and their resonances replaced with a HOD peak at 4.7 δ/ppm.

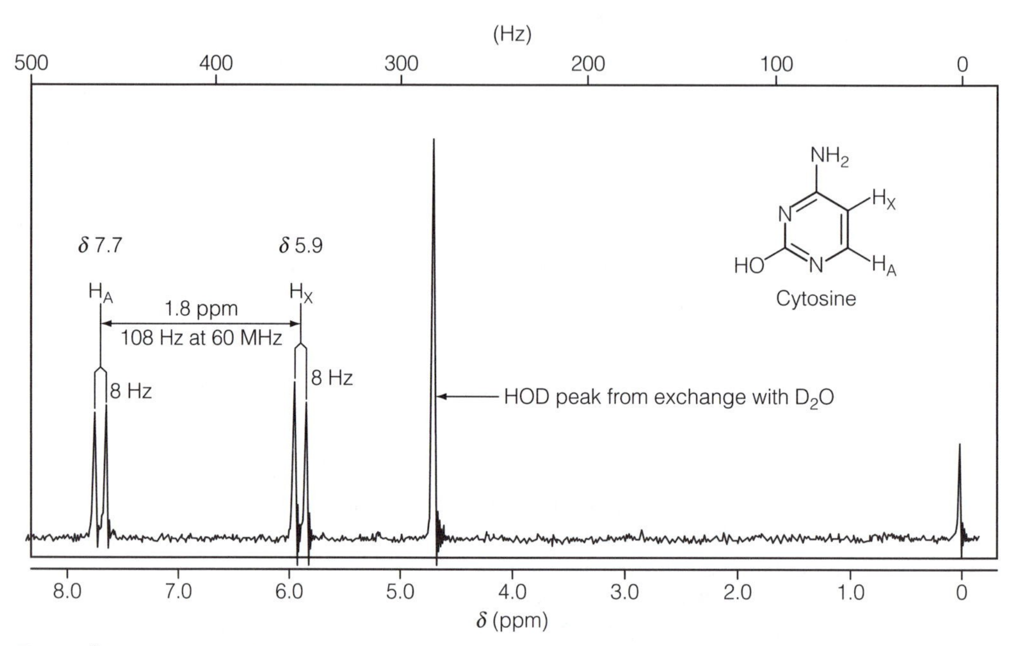

Fig. 3. [1]*H spectrum of cytosine.*

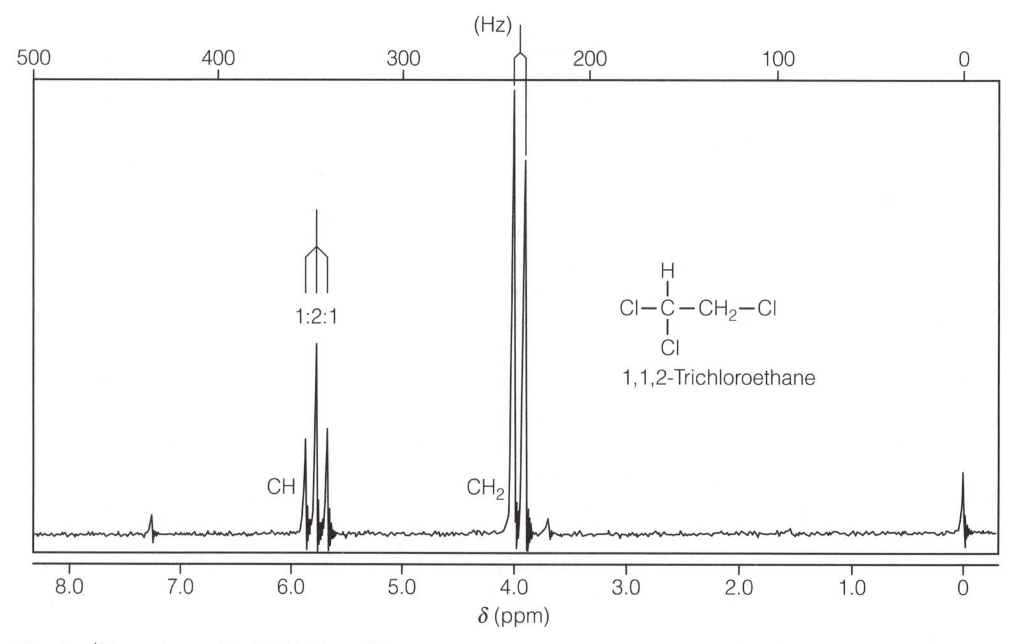

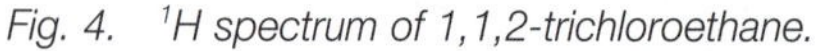

Fig. 4. [1]*H spectrum of 1,1,2-trichloroethane.*

The **CH** (methine) resonance in **1,1,2-trichloethane** (*Fig. 4*) is at a much lower field than the **CH_2** (methylene) resonance because of the very strong deshielding by two chlorines. The protons give an **AX_2** coupling pattern of a triplet and doublet, and an integral ratio of 1 : 2.

The CH_3 resonance in **2-(4-chlorophenoxy)propanoic acid** (*Fig. 5*) is well upfield as it is not affected by an adjacent electronegative group. The **CH** proton is deshielded by the oxygen, and the aromatic protons by the ring current effect. The downfield **COOH** proton is strongly deshielded by the adjacent oxygen and by H-bonding. The coupled CH_3 and **CH** protons give an AX_3 quartet and doublet, and the symmetry of the aromatic resonances is typical of a **1,4-disubstituted ring**. The integral ratios are 1:4:1:3.

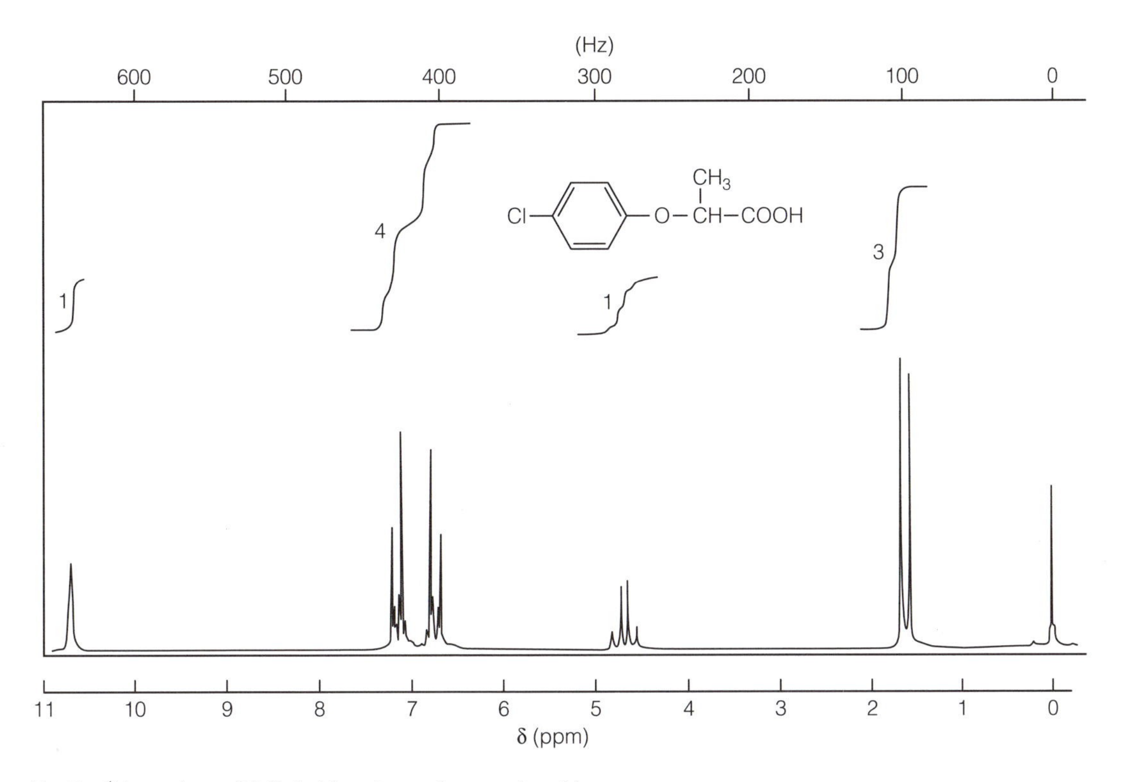

Fig. 5. [1]H spectrum of 2-(1,4-chlorophenoxy)propanoic acid.

The CH_3 resonance of phenylethyl ethanoate (*Fig. 6*) is shifted downfield by the deshielding effect of the ester group. The two CH_2 groups (methylenes) couple to give A_2X_2 triplets, the higher field one being the CH_2 adjacent to the ring, the lower field being more strongly deshielded due to the oxygen. The five ring protons are almost equally deshielded by the ring current and the integral ratios are 5:2:2:3.

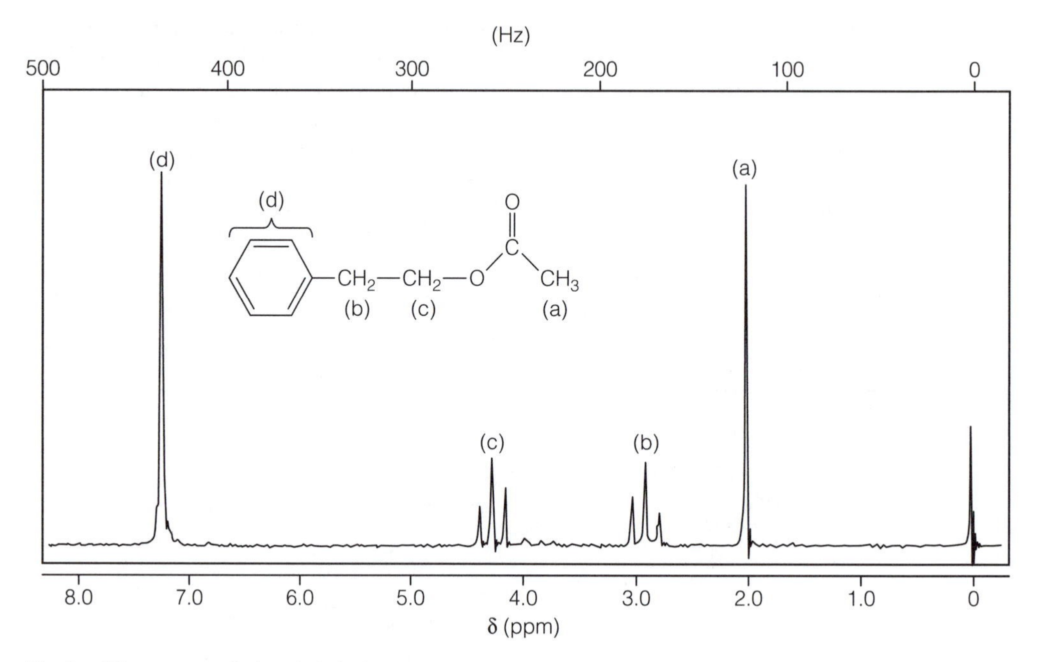

Fig. 6. *[1]H spectrum of phenylethyl ethanoate.*

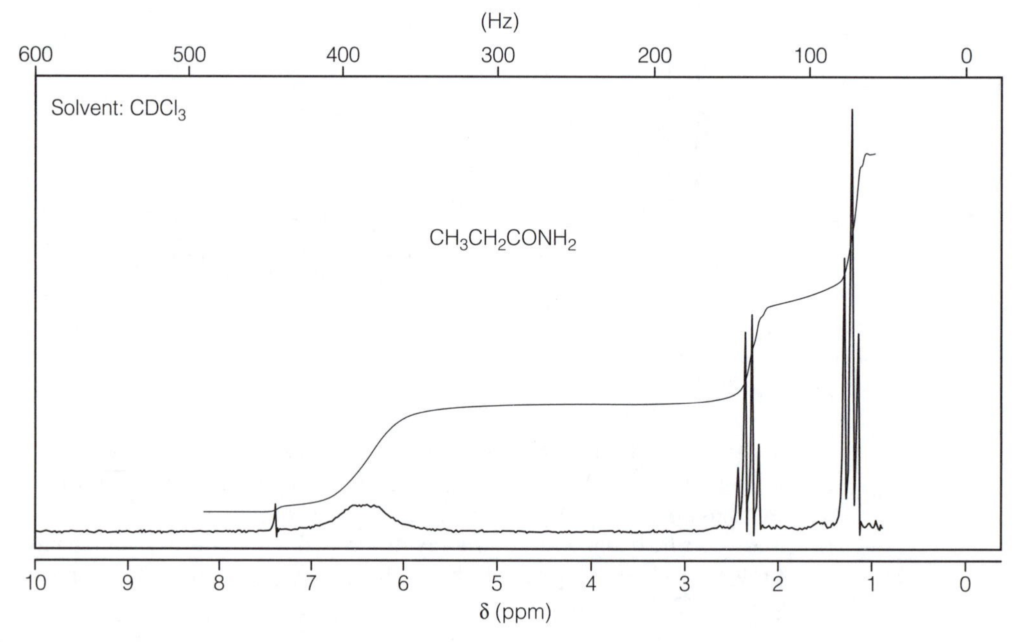

Fig. 7. *[1]H spectrum of propanamide.*

The $\mathbf{CH_2}$ protons in **propanamide** (*Fig. 7*) are deshielded by the adjacent carbonyl group and couple with the $\mathbf{CH_3}$ protons to give an $\mathbf{A_2X_3}$ quartet and triplet. The resonance of the amine protons at 6.4 δ/ppm is broadened by proton–proton exchange and could be removed by shaking with D_2O. The integral ratios are 2 : 2 : 3.

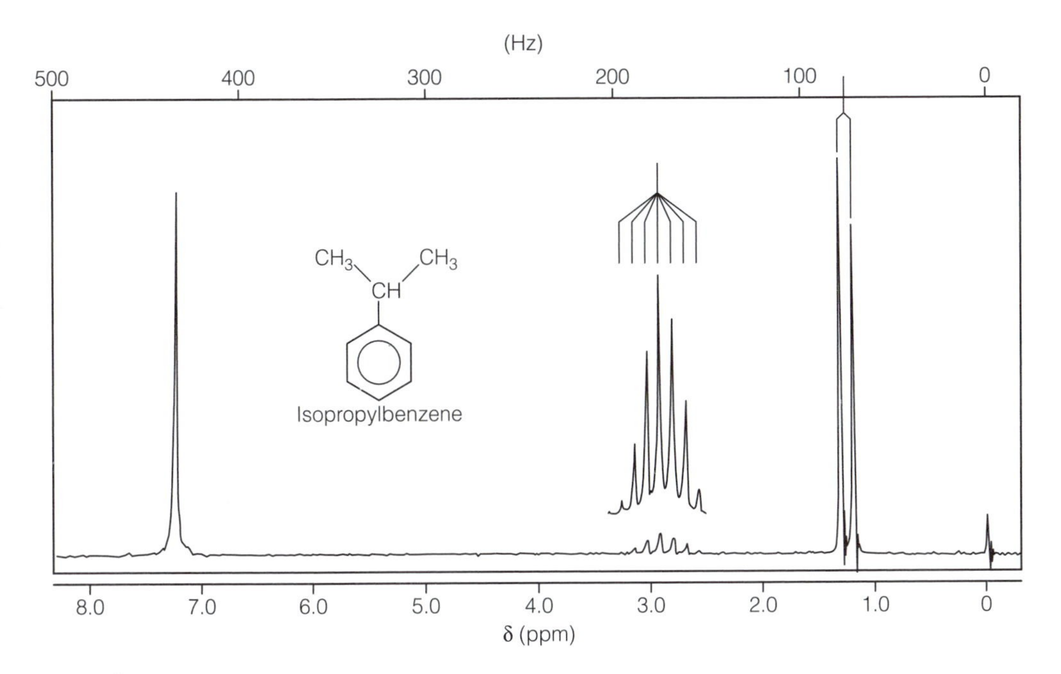

Fig. 8. [1]H spectrum of isopropyl benzene.

The two $\mathbf{CH_3}$ groups in **isopropyl benzene** (*Fig. 8*) are identical and their resonance is split into a doublet by the adjacent **CH,** which is deshielded by the ring current. The methine resonance is split into a septet by the six equivalent methyl protons. The aromatic protons are almost equivalent as there is no electronegative group present to affect their chemical shifts. The integral ratios are 5:1:6.

Carbon-13 spectra

Carbon-13 spectra cover a much wider range of chemical shifts than proton spectra, but the positions of resonances are generally determined by the same factors. However, for ease of interpretation, they are often recorded as **decoupled** spectra to eliminate the effects of coupling to adjacent protons which would otherwise split the carbon-13 resonances according to the n+1 rule and Pascal's triangle. Decoupled spectra consist of a single peak for each chemically different carbon in the molecule, and spectral interpretation is confined to the correlation of their chemical shifts with structure, augmented by reference to chemical shift data and the spectra of known compounds. Proton coupling can be observed under appropriate experimental conditions. The following is a general approach.

- Note the presence or absence of **saturated structures**, most of which give resonances between 0 and 90 δ/ppm.
- Note the presence or absence of **unsaturated structures** in the region between about 100 and 160 δ/ppm. (N.B. Alkyne carbons are an exception, appearing between 70 and 100 δ/ppm.)
- Note any **very low field resonances** (160 to 220 δ/ppm), which are associated with **carbonyl** and **ether** carbons. **Carboxylic acids**, **anhydrides**, **esters**, **amides**, **acyl halides** and **ethers** are all found in the range 160 to 180 δ/ppm, whilst **aldehydes** and **ketones** lie between 180 and about 220 δ/ppm.

Some examples of carbon-13 spectra with their resonances assigned are shown in *Figures 9–11*.

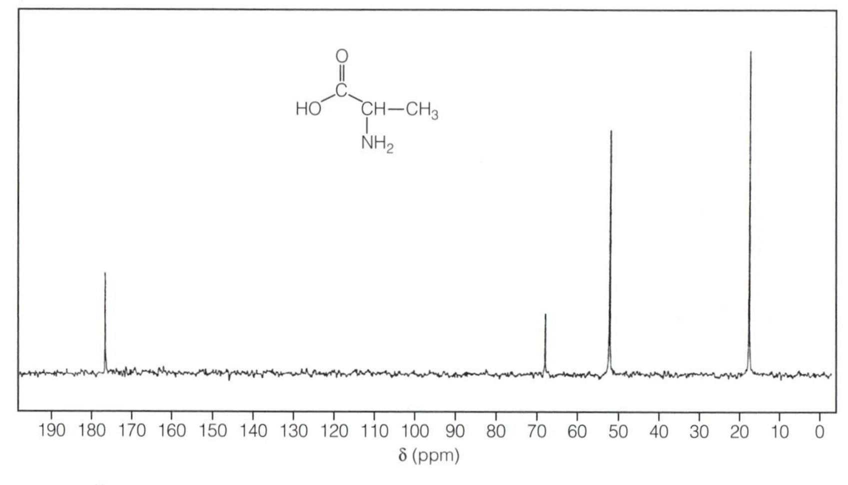

Fig. 9. ^{13}C spectrum of alanine.

The three carbons show a very wide range of chemical shifts. The lowest field resonance corresponds to the **carbonyl carbon**, which is highly deshielded by the double-bonded oxygen. The nitrogen deshields the **CH** carbon much less, and the **CH_3** carbon is the least deshielded of the three.

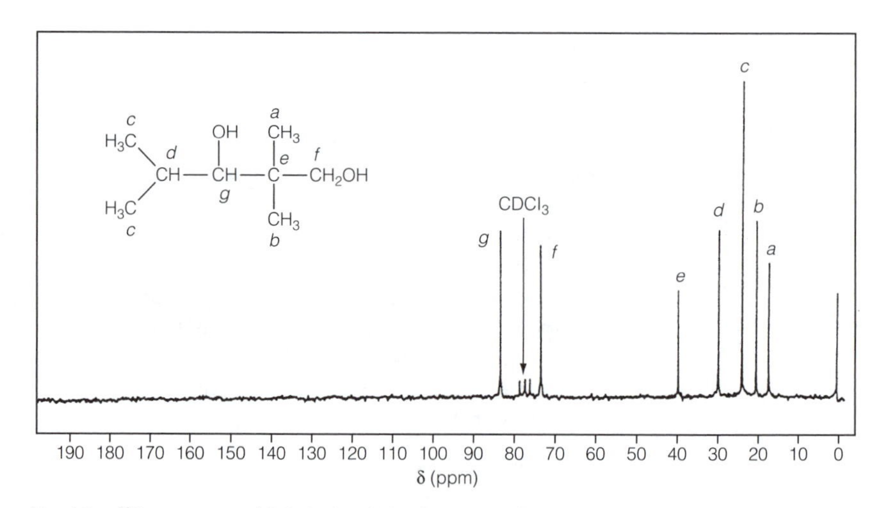

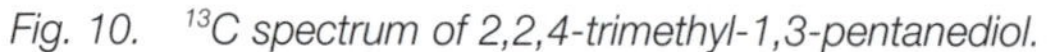

Fig. 10. ^{13}C spectrum of 2,2,4-trimethyl-1,3-pentanediol.

The carbon-13 resonances of this fully saturated compound (*Fig. 10*) are all found between 0 and 90 δ/ppm. The two carbons directly bonded to oxygens are deshielded significantly more than the **CH** carbon, which in turn is more deshielded than the **CH_3** carbons.

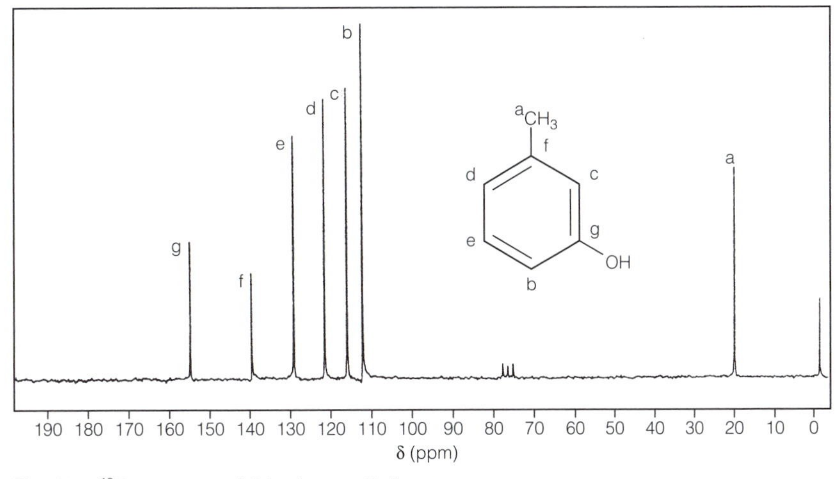

Fig. 11. ^{13}C *spectrum of 3-hydroxymethylbenzene.*

Apart from the $\mathbf{CH_3}$ carbon at 22 δ/ppm, the remaining carbon-13 resonances in **3-hydroxy methylbenzene** (*Fig. 11*) are all in the unsaturated region between 100 and 160 δ/ppm. The **3-carbon** is deshielded more than the remainder, being bonded directly to oxygen, but the chemical shifts of the remainder reflect the variations in electron density around the ring.

E14 MASS SPECTROMETRY

Key Notes

Principles — Mass spectrometry (MS) is a technique whereby materials are ionized and dissociated into fragments characteristic of the molecule(s) or element(s) present in the sample. The numbers of ions of each mass provide information for qualitative and quantitative analysis.

Mass spectrometer — A mass spectrometer, which is operated under high vacuum, incorporates a sample inlet and ion source, a mass analyzer, an ion detector and a data processing system.

Ionization techniques — Alternative ionization techniques are available differing in energy and applicability. Some produce a high degree of dissociation of molecules, while others are used primarily to establish an accurate relative molecular mass of a compound or to facilitate elemental analysis.

Fragmentation — After ionization, molecules may dissociate into fragments of smaller mass, some carrying a charge. The presence and relative abundances of the various charged fragments provide structural information and enable unknown compounds to be identified.

Isotope peaks — These are peaks in a mass spectrum arising from fragments containing naturally occurring heavier isotopes of one or more elements.

Mass spectra — Spectral data is either tabulated or shown graphically as a plot of the numbers of ions of each mass detected. For ease of interpretation, these are presented as line diagrams.

Related topics — Inductively coupled plasma spectrometry (E5); Combined techniques (Section F)

Principles

Mass spectrometry (**MS**) is an analytical technique in which gaseous ions formed from the molecules or atoms of a sample are separated in space or time and detected according to their **mass-to-charge ratio**, m/z. The numbers of ions of each mass detected constitute a **mass spectrum**, which may be represented graphically or tabulated as shown for methanol in *Figure 1*. Peak intensities are expressed as a percentage of that of the most abundant ion (m/z 31 for methanol), which is designated the **base peak**. The spectrum provides structural information and often an accurate relative molecular mass from which an unknown compound can be identified or a structure confirmed. Quantitative analysis is based on measuring the numbers of a particular ion present under closely controlled conditions.

Mass spectrometer

A block diagram of a mass spectrometer is shown in *Figure 2*. It is operated under a vacuum of 10^{-4} to 10^{-7} Nm^{-2} as the presence of air would swamp the mass spectra of samples, and damage the ion source and detector.

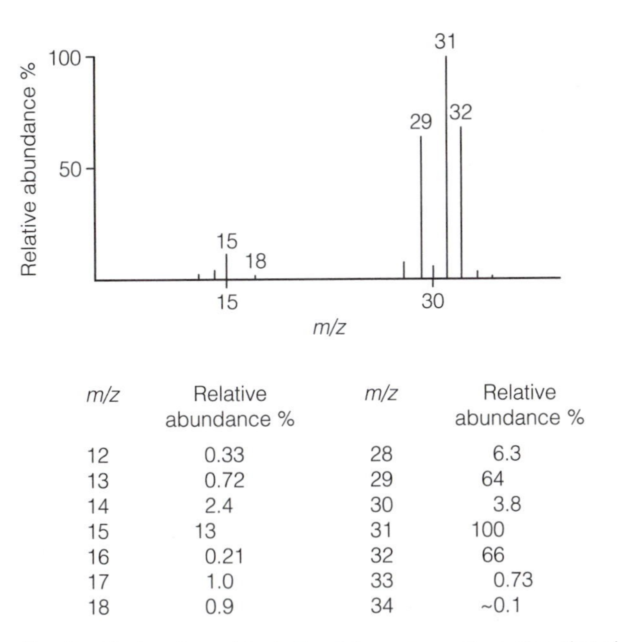

m/z	Relative abundance %	m/z	Relative abundance %
12	0.33	28	6.3
13	0.72	29	64
14	2.4	30	3.8
15	13	31	100
16	0.21	32	66
17	1.0	33	0.73
18	0.9	34	~0.1

Fig. 1. Electron impact ionization (EI) mass spectrum of methanol.

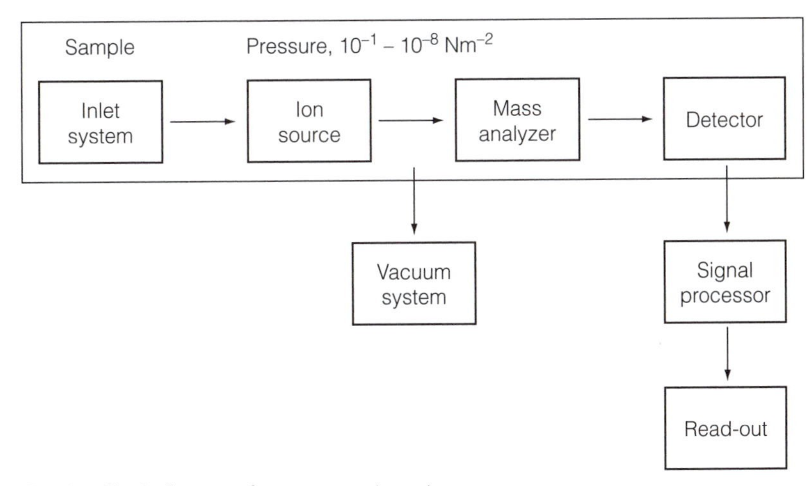

Fig. 2. Block diagram of a mass spectrometer.

The principal components are

- a **sample inlet,** which facilitates the controlled introduction of gaseous or vaporized liquid samples via a **molecular leak** (pinhole aperture) and solids via a **heated probe** inserted through a vacuum lock;
- an **ion source** to generate ions from the sample vapor;
- a **mass analyzer** which separates ions in space or time according to their mass-to-charge ratio. Ions generated in the source are accelerated into the analyzer chamber by applying increasingly negative potentials to a series of metal slits through which they pass.

There are several types of mass analyzer.

(i) A **single focusing magnetic mass analyzer** (*Fig. 3*) generates a field at right angles to the rapidly moving ions, causing them to travel in curved trajectories with radii of curvature, r, determined by their mass-to-charge ratio, m/z, the magnetic field strength, B, and the accelerating voltage, V, as given by the relation

$$\frac{m}{z} = \frac{B^2r^2}{2V}$$

The majority of ions carry a charge of +1, hence m is directly proportional to r^2. For ions of a particular mass there is a specific combination of values of B and V that allows them to pass along the center of a curved **analyzer tube** to a detector positioned at the end. Progressive variation of the field or the accelerating voltage allows ions of different mass to pass down the center of the analyzer tube, be detected and a mass spectrum recorded.

(ii) A **double focusing mass analyzer** employs an **electrostatic separator** in addition to a magnetic analyzer to improve the mass resolution. Ions of the same mass inherently acquire a range of kinetic energies when accelerated and this leads to overlapping signals from those with similar masses. Application of an electrostatic field to the moving ions allows the selection of those with the same kinetic energy so eliminating this problem.

(iii) A **quadrupole mass analyzer** consists of a set of four parallel metal rods positioned very closely together, but leaving a small space through the center (*Fig. 4*). Ions are accelerated into the space between the rods at one end and a DC potential and a high frequency RF signal is applied across

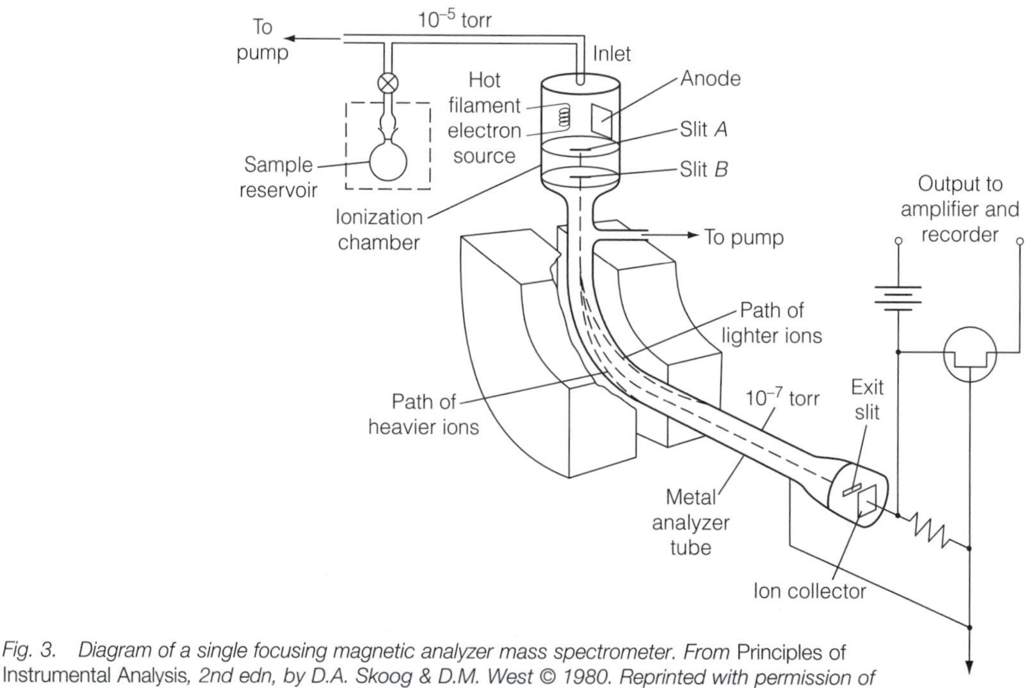

Fig. 3. Diagram of a single focusing magnetic analyzer mass spectrometer. From Principles of Instrumental Analysis, *2nd edn, by D.A. Skoog & D.M. West © 1980. Reprinted with permission of Brooks/Cole an imprint of the Wadsworth Group, a division of Thomson Learning.*

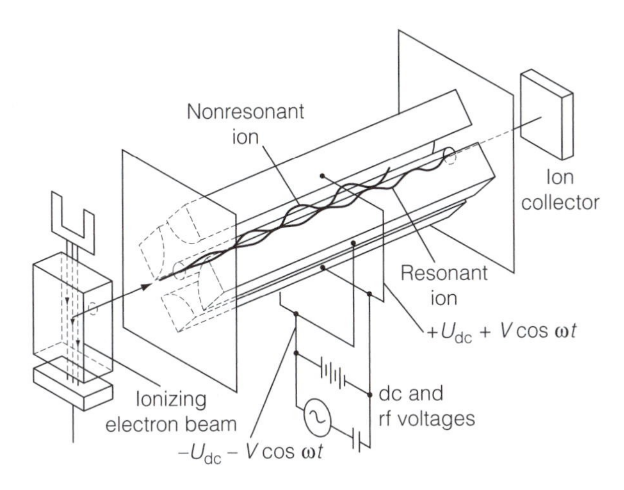

Fig. 4. Diagram of a quadrupole mass analyzer.

opposite pairs of rods. This results in ions of one particular m/z value passing straight through the space to a detector at the other end while all others spiral in unstable trajectories towards the rods. By altering the DC and RF signals applied to the rods, ions with different m/z ratios can be allowed to reach the detector in turn. The **ion trap** is a modified version with a circular polarizable rod and end caps enclosing a central cavity which is able to hold ions in stable circular trajectories before allowing them to pass to the detector in order of increasing m/z value. A particular feature of quadrupole and ion trap analyzers is their ability to scan through a wide range of masses very rapidly, making them ideal for monitoring chromatographic peaks (Section F).

(iv) **Tandem mass analyzers** incorporate several mass analyzers in series. This enables ions selected from the first analyzer to undergo **collision induced dissociation** (**CID**) with inert gas molecules contained in a collision cell producing new ions which can then be separated by the next analyzer. The technique, known as **tandem mass spectrometry**, **MS-MS** or **(MS)^n**, is used in the study of decomposition pathways, especially for molecular ions produced by soft ionization techniques (*vide infra*). Collision-induced reactions with reactive gases and various scan modes are also employed in these investigations.

Mass spectrometers are designed to give a specified **resolving power**, the minimum acceptable resolution being one mass unit. Two masses are considered to be resolved when the valley between their peaks is less than 10% of the smaller peak height. For masses m_1 and m_2, differing by Δm, resolving power is defined as $m_2/\Delta m$, and for unit mass resolution the requirement increases with the magnitudes of m_1 and m_2, for example,

Unit mass resolution for masses	Required resolving power
99 and 100	100
499 and 500	500
4999 and 5000	5000

Single magnetic analyzers have resolving powers of up to about 5000, but much higher resolution instruments are required to distinguish between ions whose masses differ by less than one in the second or third decimal place, for example,

Ion	Exact mass	Required resolving power
$C_{18}H_{36}N_2^{+\bullet}$	280.2881	
		25 000
$C_{19}H_{36}O^{+\bullet}$	280.2768	

Ionization techniques

There are numerous means of ionizing molecules or elements in a sample, the most appropriate depending on the nature of the material and the analytical requirements. In addition, mass spectrometry can be directly interfaced with other analytical techniques, such as gas or liquid chromatography (Topics D4 to D7) and emission spectrometry (Topics E3 to E5). These **hyphenated systems** are described in Section F. The more important ionization techniques are summarized below.

- **Electron impact ionization** (**EI**) employs a high-energy electron beam (~70 eV). Collisions between electrons and vaporized analyte molecules, M, initially result in the formation of molecular ions, which are radical cations:

$$M + e^- \rightarrow M^{+\bullet} + 2e^-$$

These then decompose into smaller fragments.

- **Chemical ionization** (**CI**) is a softer technique than EI, ions being produced by collisions between sample molecules and ions generated by a reagent gas such as methane or ammonia. Three stages are involved. For methane, for example:

(i) reagent gas ionized by **EI**: $CH_4 + e^- \rightarrow CH_4^{+\bullet} + 2e^-$
(ii) secondary ion formation: $CH_4^{+\bullet} + CH_4 \rightarrow CH_5^+ + CH_3^\bullet$
(iii) formation of molecular species: $CH_5^+ + M \rightarrow MH^+ + CH_4$ (pseudomolecular ion)

Compared to EI, there is much less fragmentation, but a molecular species, MH^+, which is one mass unit higher than the relative molecular mass (RMM) of the analyte is formed (*Figs 5(a)* and *(b)*).

- **Desorption techniques** are used mainly for solid samples that can be deposited on the tip of a heatable probe that is then inserted into the sample inlet through vacuum locks. Molecules are ionized by the application of a high potential gradient (**field desorption, FD**) or by focusing a pulsed laser beam onto the surface of the sample. In **matrix-assisted laser desorption, MALDI**, the sample is mixed with a compound capable of absorbing energy from the laser and which results in desorption of protonated sample molecules. These techniques are very soft, give little fragmentation and are especially useful for compounds with a high RMM.
- **Interfacing mass spectrometry** with other analytical techniques (Section F) necessitates the use of specially designed interfaces and ionizing sources. These include **thermospray**, **electrospray** and **ionspray** for **liquid chromatography-mass spectrometry** (**LC-MS**), and an **inductively coupled plasma torch** (**ICP**) for **ICP-MS** (Topic E5). For **gas chromatography-mass spectrometry** (**GC-MS**), the carrier gas flows directly into the spectrometer where EI ionization can then be used.

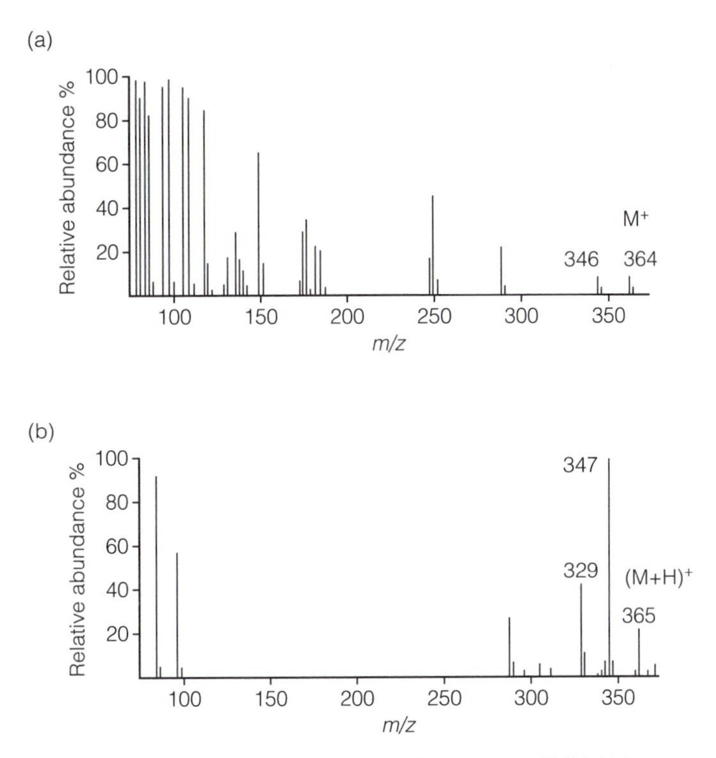

Fig. 5. EI and CI ionization spectra of dihydrocortisone, RMM 364.

Fragmentation

The production of a molecular ion is often followed by its dissociation, or **fragmentation**, into ions and neutral species of lower mass, which in turn may dissociate further. **Fragmentation patterns** are characteristic of particular molecular structures and can indicate the presence of specific functional groups, thus providing useful information on the structure and identity of the original molecule. The points of cleavage in a molecule are determined by individual bond strengths throughout the structure and, additionally, molecular rearrangements and recombinations can occur. These are illustrated below for a hypothetical molecule ABC.

$ABC + e^- \rightarrow ABC^{+\bullet} + 2e^-$	Molecular ion
$ABC^{+\bullet} \rightarrow A^+ + BC^\bullet$	Fragment ions
$ABC^{+\bullet} \rightarrow AB^+ + C^\bullet$	and radicals
$AB^+ \rightarrow A^+ + B$	
$AB^+ \rightarrow B^+ + A$	
$ABC^{+\bullet} \rightarrow CAB^{+\bullet} \rightarrow C^+ + AB^\bullet$	Rearrangements
$ABC^{+\bullet} + ABC \rightarrow (ABC)_2^{+\bullet} \rightarrow ABCA^+ + BC^\bullet$	

Fragmentation patterns are an invaluable aid in the interpretation of mass spectra and in the identification or confirmation of structural features. The fragmentation pathways of benzamide by an EI ionization source are shown in *Figure 6* and the spectrum in *Figure 13*.

Isotope peaks

Most elements occur naturally as a mixture of isotopes, all of which contribute to peaks in a mass spectrum. Examples are given in *Table 1* along with their

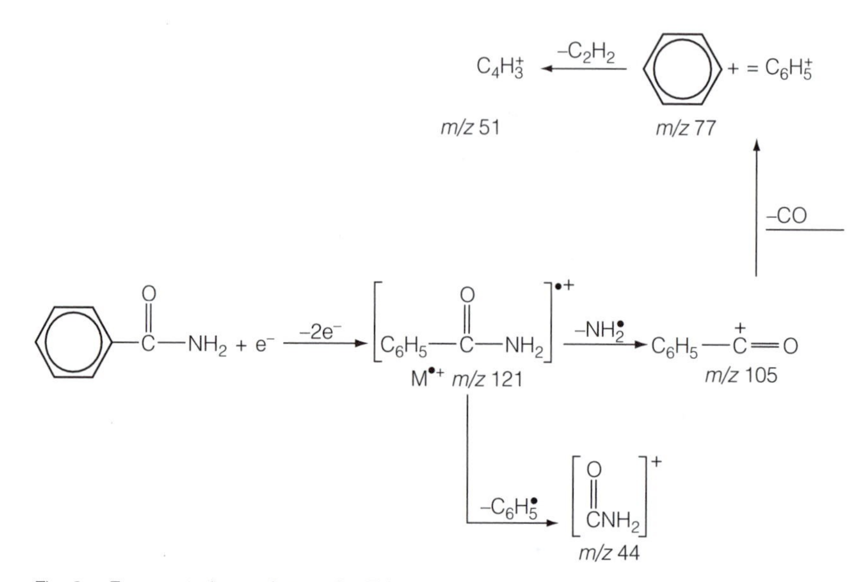

Fig. 6. Fragmentation pathways for EI ionization of benzamide (spectrum shown in Fig. 13).

percentage relative abundances. Thus, natural carbon consists of a mixture of 98.90% ^{12}C and 1.10% ^{13}C, and natural hydrogen is a mixture of 99.985% ^{1}H and 0.015% ^{2}H (deuterium).

The very small peaks above the molecular ion peak, M^+, at m/z 32 in the spectrum of methanol (*Fig. 1*) are due to the heavier isotopes of carbon, hydrogen and oxygen. These **isotope peaks**, designated **(M+1)**$^+$, **(M+2)**$^+$ etc., are of importance in the interpretation of mass spectra and can be used for two purposes.

(i) To establish an empirical formula for molecules containing C, H, O and N by comparison of the relative intensities of the M, M+1 and M+2 peaks in a recorded spectrum with tabulated values. *Table 2* is an extract of extensive

Table 1. Natural isotopic abundances of some common elements as a percentage of the most abundant isotope

Element	Most abundant isotope	Other isotopes	Abundance of heavier isotope[a]
Hydrogen	^{1}H	^{2}H	0.016
Carbon	^{12}C	^{13}C	1.11
Nitrogen	^{14}N	^{15}N	0.38
Oxygen	^{16}O	^{17}O	0.04
		^{18}O	0.20
Sulfur	^{32}S	^{33}S	0.78
		^{34}S	4.40
Fluorine	^{19}F		100
Chlorine	^{35}Cl	^{37}Cl	32.5
Bromine	^{79}Br	^{81}Br	98.0
Iodine	^{127}I		100
Phosphorus	^{31}P		100

[a]Relative to 100 for the most abundant.

Table 2. Empirical formulae and isotope peak ratios for a nominal RMM value of 70 (M = 100%)

	M+1 %	M+2 %	Exact mass
CN_3O	2.26	0.22	70.0042
CH_2N_4	2.62	0.03	70.0280
C_2NO_2	2.67	0.42	69.9929
$C_2H_2N_2O$	3.03	0.23	70.0167
$C_2H_4N_3$	3.39	0.04	70.0406
$C_3H_2O_2$	3.44	0.44	70.0054
C_3H_4NO	3.80	0.25	70.0293
$C_3H_6N_2$	4.16	0.07	70.0532
C_4H_6O	4.57	0.28	70.0419
C_4H_8N	4.93	0.09	70.0657
C_5H_{10}	5.70	0.13	70.0783

tables for all possible combinations of these elements up to an RMM of several hundred. Theoretical values for any molecular formula are readily calculated by multiplying the number of atoms of each element by the corresponding natural isotopic abundances and summing them to obtain the intensities of the M+1 and M+2 peaks as percentages of the M peak, e.g. for $C_{24}H_{22}O_7$

	C	**H**	**O**	**Total/%**
$\mathbf{M^+}$				**100.00**
$\mathbf{(M+1)^+}$	24×1.108	22×0.016	7×0.04	**27.22**
$\mathbf{(M+2)^+}$	–	–	7×0.20	**1.40**

(ii) To determine the numbers of chlorine and bromine atoms in a compound from the relative intensities of the M, M+2, M+4, M+6, etc., peaks in a recorded spectrum by comparisons with tabulated values or graphical plots. The presence of one or more atoms of either halogen will give isotope peaks two mass units apart (*Table 3*).

Mass spectra

Mass spectral data can be used to provide the following analytical information:

- an **accurate RMM** if the molecular ion can be identified;
- an **empirical molecular formula** based on isotope peak intensities;

Table 3. Isotope peak ratios for molecules containing up to three chlorine and bromine atoms (M = 100%)

Halogen atom(s)	M+2 %	M+4 %	M+6 %
Cl	32.6		
Cl_2	65.3	10.6	
Cl_3	99.8	31.9	3.47
Br	97.7		
Br_2	195	95.5	
Br_3	293	286	93.4
ClBr	130	31.9	

- the identification of unknowns and of structural features from **fragmentation patterns**, sometimes with the aid of computerized library searches;
- quantitation by **selected ion monitoring** (**SIM**) for both molecules and elements.

The interpretation of molecular mass spectra is accomplished by comparisons with the spectra of known compounds and the application of a set of empirical rules. The following are particularly useful.

(i) The **nitrogen rule** states that compounds with an even-numbered RMM must contain zero or an even number of nitrogen atoms, and those with an odd-numbered RMM must contain an odd number of nitrogen atoms.

(ii) The **unsaturated sites rule** provides a means of calculating the number of double-bond equivalents in a molecule from the formula

$$\text{No. of C atoms} + \tfrac{1}{2}(\text{no. of N atoms}) - \tfrac{1}{2}(\text{no. of H} + \text{halogen atoms}) + 1$$

For example, for C_7H_7ON, the formula gives $7 + 0.5 - 3.5 + 1 = 5$ **double bond equivalents**. This corresponds to benzamide, $C_6H_5CONH_2$, the aromatic ring being counted as three double bonds plus one for the ring.

(iii) The intensity of the molecular ion peak **decreases** with increasing chain length in the spectra of a homologous series of compounds and with increased branching of the chain.

(iv) Double bonds and cyclic structures tend to stabilize the molecular ion, saturated rings losing side chains at the α-position.

(v) Alkyl-substituted aromatic rings cleave at the β-bond to the ring giving a prominent peak at m/z 91, which corresponds to the tropylium ion, $C_7H_7^+$.

(vi) Small neutral molecules, such as CO, C_2H_4, C_2H_2, H_2O and NH_3 are often lost during fragmentation.

(vii) The C–C bond adjacent to a heteroatom (N,O,S) is frequently cleaved leaving the charge on the fragment containing the heteroatom, whose nonbonding electrons provide resonance stabilization, e.g.

$$CH_3\text{–}CH_2\text{–}Y^{+\bullet}\text{–}R \xrightarrow{-CH_3^{\bullet}} CH_2{=}Y^+\text{–}R \leftrightarrow {}^+CH_2\text{–}Y\text{–}R$$

McLafferty rearrangements in carbonyl compounds are common, e.g.

$$HCH_2\text{–}CH_2\text{–}CH_2\text{–}CO^{+\bullet}\text{–}OR \leftrightarrow C_2H_4 + CH_2{=}C(O^{+\bullet}H)\text{–}OR$$

A neutral molecule of ethene is eliminated in the process (Rule (vi)).

Some examples of fragmentation patterns in EI mass spectra are shown in *Figures 7–15*. The relative intensities of the peaks within a spectrum are a reflection of the statistical probability of particular bonds in the molecule being cleaved and are closely related to the bond energies.

Spectra of saturated straight chain compounds are characterized by clusters of peaks 14 mass units (CH_2 groups) apart, as successive C–C bonds along the chain are cleaved in different molecules. **Octane** (*Fig. 7*) has a base peak at m/z 43 due to the $CH_3CH_2CH_2^+$ fragment ion, and a small molecular ion peak at m/z 114 (Rule (iii)).

Branching of the chain alters the relative intensities of the clusters, as shown by the spectrum of the isomeric **2,2,4-trimethylpentane** (*Fig. 8*), which has a base peak at m/z 57 due to the $(CH_3)_3C^+$ fragment ion, and no significant m/z 71, 85 or molecular ion peaks (Rule (iii)).

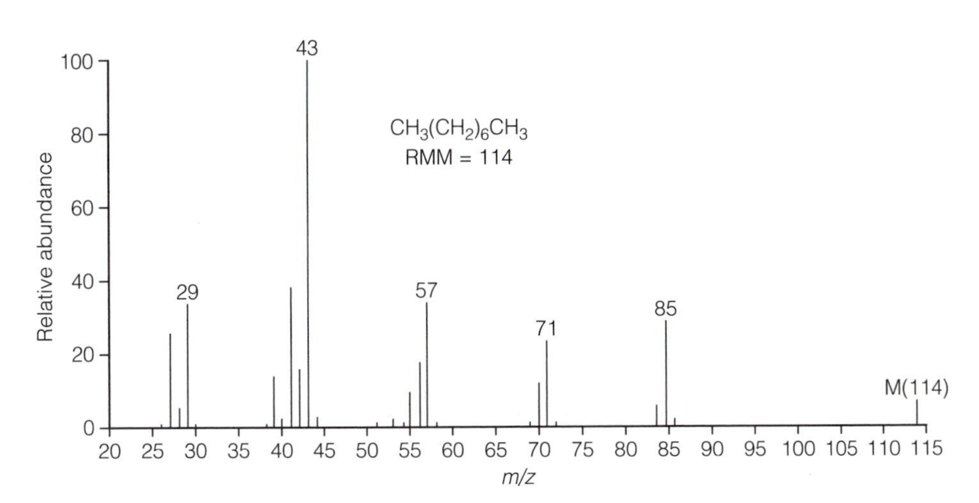

Fig. 7. Mass spectrum of octane.

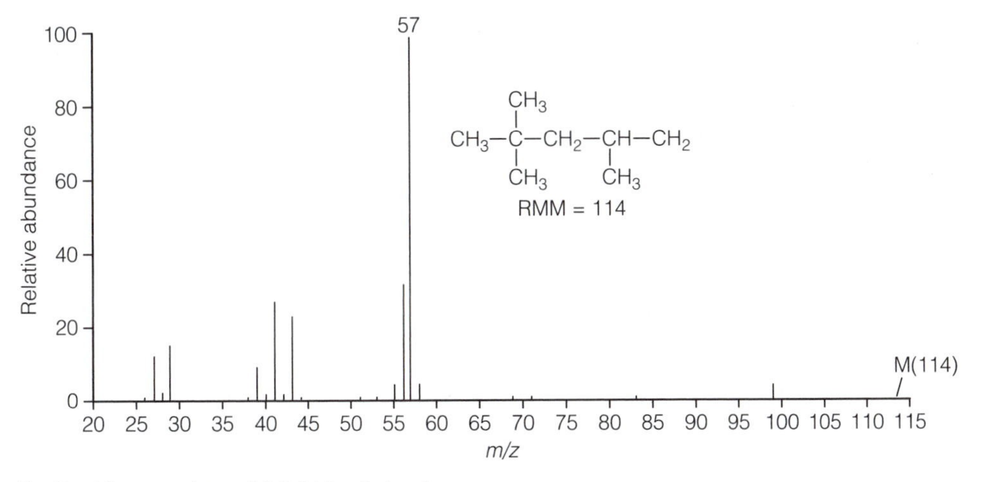

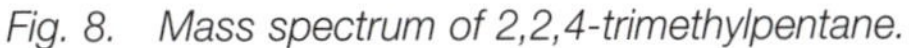
Fig. 8. Mass spectrum of 2,2,4-trimethylpentane.

The spectrum of **methylbenzene** (*Fig. 9*) typifies alkyl-substituted aromatic compounds, with a base peak corresponding to the **tropylium ion**, $C_7H_7^+$, at *m*/*z* 91 and a large molecular ion peak *at m*/*z* 92 (Rules (iv) and (v)).

The base peak of **butanone** (*Fig. 10*) is due to the resonance-stabilized CH_3CO^+ fragment ion, *m*/*z* 43, originating from cleavage of the C–C bond adjacent to the carbonyl group (Rule (vii)). The loss of a methyl radical gives a prominent peak at *m*/*z* 57 from the $CH_3CH_2CO^+$ fragment ion.

Alcohols readily lose water to give $(M\text{-}18)^+$ fragment ions (Rule (vi)). In the case of **1-butanol** (*Fig. 11*), this occurs at *m*/*z* 56. The molecular ion peak is correspondingly small, and the base peak at *m*/*z* 31 is due to CH_2OH^+.

Methyl butyrate provides a good example of a McLafferty rearrangement peak in a carbonyl compound (Rule (vii)). In this spectrum (*Fig. 12*), the peak, which is due to the $CH_2{=}CO^+HOCH_3$ fragment ion, is at *m*/*z* 74, while the base

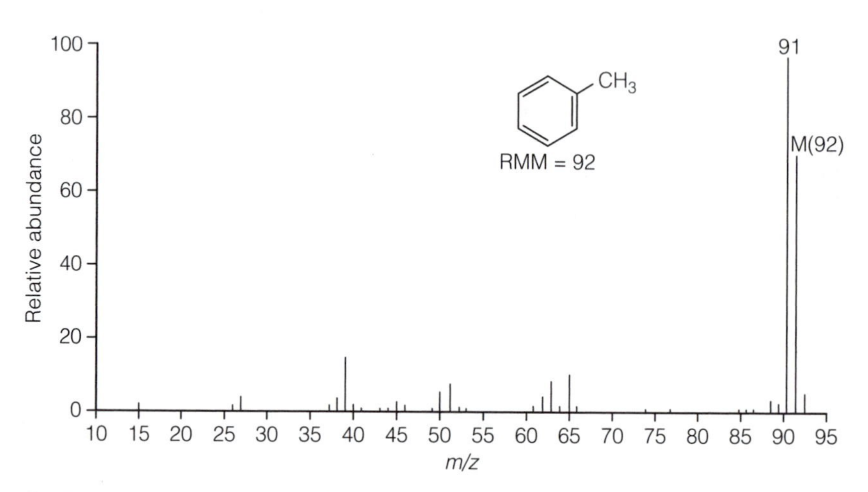

Fig. 9. Mass spectrum of methylbenzene.

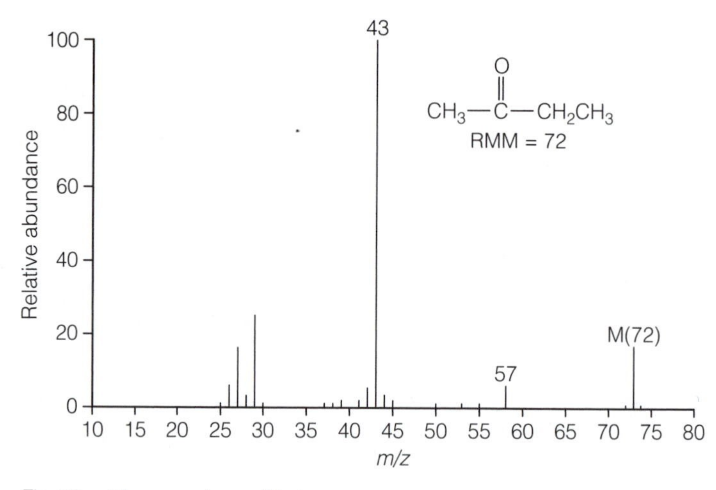

Fig. 10. Mass spectrum of butanone.

peak at m/z 43 arises from the $CH_3CH_2CH_2^+$ fragment ion. Peaks at m/z 59 and 71 are from the CH_3OCO^+ and the $CH_3CH_2CH_2CO^+$ fragment ions respectively.

The spectrum of **benzamide** (*Fig. 13*) demonstrates the stability of an aromatic ring structure (Rule 4). After the base peak at m/z 77, due to the $C_6H_5^+$ fragment ion, the next two most intense peaks (m/z 105 and 121) are from the $C_6H_5CO^+$ fragment ion and the molecular ion, respectively. The former loses a neutral CO molecule to give the base peak, and then a neutral C_2H_2 molecule to give the $C_4H_3^+$ fragment ion at m/z 51.

In the spectrum of **dichloromethane** (*Fig. 14*), the relative intensities of the m/z 84 (M), 86 (M+2) and 88 (M+4) peaks, correspond to the pattern expected from a molecule with two chlorine atoms whose isotopes are ^{35}Cl and ^{37}Cl (*Table 3*). The base peak is at m/z 49, from the $CH_2{}^{35}Cl^+$ fragment ion after the loss of one chlorine from the molecular ion. This is accompanied by a $CH_2{}^{37}Cl^+$ ion

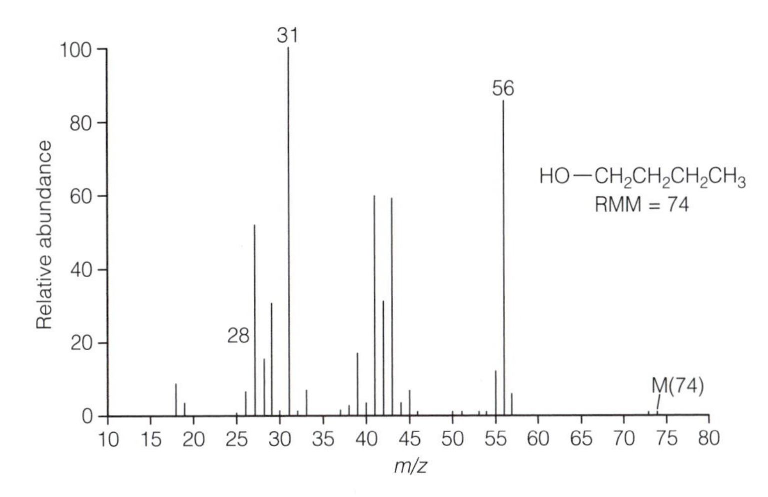

Fig. 11. Mass spectrum of 1-butanol.

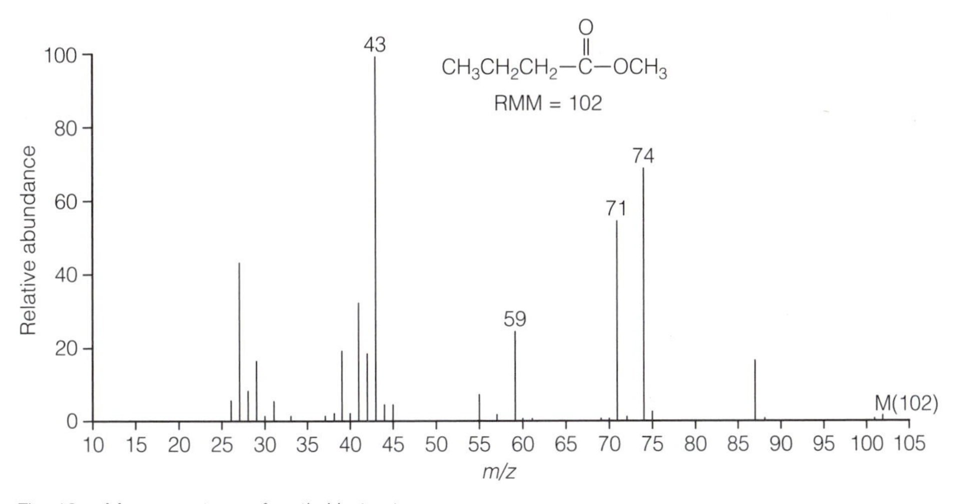

Fig. 12. Mass spectrum of methyl butyrate.

of about one-third the intensity at *m/z* 51, the expected ratio for a species with only one chlorine atom.

For **1-bromohexane**, the almost equal abundances of the ^{79}Br and ^{81}Br isotopes result in two pairs of nearly equal sized peaks at *m/z* 164 and 166 (M and M+2) and at *m/z* 135 and 137 from the $CH_2CH_2CH_2CH_2Br^+$ fragment ion (*Fig. 15*). Both the base peak at *m/z* 43, due to the $CH_3CH_2CH_2^+$ fragment ion, and that at *m/z* 85, due to the intact alkyl chain fragment ion, have no bromines to give associated isotope peaks two mass units higher.

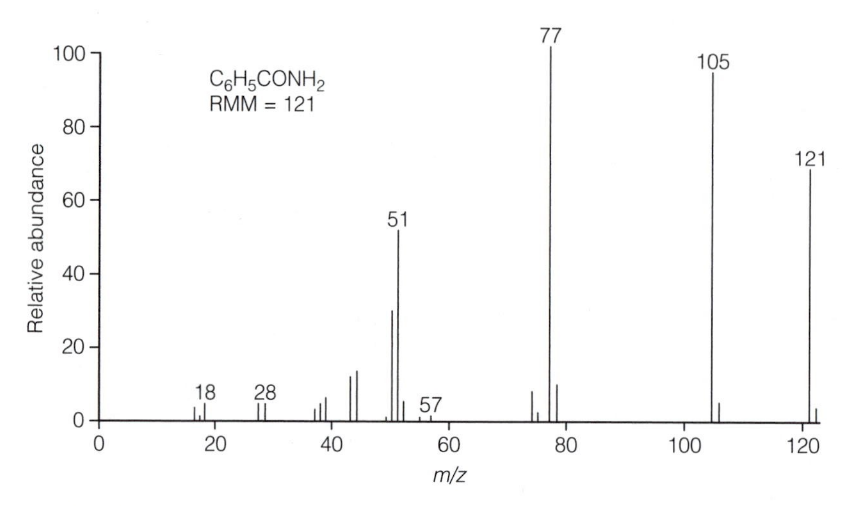

Fig. 13. Mass spectrum of benzamide.

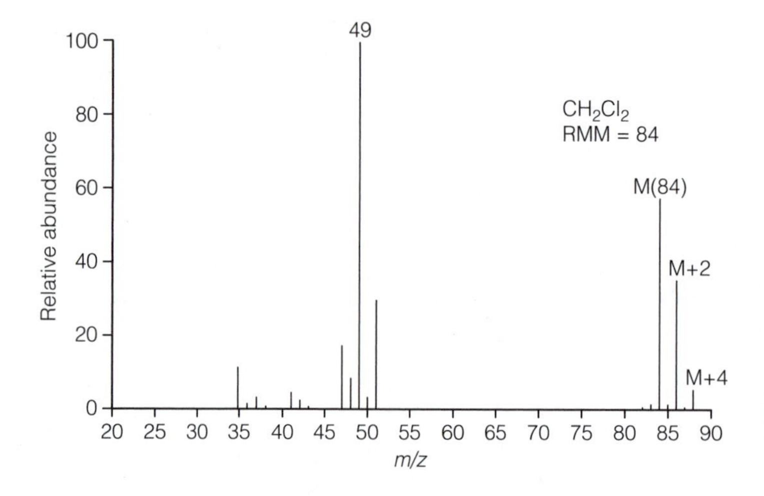

Fig. 14. Mass spectrum of dichloromethane. From Instrumental Methods of Analysis, *2nd edn, by H.H. Willard, L.L. Merritt, J.A. Dean & F.A. Settle © 1988. Reprinted with permission of Brooks/Cole an imprint of the Wadsworth Group, a division of Thomson Learning.*

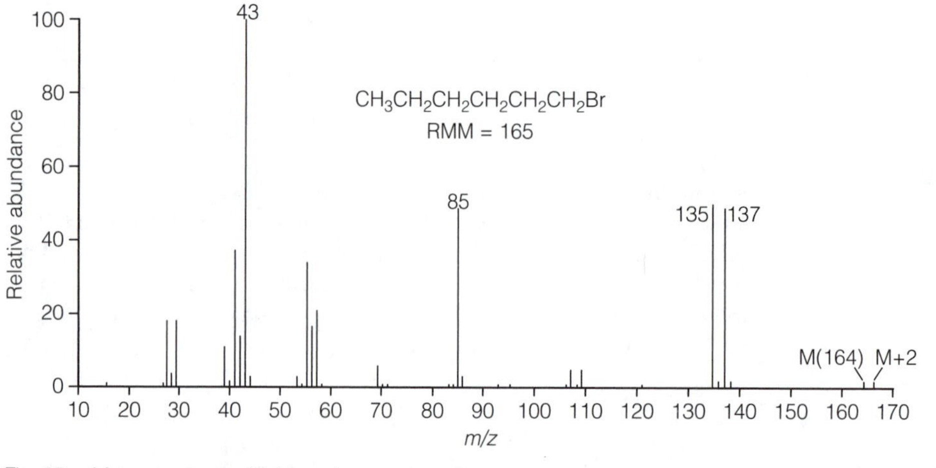

Fig. 15. Mass spectrum of 1-bromohexane.

F1 ADVANTAGES OF COMBINED TECHNIQUES

Key Notes

Combined approach

The strategy for analyzing a sample to determine its composition, structure and properties often requires the application of more than one of the analytical techniques described in this text. The use of multiple techniques and instruments, which allow more than one analysis to be performed on the same sample at the same time, provide powerful methods for analyzing complex samples.

Problem solving

In order to solve many analytical problems, it is necessary to use a number of techniques and methods. The components of naturally occurring substances may need to be separated and identified. Problem solving is aided if several analytical techniques are used, and time may be saved if the analyses are performed simultaneously.

Advantages

By using many techniques either separately or in combinations, the advantages to the analyst in the additional information, time saving and sample throughput are considerable.

Related topics

Separation techniques (Section D)
Spectrometric techniques (Section E)
Sensors, automation and computing (Section H)

Combined approach

Other sections of this text describe some of the many techniques and methods of qualitative, quantitative and structural analysis. In the case of samples originating in the real world, that is from man-made materials, environmental sampling or complex natural mixtures, each of the techniques has a place, and often several must be used in order to obtain a complete overview of the nature of the sample.

Particular emphasis has been placed on separation and spectrometric techniques in Sections D and E. Together with the electrochemical techniques of section C, they represent the major tools of the analytical chemist. Other special techniques, such as thermal analysis, may also be combined with them to reveal precise details of the processes occurring (see Topic G4).

If an unknown material is presented for analysis, it should first be determined whether the sample is a single substance, or a mixture. The purity of substances such as pharmaceuticals is very important. Separation by an appropriate technique (Section D) should reveal the number of components in the sample.

Even single substances can be complex chemical molecules. If these are to be assigned a unique identity, then several of the molecular spectrometric techniques may need to be used on the sample.

Instrumentation may need to be specially adapted if two or more techniques are to be interfaced successfully; for example, passing the effluent gas emerging from a gas chromatograph at atmospheric pressure into the vacuum inlet of a mass spectrometer.

There are two alternative strategies for analyzing samples by multiple techniques.

- Several representative samples are taken from the specimen and each is analyzed using optimum conditions for the techniques selected. This is referred to as a **multidisciplinary approach**. For example, a pharmaceutical sample may need to be analyzed by UV, IR, NMR, MS and other spectrometric techniques.
- **A single** representative sample is taken and analyzed by an instrument designed to perform more than one analytical technique at the same time. For example, a sample that is chemically or thermally unstable could be studied by gas chromatography linked to an infrared spectrometer. If the instruments for GC and IR spectrometry are combined so that the analyses are done essentially at the same time, this is called a **simultaneous approach** and is often written with a hyphen, so that they may be referred to as **hyphenated techniques,** for example **GC-MS** and **GC-IR**.

Problem solving

The procedure for the analysis of each sample must be optimized at all stages to take account of the information required, as detailed in Topic A1.

The sample handling and the amount of sample available may determine which techniques are possible. The quantitative accuracy and the components sought govern the choice of method. Whether the exact structure of an unknown component is required, or merely an indication of its chemical class, may control the time required for the analysis and preparation of the final analytical report.

The time factor is of vital importance in a busy analytical laboratory schedule. Any method that saves time, or that can be operated automatically under computer control on a sequence of different samples, as described in Section H, is of value. Similarly, if this can be linked to a computer search facility to aid structure determination, then this offers a further time saving.

Advantages

The advantages are summarized below.

- The combined approach often provides more information than could be obtained by using the individual techniques in isolation.
- Multidisciplinary analysis is a necessary tool in the work of the analytical chemist, otherwise misleading or incorrect results may be obtained.
- Hyphenated techniques provide a time saving because two or more analyses can be run at the same time, since the instrumentation combines the features of both, and sample preparation time is reduced.
- Hyphenated techniques provide analytical information on exactly the same sample, provided that the sample is truly representative (Topic A4) of the material to be analyzed. If comparisons between samples are required, then it is essential that cross-contamination does not occur.

Another advantage of using a combined approach is that of obtaining **time-resolved** (or **temperature-resolved**) **spectra** or **chromatograms**. If changes occur, due to reaction, evaporation or phase separation, these changes may be studied by recording the analytical data at specific time (or temperature) intervals.

F2 SAMPLE IDENTIFICATION USING MULTIPLE SPECTROMETRIC TECHNIQUES

Key Notes

Information from each spectrum

The information that may be obtained from ultraviolet-visible, infrared, proton and carbon-13 nuclear magnetic resonance and mass spectra is complementary, and it is much easier to identify the structure of a compound if all the spectra are considered.

Spectrometric identification

Each spectrometric technique provides characteristic data to assist in the eventual identification of the sample. These have been considered in the individual topics, but need to be combined to extract the maximum information.

Applications

Examples of the use of the multidisciplinary approach to sample identification exemplify a general scheme of interpretation.

Related topics

Spectrometric techniques (Section E)

Information from each spectrum

In order to study and identify any unknown analytical sample using spectrometric techniques, the analytical chemist must first obtain good quality spectra and then use these to select the information from each technique that is of most value. It is also important to recognize that other analytical observations should be taken into account. For example, if the sample is a volatile liquid and the spectral information suggests that it is an involatile solid, clearly, there is conflicting evidence. *Table 1* summarizes the information of most value that may be obtained from the common spectrometric techniques.

The use of computerized library databases can assist in the matching of spectra to recorded examples (Topic H4). If difficulties are found in distinguishing between two possibilities for the sample identity, then it may be necessary to consult reference texts or additional computer databases so that an exact match is found. Some databases give information that helps when working with samples that are new

Table 1. Information from common spectrometric techniques

Information	Techniques			
	IR	MS	NMR	UV
Molecular formula		h		
Functional groups	h	m	m	m
Connectivity	m	m	h	m
Geometry/stereochemistry	m		h	

h, high value; m, moderate value.

compounds or whose spectra are not present in the database. For example, the presence of a strong peak in an IR spectrum near 1700 cm^{-1} should suggest a high probability that the sample might be a carbonyl compound.

Spectrometric identification

The conditions under which each spectrum has been obtained must be taken into consideration. For example, if the UV, IR and NMR spectra were run in solution, what was the solvent? The instrumental parameters also need to be considered. In MS, the type of ionization used will affect the spectrum obtained.

Sometimes the source of the analytical sample is known, and this can be a great help in elucidating the identity of the material.

It is a worthwhile exercise to follow the same general scheme and to note down the information that is deduced from the study of each spectrum. One suggested scheme is given below, but the value of 'feedback' in checking the deductions must not be overlooked.

(i) **Empirical formula.** Occasionally, if the sample has been analyzed to find the percentage of carbon, hydrogen, nitrogen, sulfur and other elements, and to deduce the percentage of oxygen by difference, this can be a useful first step. If this information is not available, it may be found from the MS if an accurate relative molecular mass has been measured.

Example: A solid sample contained C 75.5%, H 7.5%, and N 8.1% by weight. What is the empirical formula of the sample?

Dividing by the relative atomic masses gives the ratio of numbers of atoms, noting that there must be (100 – 75.5 – 7.5 – 8.1) = 8.9% oxygen.

C = 75.5/12 = 6.292
H = 7.5/1 = 7.5
N = 8.1/14 = 0.578
O = 8.9/16 = 0.556

This corresponds (roughly) to $C_{11}H_{13}NO$, with an RMM of 175, which may give a molecular ion in the mass spectrum.

(ii) **Double bond equivalents.** The presence of unsaturation in a structure should be considered. Since a saturated hydrocarbon has the formula C_nH_{2n+2}, and since a single-bonded oxygen can be thought of as equivalent to $-CH_2-$, and a single-bonded nitrogen as –CH<, the number of double bonds, or rings, called the **double bond equivalents (DBE)** for the compound is given by:

$$DBE = (2n_4 + 2 + n_3 - n_1)/2$$

where n_4 is the number of tetravalent atoms (e.g., carbon), n_3 is the number of trivalent atoms (e.g., nitrogen), n_1 is the number of monovalent atoms (e.g., hydrogen or halogen).

Therefore, for benzene, C_6H_6, the DBE is (14 – 6)/2 = 4, that is, three double bonds and one ring.

For the example in (i) above, $C_{11}H_{13}NO$, the DBE is (24 + 1 – 13)/2 = 6, which would correspond to one benzene ring (4) plus one –C=C- plus one >C=O. Note that other spectra must be used to distinguish between a ring and a double bond or between a –C=C- and a >C=O.

(iii) The **IR** spectrum gives evidence about the presence or absence of functional groups as discussed in Topics E10 and E11. The example in (ii) above would be solved if the infrared spectrum showed no carbonyl to be present. The presence of aliphatic groups, or unsaturated or aromatic structures may be inferred from the position of the –C-H stretching bands around 3000 cm^{-1}, and confirmed by the presence of other bands. A useful application of Raman spectrometry (Topic E10) is the detection of groups that have very weak absorbances in the infrared region, such as substituted alkynes.

(iv) The **UV** spectrum does give some structural information, even when there is little or no absorbance, which would suggest the absence of any aromatic, conjugated or ketonic structures. If there are double bonds or unsaturated rings present, the UV spectrum should provide further information.

(v) Much useful information may be derived from the mass spectrum as discussed fully in Topic E14. A brief summary of what to look for should include:

- the *m*/*z* of the molecular ion. This corresponds to the **molecular formula**, which may be a multiple of the empirical formula derived in (i). An odd value for the *m*/*z* of the molecular ion requires that an odd number of nitrogen atoms are present, as in the example in (i) above. Prominent isotope peaks indicate the presence of Cl, Br or S.
- the **exact** value of *m*/*z* for the molecular ion. For example, the nominal RMM of the example in (i) is 175, and the formula might have been deduced if the exact mass was determined as 175.0998, since, excluding some impossible formulae, some others are:

$C_8H_5N_3O_2$	175.0382
$C_7H_{13}NO_4$	175.0845
$\mathbf{C_{11}H_{13}NO}$	**175.0998**
$C_{10}H_{13}N_3$	175.1111

- the fragments present and the fragments lost.

(vi) Both the **1-H NMR** and the **13-C NMR** give essential information about the types of protons and carbons present, their environment and their connections to neighboring atoms. This is discussed in detail in Topics E12 and E13.

(vii) Before the final report is given, it is always a good idea to retrace the steps above to check whether the data is self-consistent. For example, if there is no evidence for aromatic structures in the IR spectrum, is this consistent with the NMR spectrum? If an isomer must be identified, do the positions of the peaks in the IR and NMR spectra correspond, and does the fragmentation in the mass spectrum provide confirmation?

Applications

Example 1

The spectra shown in *Figure 1 (a)–(d)* were obtained for a compound of composition C 67.0%, H 7.3%, N 7.8%; melting at 135°C.

(i) **Empirical formula**: $C_{10}H_{13}NO_2$; RMM = 179

(ii) **DBE** = 5

(iii) 1a: **IR** (KBr disk)

3300 cm^{-1}	H-N- stretch
3000+	H-C- aromatic stretch
3000	H-C- aliphatic
1670	C=O stretch (amide or aromatic links?)
1650, 1510, etc.	aromatic ring vibrations

This suggests a substituted aromatic amide.

(iv) **UV** (methanol solution): major peaks at 243 and 280 nm also suggest an aromatic compound

(v) 1b: **MS** (EI)

m/z	
179	$M^{+\bullet}$ Must be odd number of nitrogens
137	M- 42: loss of CH_2CO; CH_3CO- compound?
43	CH_3CO^+
27 and 29	$C_2H_3^+$ and $C_2H_5^+$
108/109	$(HO–C_6H_4–NH_2)^+$ and less 1 H

(vi) 1c: **1-H NMR** (80 MHz, CCl_4 solution)

δ/ppm	Relative integral	Multiplicity	Assignment
1.3	3	3	**CH_3**–CH_2–
2.1	3	1	**CH_3**–CO–
4.0	2	4	O–**CH_2**–CH_3
6.8, 7.3	4	~ 2 doublets	1,4–Ar**H**–
7.6	1	1, broad	Ar–N**H**–CO

1d: **13-C NMR** (20.15 MHz, $CDCl_3$ solution)

δ/ppm	Multiplicity	Assignment
14.8	4	**C**H_3–CH_2-
24.2	4	**C**H_3–CO-
63.7	3	O–**C**H_2–CH_3
114.7	2	**ArC**H–
122.0	2	**ArC**H
131.0	1	**ArC**–CO–
155.8	1	**ArC**–N–
168.5	1	Ar–**C**O–

(vii) The pair of doublets in the proton NMR suggests a 1,4-disubstituted aromatic compound. Evidence for an **ethyl** group and for an **amide** suggest the structure **$C_2H_5O–C_6H_4–NHCOCH_3$**, **4-ethoxyacetanilide (phenacetin)**.

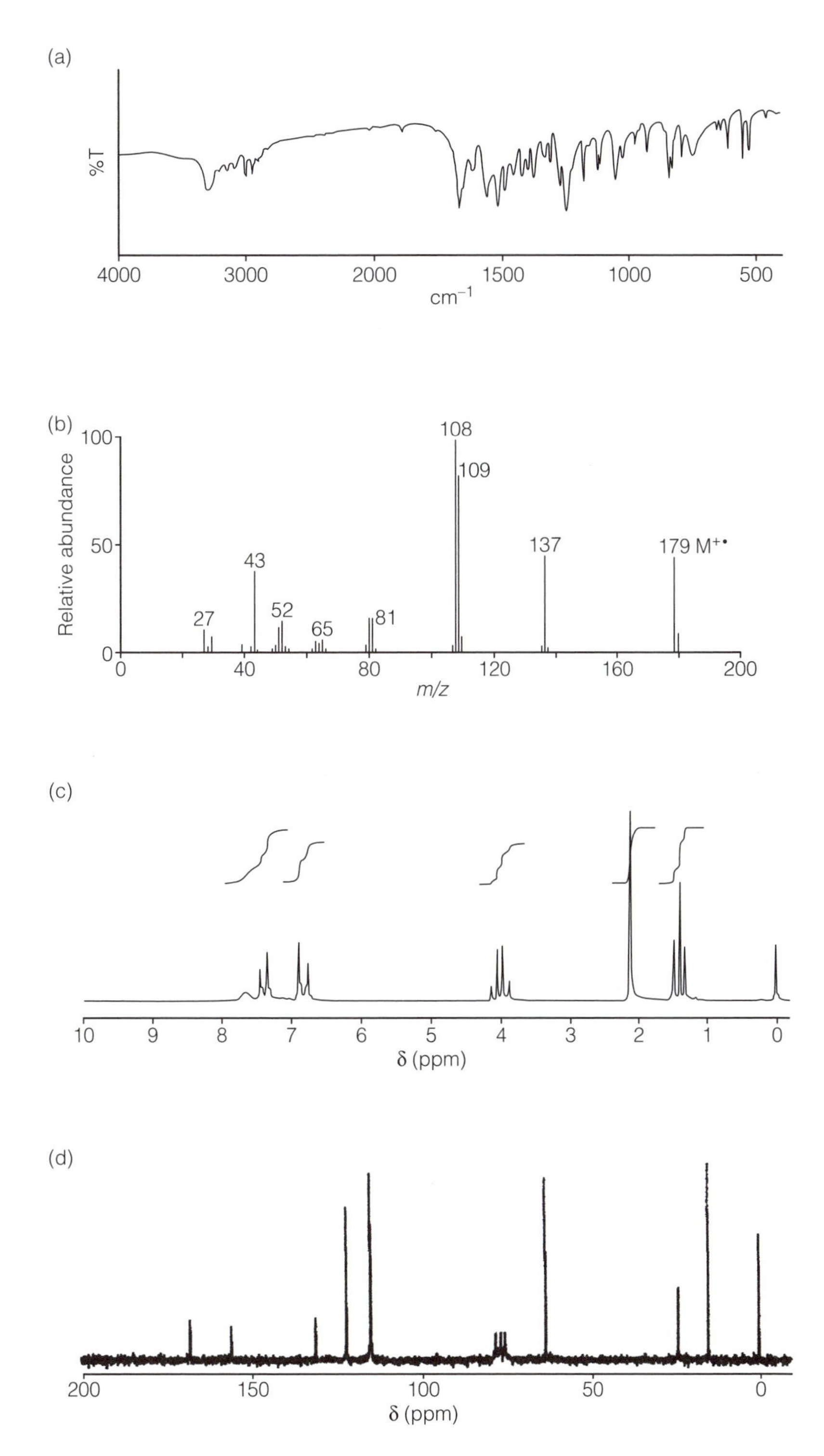

Fig. 1. Example 1.

Example 2

The spectra shown in *Figs 2(a)–(d)* are for a liquid, boiling at 141°C and soluble in water. The elemental composition was C 48.6%, H 8.1%.

(i) **Empirical formula:** $C_3H_6O_2$; RMM = 74

(ii) **DBE** = 1

(iii) 2a: **IR** (liquid film) The most notable features of the spectrum are the broad band around 3000 cm^{-1} and the strong carbonyl band at 1715 cm^{-1}.

3000 cm^{-1}	H-O-, hydrogen bonded stretch of acid
2900	H-C- aliphatic str
1715	C=O str of acid
1450	CH_2 and CH_3 bend
1380	CH_3 bend
1270	C-O- str

This strongly suggests a **carboxylic acid.**

(iv) **UV**: No significant UV absorption above 220 nm, therefore **aliphatic**.

(v) 2b: **MS** (EI)

m/z	
74	$M^{+\bullet}$
57	M- 17 possibly M–OH
45	–COOH
27, 29	C_2H_5 present?

The fragment ions suggest an **aliphatic carboxylic acid**

(vi) 2c: **1-H NMR** (200 MHz, $CDCl_3$ solution)

δ/ppm	Relative integral	Multiplicity	Assignment
1.2	3	3	$\mathbf{CH_3}$–CH_2–
2.4	2	4	–CO–$\mathbf{CH_2}$–CH_3
11.1	1	1, broad	**H**OOC–

The fact that the proton at δ = 11.1 ppm exchanges with D_2O suggests an **acid**.

2d: **13-C NMR** (50.0 MHz, $CDCl_3$ solution)

δ/ppm	Multiplicity	Assignment
9.5	4	$\mathbf{CH_3}$–C
28.2	3	–CO–CH_2–C
180.0	1	–**C**O–

(vii) The compound is **propanoic acid, CH_3-CH_2-COOH**. This is in agreement with all the spectrometric data, and with the boiling point.

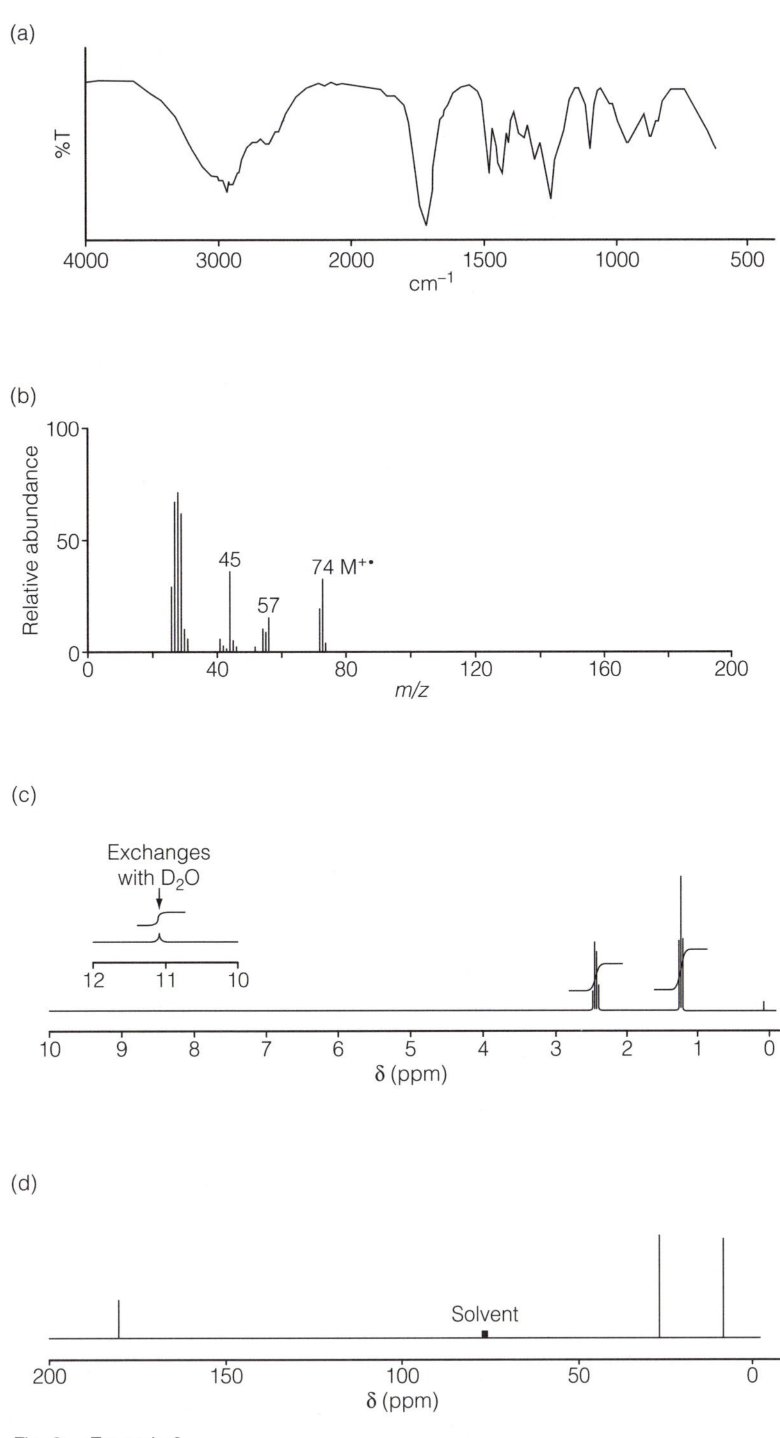

Fig. 2. Example 2.

Example 3

The spectra shown in *Figures 3(a)–(d)* are for a compound boiling at 205°C and insoluble in water. The composition is C 52.2%, H 3.7%, Cl 44.1%.

(i) **Empirical formula:** $C_7H_6Cl_2$; RMM = 161

(ii) **DBE** = 4

(iii) 3a: **IR** (liquid film)

3000–3050 cm^{-1}	H–C aromatic
2000–1600	Monosubstituted aromatic bands
1700	**not** a carbonyl, since no oxygen: probably aromatic
1500, 1450	aromatic ring vibrations
696	C–Cl stretch

(iv) **UV**: The weak absorbance at 270 nm may indicate an aromatic compound.

(v) 3b: **MS** (EI). The multiple molecular ion at *m/z* 160, 162 and 164 and the doubled fragment ions at 125 and 127 would strongly suggest 2 Cl atoms, even without the analytical information.

m/z			Relative abundance
164	$M^{+\bullet}$	$C_7H_6{}^{37}Cl_2$	9
162	$M^{+\bullet}$	$C_7H_6{}^{35}Cl^{37}Cl$	6
160	$M^{+\bullet}$	$C_7H_6{}^{35}Cl_2$	1
127	M–Cl	$C_7H_6{}^{37}Cl$	33
125	M–Cl	$C_7H_6{}^{35}Cl$	100

Other peaks suggest **aromatic ring residues**.

(vi) 3c: **1-H NMR** (200 MHz, $CDCl_3$ solution)

δ/ppm	Relative integral	Multiplicity	Assignment
6.7	1	1	**H-**C(Cl)–Ar
7.3	3	m	*m-*, *p-*Ar**H**
7.5	2	m	*o-*Ar**H**

(m = multiplet)

This suggests a **monosubstituted aromatic compound**.

3d: **13-C NMR** (50.0 MHz, $CDCl_3$ solution)

The presence of 5 carbon resonances when 7 carbons are present means that at least two pairs are equivalent, which would correspond to a **monosubstituted aromatic compound**.

δ/ppm	Multiplicity	Assign
73.0	2	H–**C**(Cl)–Ar
128.0	2	**ArC**–H
130.5	2	**ArC**–H
132.0	2	**ArC**–H
142.0	1	**ArC**–CH(Cl)

(vii) The deductions above indicate that this is an aromatic compound, which is consistent with a DBE of 4. Monosubstitution is confirmed by the IR spectrum and by both NMR spectra. This means that the two chlorines cannot be substituted on the ring. Since the carbon external to the ring has only a single hydrogen, both the chlorines must also be attached to it. The compound is therefore **dichloromethyl benzene, Cl_2CH-C_6H_5 (benzylidene chloride)**.

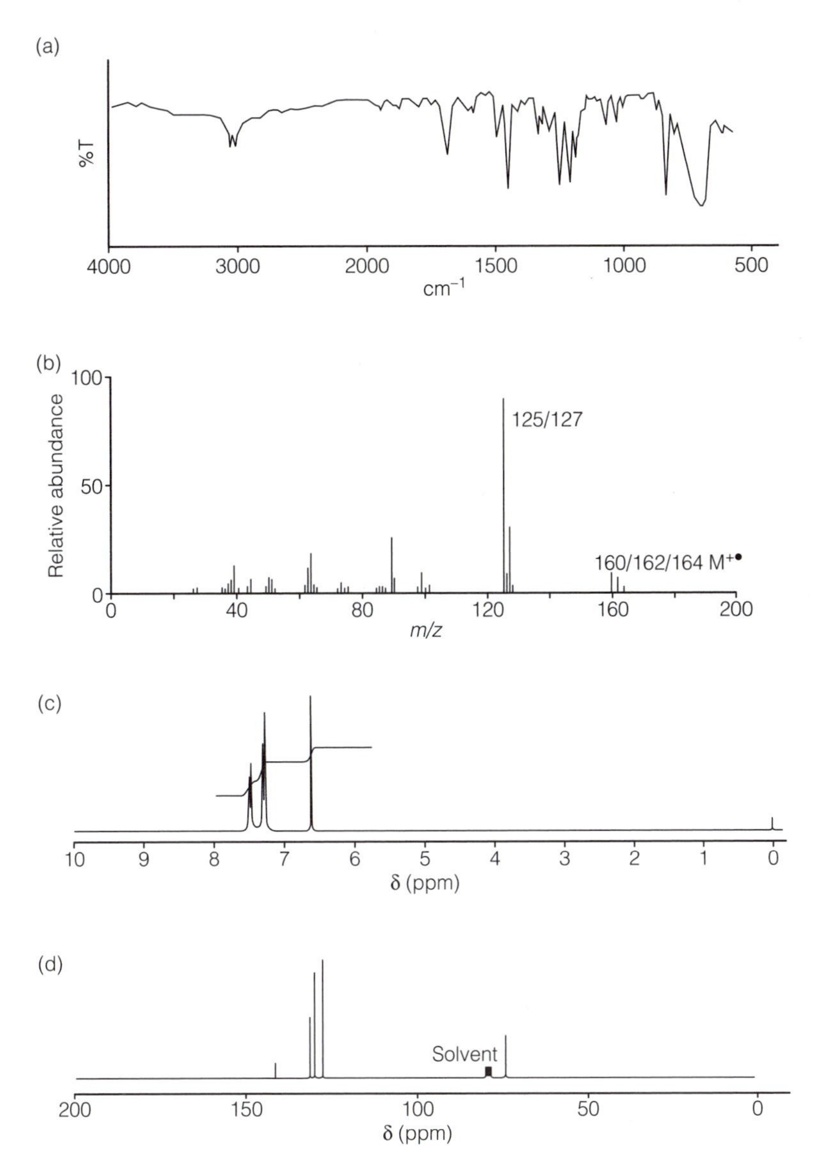

Fig. 3. Example 3.

While these three examples are relatively simple, they illustrate the advantages of the multidisciplinary approach in the analysis of molecular spectra. However, caution should always be excercised for two important reasons.

- Aromatic systems other than benzene are widely distributed. Many natural products contain heterocyclic aromatic rings containing nitrogen, oxygen or sulfur and these should be included as possibilities in the interpretation of spectra. Correlation tables for such compounds are contained in the further reading.
- While spectrometry is very powerful in identifying pure compounds, it often fails when the sample is a mixture. The following topics in this section discuss how mixtures may be separated prior to spectrometric analysis.

F3 GAS CHROMATOGRAPHY-MASS SPECTROMETRY

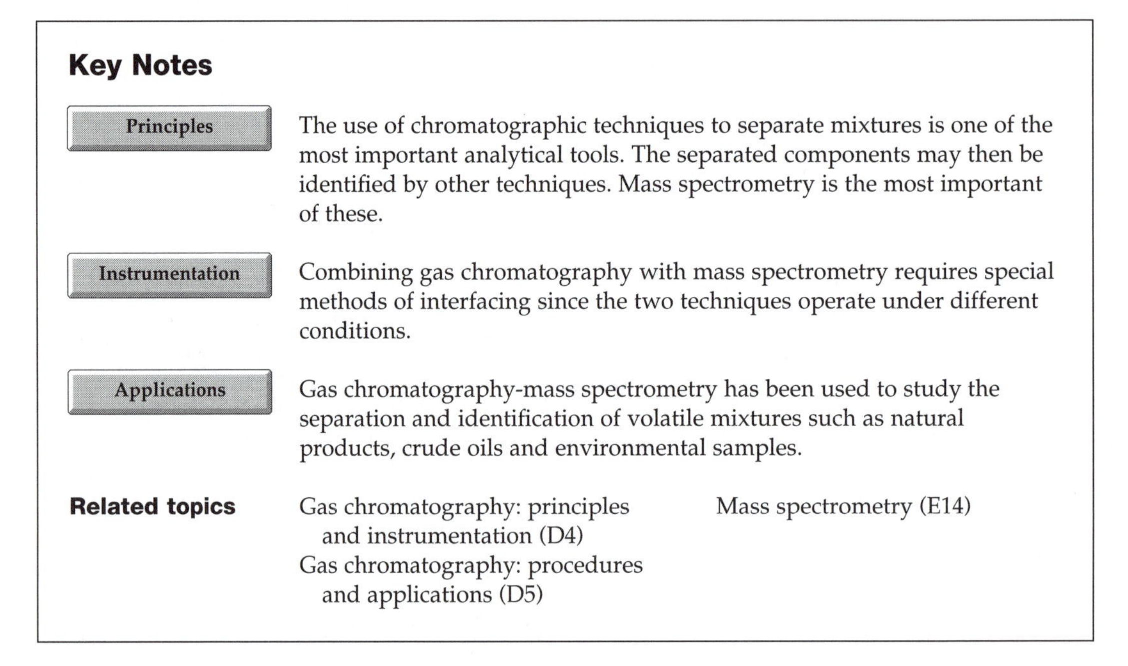

Key Notes

Principles — The use of chromatographic techniques to separate mixtures is one of the most important analytical tools. The separated components may then be identified by other techniques. Mass spectrometry is the most important of these.

Instrumentation — Combining gas chromatography with mass spectrometry requires special methods of interfacing since the two techniques operate under different conditions.

Applications — Gas chromatography-mass spectrometry has been used to study the separation and identification of volatile mixtures such as natural products, crude oils and environmental samples.

Related topics — Gas chromatography: principles and instrumentation (D4); Gas chromatography: procedures and applications (D5); Mass spectrometry (E14)

Principles

The separation of mixtures by gas chromatography requires that they are volatile within the operating temperature range of the instrument. Since stable stationary phases are available for use up to 400°C and ovens may be temperature-programmed to operate from ambient to high temperatures, this allows the separation of many samples, provided they do not decompose in the system. The separated components may be classified according to their retention times or by chromatographing spiked samples, but for unambiguous identification, other techniques are required. Mass spectrometry, which is fully described in Topic E14, is an important identification tool. Solutes may be ionized by electron impact, or by softer techniques such as chemical ionization. This is very useful in the identification of biological and less stable species.

Instrumentation

The effluent gases from the gas chromatograph contain both the carrier gas and the separated components at a pressure close to atmospheric. Detection by a flame ionization detector (FID) or one of the other GC detectors is possible, and effected by splitting the effluent stream at the column exit, allowing most of the sample to enter the mass spectrometer. It is then necessary to reduce the pressure to the operating pressure of the mass spectrometer, which is around 10^{-8} Nm^{-2}. With a capillary GC column, the flow of carrier gas is small, and the effluent can be fed through a fine capillary directly into the mass spectrometer. For a packed GC column, an interface between the GC and the MS is required. This may either be a porous tube separator, or a jet separator, shown in *Figure 1(b)*. In both of these, the low-mass carrier gas, usually helium, diffuses away

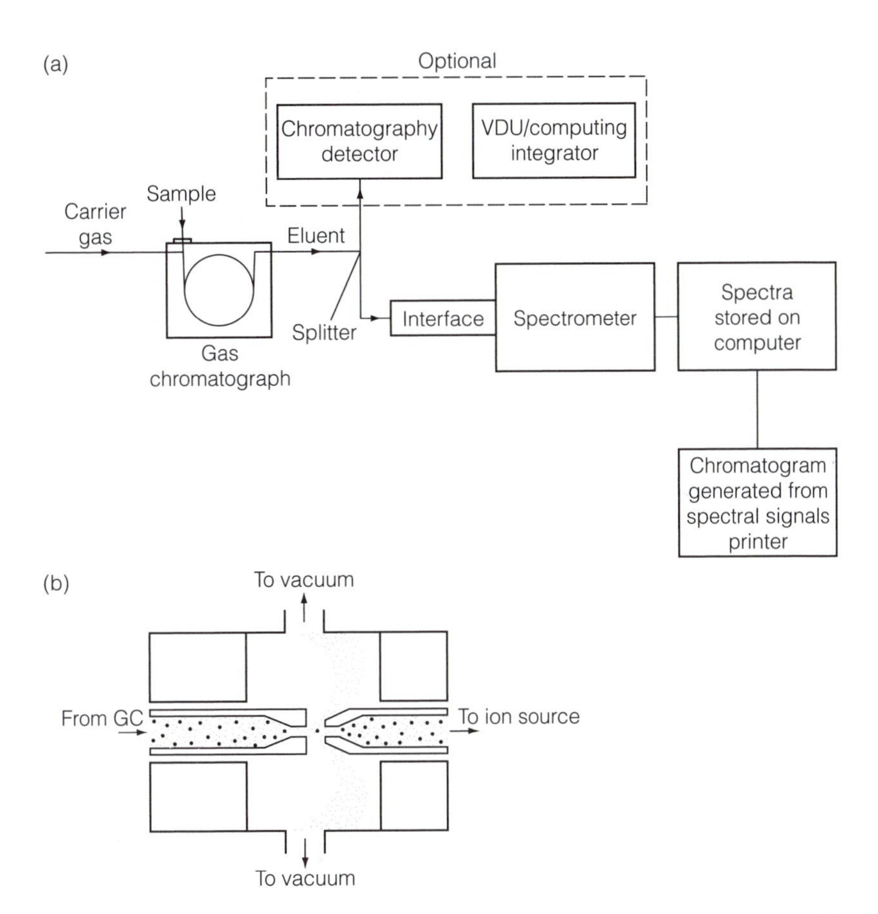

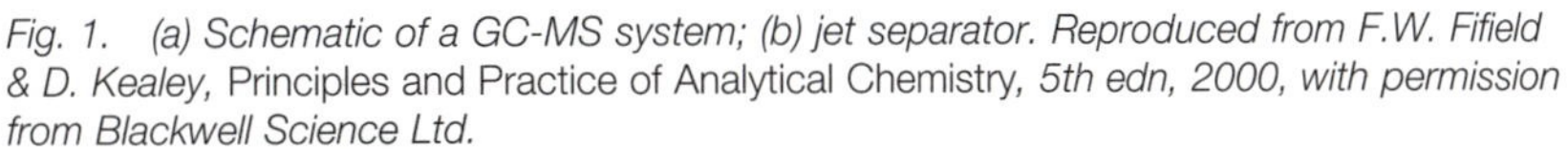

Fig. 1. (a) Schematic of a GC-MS system; (b) jet separator. Reproduced from F.W. Fifield & D. Kealey, Principles and Practice of Analytical Chemistry, *5th edn, 2000, with permission from Blackwell Science Ltd.*

and is removed by pumps, while the larger sample molecules continue through into the MS. A complete GC-MS system is shown in *Figure 1(a)*. The interface should be maintained at the temperature of the GC outlet, usually by enclosing it in the GC oven.

Most types of mass spectrometer are suitable for GC-MS work, although those with a quadrupole analyzer (Topic E14) are very often used because of their ability to scan rapidly.

The chromatogram will be recorded by the GC detector and data system, but can also be derived by continuously measuring the **total ion current (TIC)** for the ions generated as a function of the elapsed time. This total ion current chromatogram matches that from the GC detector, and may also detect other solutes. By selecting a particular mass/charge (m/z) ratio, **selected ion monitoring (SIM)** may be used to detect a particular ion; for example, m/z 320, 322 and 324 may be studied in analyzing mixed dioxins, since these ions are characteristic of a particular tetrachlorodibenzo-*p*-dioxin. Detection down to 10^{-15} g of a solute is possible.

Applications

Both qualitative and quantitative analysis can be achieved with GC-MS and it is now used widely on complex samples of all types.

Mixtures of fragrances are readily analyzed. *Figure 2* shows the TIC chromatogram of a mixture of six fragrance components, injected as 5 μl of a solution in ether.

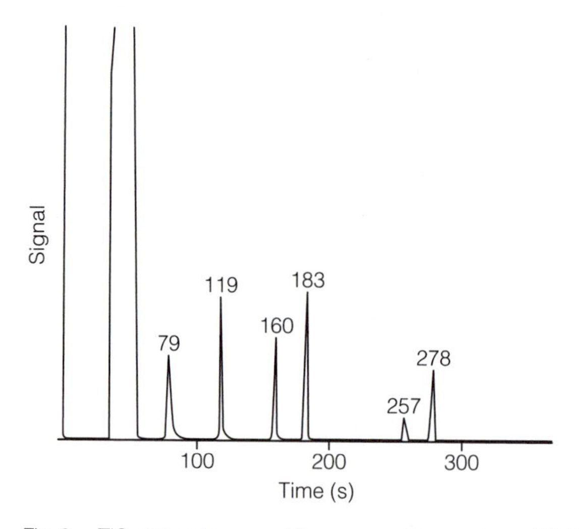

Fig. 2. TIC chromatogram of fragrance mixture using a BP-1, nonpolar capillary column, direct injection and MS.

The large peak at low retention time is the solvent. The solute peaks are readily identified, either by consideration of their retention index and mass spectral fragmentation patterns, or directly by computer searching.

The mass spectrometer scan at 160 seconds, which corresponds to a boiling point of about 220°C, gave the major peaks listed in *Table 1*.

Table 1. Mass spectrometric peaks for peak at 160 s

m/z	Relative intensity %	Assignment
39	55	$C_3H_3^+$
65	30	$C_5H_5^+$
92	90	$(M^- CH_3COOH)^+$
120	100	$(M^- CH_3OH)^+$
152	50	M^+, $C_8H_8O_3$

This suggests a methyl ester of an aromatic acid, and the spectrum matches the mass spectrum of **methyl salicylate**, present in oil of wintergreen (*Fig. 3*). Peaks 5 and 6 give almost identical mass spectra and may be identified as isomers of **eugenol**, present in oil of cloves. The higher boiling **isoeugenol** is probably the peak at 278 s.

This spectrum does not correspond to any likely essential oil or fragrance, but may be identified as a **contaminant**. Many of the fragment ions are separated by a constant mass difference of 14. This suggests a long-chain aliphatic compound, possibly a hydrocarbon. A molecular ion at *m/z* 156 would correspond to $C_{11}H_{24}$, and this component matches the spectrum of *n*-**undecane**.

Other applications include the analysis of alcoholic drinks, such as whisky, of pesticides from environmental samples (e.g., chlorinated pesticides from marine

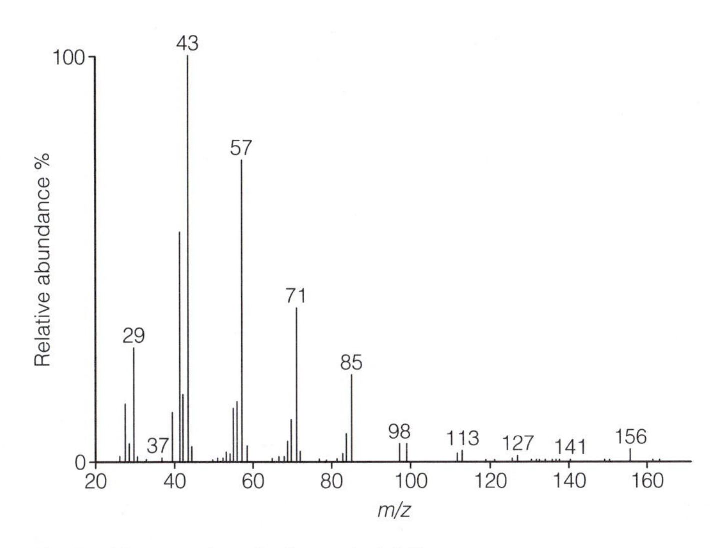

Fig. 3. Mass spectrum for the peak at 119 s.

animals) and of pyrethrin insecticides using field ionization MS. Pharmaceutical samples such as antibiotics can be studied, but soft ionization techniques may be needed.

One example of the ability of this combined technique to analyze complex mixtures is provided by the use of GC-MS for a sample of mineral oil with over 50 components. Accurate identification of peaks from amounts of solutes ranging from 5 ng to 200 ng was possible. The pyrolysis of polymers, followed by separation by gas chromatography and identification by mass spectrometry (**Py-GC-MS**) extends the combination to include three distinct techniques.

F4 GAS CHROMATOGRAPHY-INFRARED SPECTROMETRY

Key Notes

Principles

The use of chromatographic techniques to separate mixtures is one of the most important analytical tools. The separated components may be classified by their retention times, but other techniques should be used to aid identification. Infrared spectrometry is capable of establishing which functional groups are present in the separated components.

Instrumentation

Combining gas chromatography with infrared spectrometry involves passing the solutes in the carrier gas stream through a heated infrared gas cell positioned in a rapid scanning Fourier transform spectrometer.

Applications

Gas chromatography-Fourier transform infrared spectrometry has been employed in the analysis of biological materials such as fragrances, to determine the proportions and nature of each component, of solvents to determine their purity and composition, and to identify the evolved products when substances are degraded by heating.

Related topics

Gas chromatography: principles and instrumentation (D4)
Gas chromatography: procedures and applications (D5)
Infrared and Raman spectrometry: principles and instrumentation (E10)
Infrared and Raman spectrometry: applications (E11)

Principles

Chromatographic techniques (Section D) are used for the separation of multi-component mixtures. However, identification of the separated solutes by their retention times alone is often ambiguous. Infrared spectrometry (Topics E10 and E11) is a very powerful and versatile technique for the identification of functional groups. The newer instruments, using computerized Fourier transform (FT) processing of spectral information, are readily interfaced with gas chromatographs using heated transfer lines and gas cells which facilitate the rapid detection and identification of unknown compounds.

Instrumentation

The choice of **gas chromatography** requires that the components must be volatile enough to pass through the chromatographic column and sufficiently different in properties to be resolved by the stationary phase. In a simple case, a difference of a few degrees in the boiling point is enough to give separation. All the many columns, stationary phases, gases and temperature programs described for GC can be adapted for use with GC-FTIR.

The choice of column is important. The greater sensitivity of modern FTIR systems and improvements in design have allowed the use of capillary columns, which produce sharper peaks. Typically, a 30 m long, 0.3 mm diameter fused silica column coated with a 1-μm thickness of stationary phase may be employed. A nonpolar stationary phase (e.g. a silicone), which separates compounds chiefly according to their boiling points, is particularly useful.

Since the GC oven is often programmed to fairly high temperatures, sometimes to 400°C, it is essential that the stationary phase is stable and does not 'bleed' off into the detector or spectrometer. Columns where the stationary phase has been chemically bonded to the column wall minimizes this problem.

Infrared spectrometry is usually carried out at atmospheric pressure, unlike mass spectrometry. This allows direct transfer of the separated components between the gas chromatograph and the IR spectrometer through a simple heated tube as an **interface**. This is shown schematically in *Figure 1*.

GC detectors, such as the flame ionization detector (FID) are extremely sensitive. By contrast, the infrared spectrometer is far less sensitive. The gas stream is therefore usually split at the column exit so that about 90% goes on to the spectrometer, and only 10% to the FID. The major portion is transferred through the interface, usually a short heated tube, often an inert piece of GC capillary, into the FTIR spectrometer. To obtain the maximum response it is necessary to do one of two things: either (i) the sample must be concentrated by condensation onto a cooled surface, or by absorption onto an active solid, or (ii) if the sample is to remain gaseous, it must be passed through a small volume cell, or **light pipe**, which gives a long path length to maximize the absorption of IR radiation.

Both these alternatives have their advantages. A condensed sample gives the spectrum normally obtained by conventional IR sampling (mulls, KBr disks, melts). The vapor phase spectrum will be rather different, since less intermolecular interactions, especially hydrogen bonding, can occur. However, the vapor sample is readily and rapidly removed from the cell, and may even be collected for further investigation.

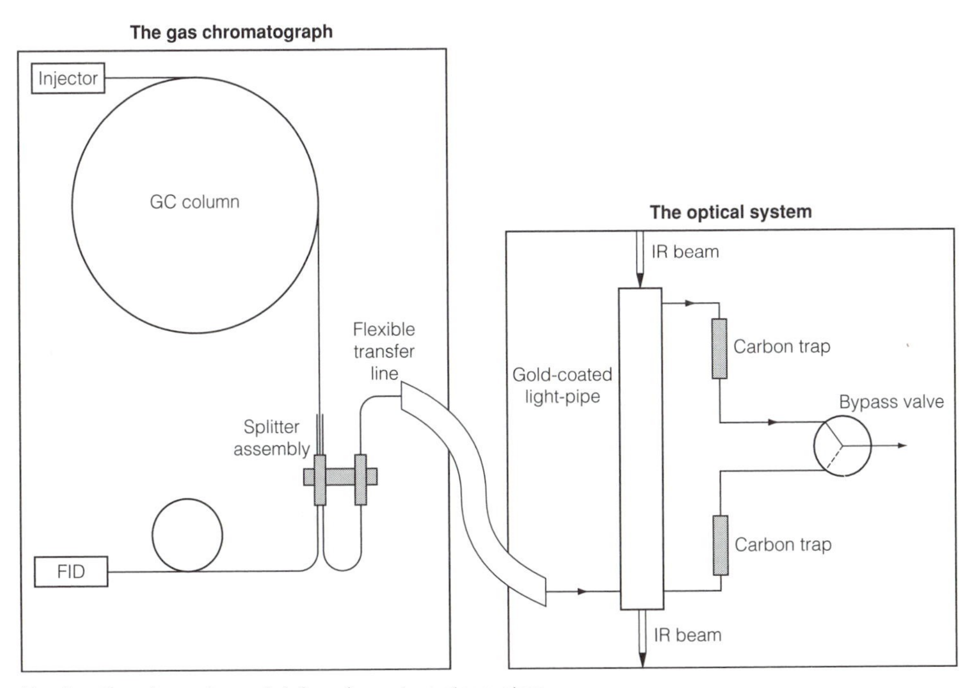

Fig. 1. Gas chromatograph-infrared spectrometer system.

Applications

A gas chromatogram obtained from a few microliters of a 12-component mixture is shown in *Figure 2*.

As with infrared spectra obtained by conventional sampling methods, the IR absorption bands correspond to characteristic vibrational frequencies of particular functional groups. Vapor-phase spectra do not show the effects of strong hydrogen bonding, such as the broadening of the –OH stretching peak around 3600 cm^{-1}.

Peak 1, representing less than 1% of the total sample (*Fig. 3*), shows the characteristic group frequency bands for

(i) free hydroxyl stretch at 3600 cm^{-1}, indicating an alcohol;
(ii) aliphatic CH stretch at 2900–3000 cm^{-1}, suggesting an aliphatic compound;
(iii) C–O stretch at around 1050 cm^{-1} confirming the alcohol.

Computer matching shows that this compound is 2-methyl propanol (*iso*-butanol).

A mixture of esters might be characterized by monitoring the absorbance due to the -CO- peak around 1700 cm^{-1}, and a mixture of aromatic compounds by selecting the aromatic CH- stretch peaks just above 3000 cm^{-1}. By plotting the

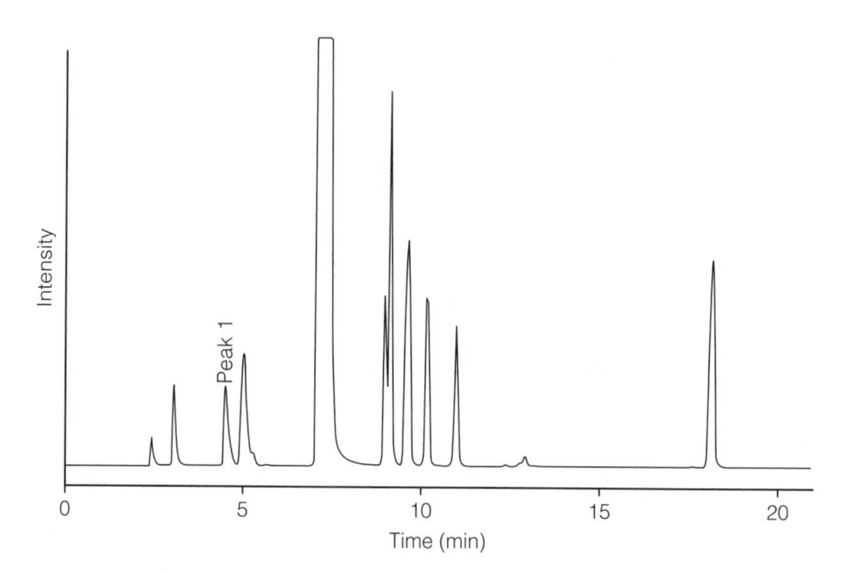

Fig. 2. Gas chromatogram of an unknown mixture.

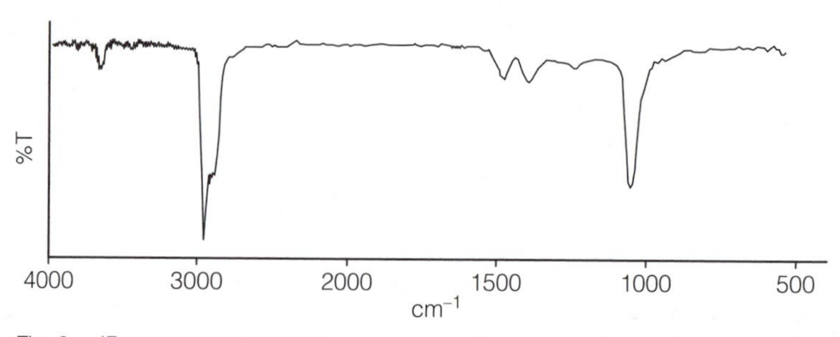

Fig. 3. IR spectrum of peak 1.

absorbance of a particular peak in the spectrum, or integrating the total IR absorbance over a range, an **infrared chromatogram** may be generated, showing how each component relates to the IR sensitivity.

Example

A sample of the herb basil was extracted with a solvent. The extract was then injected into the GC, which had a 50 m capillary column with a methyl silicone coating maintained at 40°C. The chromatogram showed 7 major peaks and over 20 minor peaks. Each major peak gave a good IR spectrum, and these were identified as eucalyptol, estragol, eugenol, and linalool, plus various terpenes and cinnamates. Many of the minor components were also identified.

F5 LIQUID CHROMATOGRAPHY-MASS SPECTROMETRY

Key Notes

Principles

The components of a mixture, after separation by liquid chromatography, may be identified and quantified by mass spectrometry.

Instrumentation

The removal of the liquid mobile phase, while allowing the analytes to be transferred to the mass spectrometer has presented difficulties, and the design of the interface is critical.

Applications

The analysis of mixtures of pharmaceuticals and drugs, the detection of degradation pathways using isotopic labeling, and the separation and analysis of peptides using soft ionization methods are typical of the application of LC-MS.

Related topics

High-performance liquid chromatography: modes, procedures and applications (D7)

Mass spectrometry (E14)

Principles

The wide variety of modes of liquid chromatography available and the separations that may be achieved have been described in Topic D7. Since these employ liquid mobile phases, sometimes containing inorganic salts, the most difficult problem is how to transfer the separated component of the analyte to the mass spectrometer without interference from the solvent.

Materials of high relative molecular mass are readily separated by liquid chromatography and, consequently, ionization methods that produce less fragmentation in the mass spectrometer may have to be employed.

Instrumentation

The **interface** between the liquid chromatograph and the mass spectrometer is the most vital part of the combined instrument. Early systems using a moving belt interface have been superseded by spray devices and interfaces, which operate near atmospheric pressure.

In **atmospheric pressure chemical ionization (APCI)** interfaces, nitrogen is introduced to nebulize the mobile phase producing an aerosol of nitrogen and solvent droplets, which are passed into a heated region. Desolvation occurs, and ionization is achieved by gas-phase ion-molecule reactions at atmospheric pressure, electrons and the primary ions being produced by a corona discharge.

Since the pressure is close to atmospheric, the collision frequency is high and **pseudomolecular ions**, $(M + H)^+$ and $(M - H)^+$, are formed with high efficiency by chemical ionization. The analyte ions are accelerated into the mass spectrometer and the uncharged solvent molecules are removed by vacuum pumps.

In the **electrospray (ES)** interface, also operating at atmospheric pressure, the liquid mobile phase is ejected from a metal capillary tube into an electric field obtained by applying a potential difference of 3–6 kV between the tube and a counter electrode. The drops accumulate charge on their surface, and as they shrink by evaporation they break into ever smaller charged droplets. The uncharged solvent molecules are pumped away and the charged ions pass into the mass spectrometer. A schematic is shown in *Figure 1*.

These interfaces can deal with a wide range of solvent polarities, although for ionic mobile phases the electrospray is preferable.

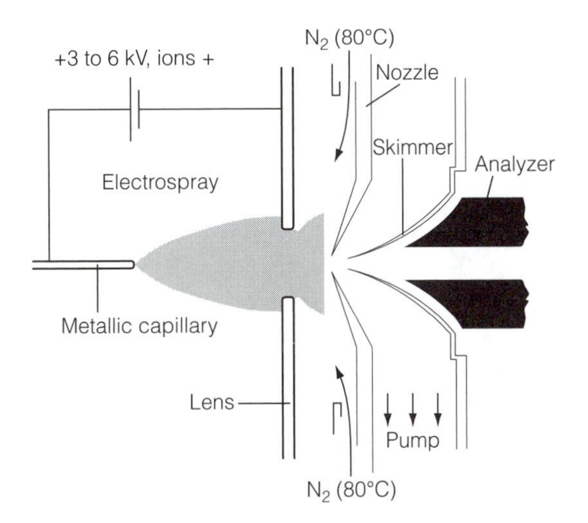

Fig. 1. Electrospray (ES) interface. Reproduced from J. Barker Mass Spectroscopy, *2nd edn, 1999, with permission from Her Majesty's Stationery Office. Crown copyright.*

Applications

A typical example of LC-MS is the detection of impurities in synthesized drugs using an acetonitrile-aqueous ammonium ethanoate mobile phase and a reversed-phase column, coupled to a quadrupole mass spectrometer using an electrospray interface. The total ion current (TIC) chromatogram in *Figure 2(a)* shows that the parent compound elutes at 17.7 minutes, and impurity peaks occur before and after this time. The compound eluting at 8.35 minutes gave the mass spectrum shown in *Figure 2(b)*. The peaks due to $[M+H]^+$ occur at m/z = 225 and 227 in the ratio 3:1, indicating the presence of one chlorine atom. The compound was identified as a trisubstituted quinazoline.

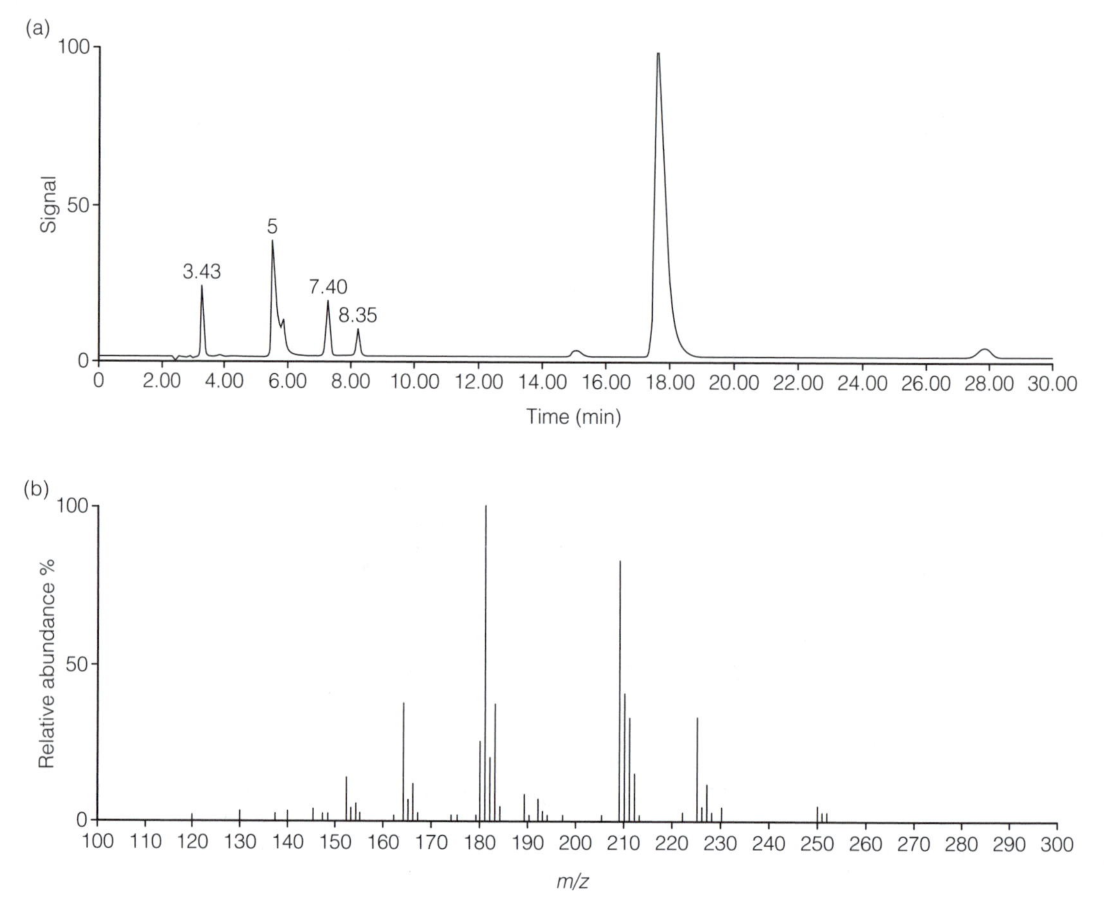

Fig. 2. (a) TIC chromatogram of drug sample; (b) mass spectrum of drug impurity.

G1 THERMOGRAVIMETRY

Key Notes

Principles

Thermal methods investigate changes that occur upon heating a sample. Thermogravimetry measures changes in the mass of a sample that occur when it is heated. These changes relate to the reactions during decomposition, the loss of volatile material and the reactions with the surrounding atmosphere.

Instrumentation

A crucible containing the sample is heated in a furnace at a controlled rate and weighed continuously on a balance. Temperature and mass data are collected and processed by a computer dedicated to the system. Control of the atmosphere surrounding the sample is important.

Proper practice

There are many factors that affect the results obtained by thermal methods. These must be carefully controlled and recorded.

Applications

Any physical or chemical change involving mass may be studied. Evaporation of volatile material, oxidation, and particularly decompositions of inorganic salts, organic and polymeric samples are investigated analytically.

Related topics

Gravimetry (C8)
Differential thermal analysis and differential scanning calorimetry (G2)
Thermomechanical analysis (G3)
Evolved gas analysis (G4)

Principles

One of the simplest tests that may be applied to an analytical sample is to heat it and observe the changes that occur. These may be color changes, burning, melting or a variety of other reactions. The group of techniques that has been developed to make analytical measurements during heating is given the general name **thermal analysis**. Any property change may be monitored, and *Table 1* lists the more important thermal methods.

While some analytical methods, such as spectrometry, give results that are very specific for the particular sample, thermal methods will respond to the totality of the effects. Anything that changes the mass at a particular temperature: evaporation, reaction or oxidation will affect the thermogravimetric measurement. It is sometimes an advantage to combine techniques, or to run two simultaneously (see Section F) to extract the maximum benefit from the analysis.

In thermogravimetry, the sample is heated, often at about 10°C min^{-1} in a thermobalance instrument as described below. Only those changes that affect the **mass** of the sample will affect the measurements, so that condensed phase changes such as melting or crystalline transitions cause no mass change. The **rate of change** of mass, dm/dt depends on the amount of sample present, and the **reaction rate constant** at the experimental temperature. With solids it is

Table 1. Principal thermal analysis techniques

Technique	Property	Uses
Thermogravimetry (TG) (thermogravimetric analysis, TGA)	Mass	Decompositions Oxidation
Differential thermal analysis (DTA)	Temperature difference	Phase changes Reactions
Differential scanning calorimetry (DSC)	Heat flow	Heat capacity Phase changes Reactions
Thermomechanical analysis (TMA)	Deformations	Softening Expansion
Dynamic mechanical analysis (DMA)	Moduli	Phase changes Polymer cure
Dielectric thermal analysis (DETA)	Electrical	Phase changes Polymer cure
Evolved gas analysis (EGA)	Gases	Decompositions
Thermoptometry	Optical	Phase changes Surface reactions

better to use the **fraction reacted, α,** instead of concentration, and for a simple (first order) reaction the rate may be written in terms of α:

$$d\alpha/dt = k(1 - \alpha) \quad (1)$$

where k is the rate constant at the experimental temperature and $(1 - \alpha)$ is the amount of sample remaining. It should be pointed out that many solid state reactions follow very complex kinetic mechanisms, and their rate equations are much more complex than this.

Any chemical change speeds up as the temperature is raised. This is most simply represented by the Arrhenius equation:

$$k = A \exp(-E/RT) \quad (2)$$

where A is called the **pre-exponential factor**, E is the **activation energy**, R is the molar gas constant and T the thermodynamic temperature in Kelvin.

This expression shows that as the temperature increases, the rate constant also increases exponentially. Combining the effects of equations (1) and (2), the result is that a solid state reaction will start very slowly at low temperature, speed up as the temperature is raised, and then slow down again as the reactant is used up. This produces a **sigmoid** curve as shown in *Figure 1*. It also means that it is difficult to quote with accuracy a single **decomposition temperature**. It is probably better to give the range of temperature over which a reaction occurs or even to quote the temperature at which only a small fraction, say 0.05%, has decomposed.

If only a single mass loss occurs, for example, when the amount of moisture in a soil or a polymer is measured, then the evaluation is simple. A mass loss of 1.5% when a sample of nylon was heated from room temperature to 130°C at 10°C min^{-1} measures the percentage moisture in the nylon, assuming that the loss is due only to moisture, not to any other solvent or other volatile material.

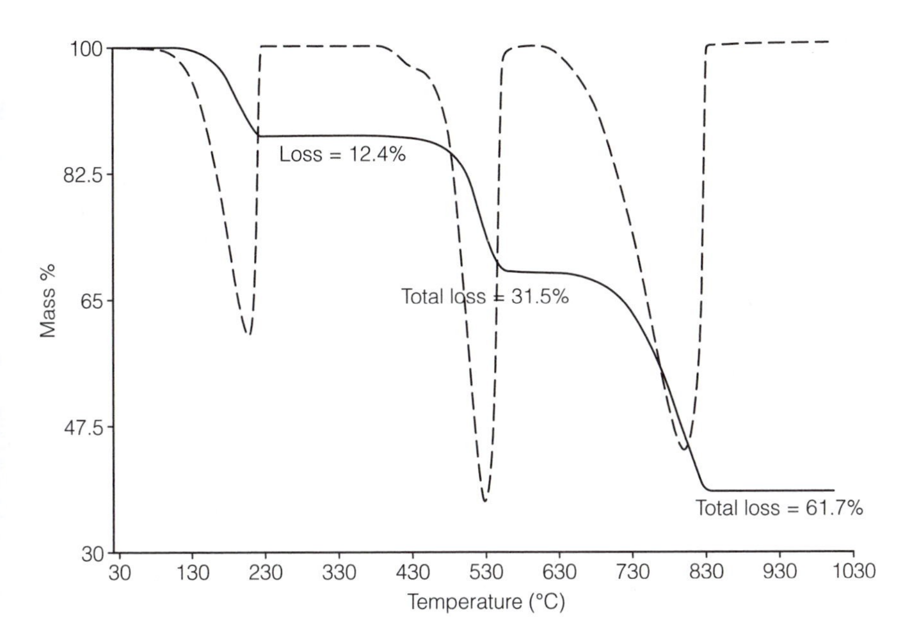

Fig. 1. TG and DTG curves for calcium oxalate monohydrate, 12.85 mg, platinum crucible, 20°C min^{-1}, nitrogen, 30 cm^3 min^{-1}.

When a complex mass loss occurs, involving several reactions, matters become more complicated.

Calcium oxalate monohydrate, $CaC_2O_4.H_2O$, a well-studied example, is shown in *Figure 1*. Three separate stages occur, around 150, 500 and 750°C. The mass losses are about 12, 19 and 30% of the original mass in the three stages. This can be explained as follows:

$CaC_2O_4.H_2O(s)$	=	CaC_2O_4 (s)	+	H_2O (v)	
146.1		128.1		18	loss = 12.3%
CaC_2O_4 (s)	=	$CaCO_3$ (s)	+	CO (g)	
		100.1		28	loss = 19.2%
$CaCO_3$(s)	=	CaO (s)	+	CO_2 (g)	
		56.1		44	loss = 30.1%

These reactions have been confirmed by analyzing the gases evolved (see Topic G4). Note that, in this example, the mass losses are calculated as a percentage of the **original** sample mass.

In more difficult cases, the reactions may overlap, and then it is difficult to assess the separate temperature ranges and mass losses. An aid to this is the **derivative thermogravimetric (DTG)** curve shown as a dashed line in *Figure 1*. This is produced electronically from the TG trace by the computer and represents the dm/dt, or occasionally the dm/dT, as a function of time or temperature.

Instrumentation

In general, thermal analysis instrumentation consists of four components:

- the furnace, controlled by the computer and a temperature sensor, often with controlled atmosphere as well;

- the sample and its container;
- the sensors for measuring temperature and the sample property;
- the computer, data collection and processing equipment and a display device for the results.

An electrical furnace is often used and the furnace enclosure is purged with a suitable gas, either to provide an inert atmosphere, or to provide a reactive environment where the sample may be burnt, or reacted. The temperature is measured by a thermocouple or resistance sensor and the heating is controlled by the system computer (*Fig. 2*).

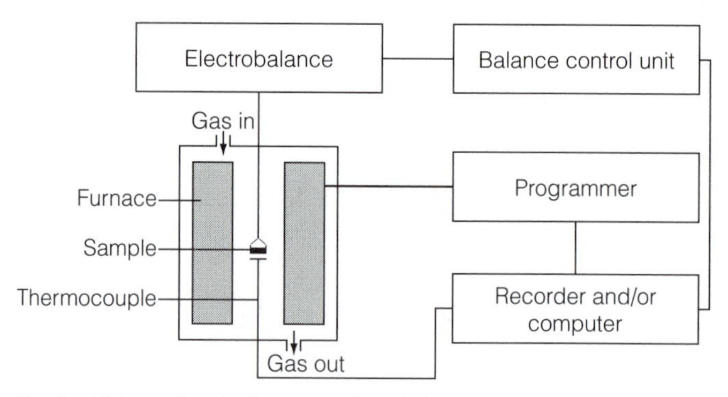

Fig. 2. Schematic of a thermogravimetric instrument.

The analytical sample is contained in a suitably inert crucible. This may be made of alumina, platinum or ceramic. A sample size of around 10 mg is fairly typical. The sensor for measuring temperature is most often a thermocouple suitable for the temperature range to be studied. For measuring the mass, a **thermobalance** is used. This is a sensitive electronic balance sufficiently far away from the furnace to avoid all effects of heat and any corrosive gases produced by the sample and capable of detecting changes as small as 1 μg and of weighing samples of 10–100 mg. The balance is purged with dry nitrogen for protection.

Temperature calibration cannot be carried out by the normal IPTS standards since these involve no mass change. A method has been devised using the Curie point (T_{Ci}) of metals. This is the temperature above which they have no ferromagnetism. For example, for nickel metal, the Curie point temperature is 353°C. By placing a magnet near a calibration sample of nickel, the mass is increased at temperatures below T_{Ci}, but not above that temperature. A step occurs in the TG trace that allows calibration.

Proper practice

Since the experimental conditions have a profound effect on the results obtained by thermogravimetry, and also other thermal methods, it is as well to establish a set of rules to follow in order to obtain the most reproducible results, or to recognize why runs differ.

The acronym **SCRAM,** standing for **S**ample, **C**rucible, **R**ate of heating, **A**tmosphere, and **M**ass of sample is a useful reminder of the things to remember and report:

- the **sample**, its source, history and chemical nature;

- the **crucible** or container, its size, shape and material;
- the **rate of heating** and any special program;
- the **atmosphere**, and whether static or flowing;
- the **mass** of the sample, and its properties, e.g. particle size.

One example may illustrate this. Several thermogravimetric experiments were carried out on a polymeric sample. One was very different from the others. This was traced to the use of a copper crucible in an oxidizing atmosphere and the consequent formation of copper oxide above 500°C.

Applications

The analysis of the thermal decomposition of inorganic salts and complexes is an important part of the study of catalysts, semiconductors and fine chemicals. The decomposition of barium perchlorate, $Ba(ClO_4)_2 \cdot 3H_2O$ was investigated by thermogravimetry and other techniques. The TG curve is shown in *Figure 3*.

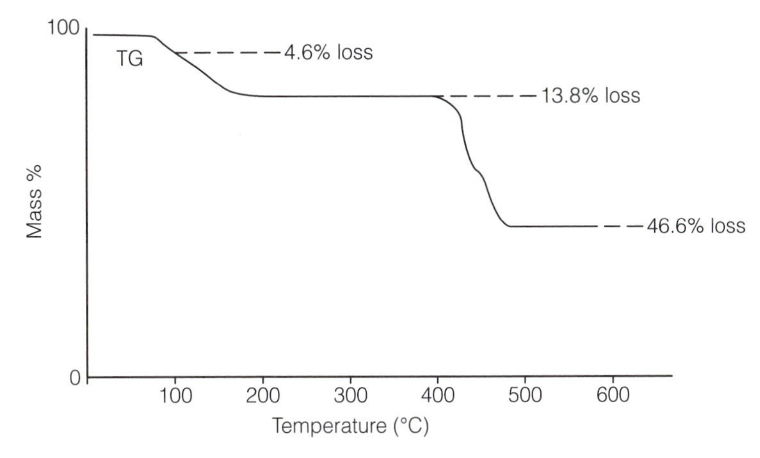

Fig. 3. Thermal decomposition of barium perchlorate, powder 30 mg, platinum dish, 10°C min^{-1}, nitrogen, 10 cm^3 min^{-1}.

There are clearly two losses, both occurring in two steps. The first, around 100°C, might well be loss of hydrate water, which is substantiated by the calculated loss of 13.8%, agreeing with that found by experiment. The major loss of a further 32.8% (of the original mass) near 450°C agrees with the formation of barium chloride:

$$Ba(ClO_4)_2 = BaCl_2 + 4O_2$$

Polymer stability is very important, and the decomposition temperatures of commodity plastics are often investigated by TG. The mechanism of decomposition varies between different polymers, as does the temperature at which this occurs. For example, polyethylene and polypropylene degrade completely in a single step between 150 and 450°C, but polyvinyl chloride (PVC) shows two steps, the first with a loss of 60%, by about 400°C. Cellulose and polyacrylamide both show several steps.

The analysis of rubber may be carried out by TG, as shown in *Figure 4*. The analysis is started in a nitrogen atmosphere. The low temperature loss to 200°C is due to the oil content of the rubber, whereas the polymer rubber is lost before

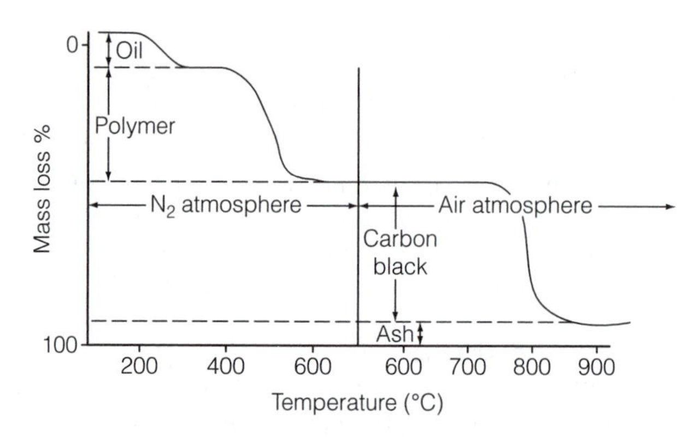

Fig. 4. TG curve for carbon-black filled rubber.

600°C leaving carbon black. The atmosphere is now changed to air, when the carbon black is oxidized away leaving a residue of ash.

The analysis of other materials such as pharmaceuticals, coal and minerals by thermogravimetry has proved a valuable tool in the study of complex thermal events.

The DTG curve permits the analyst to obtain additional information since:

- the magnitude of the DTG signal is directly proportional to the rate of reaction, it allows the comparison of reaction rates; and
- the DTG curve separates overlapping reactions more clearly than the TG curve.

G2 DIFFERENTIAL THERMAL ANALYSIS AND DIFFERENTIAL SCANNING CALORIMETRY

Key Notes

Principles

Both of these methods relate to the monitoring of the heat absorbed or evolved during the heating of a sample and a reference in equivalent environments. Differential thermal analysis (DTA) monitors temperature difference, while differential scanning calorimetry (DSC) measures the power supplied.

Instrumentation

The measurement unit has a matched pair of temperature sensors placed in or near the sample and reference pans and is heated in a temperature-controlled furnace.

Physical properties and changes

Since heat capacity relates to the quantity of heat required to raise the sample temperature by one Kelvin it may be studied by DTA and DSC. Physical changes such as melting and vaporization as well as crystal structure changes give peaks, and some may be used to calibrate the system.

Chemical reactions

Heating chemical substances may cause decompositions, oxidations or other reactions. The temperatures at which these occur and the nature and rates of the reactions are studied by these methods.

Applications

Both DTA and DSC are used to study pure chemicals, mixtures such as clay minerals and coal, biological samples, pharmaceuticals and especially polymers and materials.

Related topics

Thermogravimetry (G1)
Thermomechanical analysis (G3)
Evolved gas analysis (G4)

Principles

If an inert sample, such as alumina, is heated at a constant rate of 10°C min^{-1}, the temperature-against-time curve is practically a straight line. A sample that reacts or melts within the temperature range studied will give small changes on its temperature-time curve. By heating both a reactive sample and an inert reference together at the same rate, these small differences may be detected and amplified as a function of temperature. The simplest example is the melting of a crystalline solid. If 10 mg of metallic indium are heated as sample and a similar amount of alumina as reference, both heat at nearly the same rate until around 156°C the indium starts to melt. This absorbs energy and the temperature of the indium rises less fast. This goes on until all the indium has melted when the temperatures of the liquid indium and alumina again rise at the same rate.

Two alternative strategies can now be adopted. If the temperatures of sample S and reference R are measured and the temperature difference recorded (the **differential thermal analysis** or **DTA** strategy),

$$\Delta T = T_S - T_R$$

a downward peak (i.e. a minimum) is recorded. Under carefully controlled instrumental conditions, this may be related to the enthalpy change for the thermal event:

$$\Delta H = -K\int_i^f \Delta T dt = -KA$$

where A is the area of the temperature-time peak from initial (i) to final (f) point.

This leads to quantitative or **heat-flux differential scanning calorimetry** (heat-flux DSC). The negative sign is required since the enthalpy change on melting is positive, but ΔT for melting is negative.

The second strategy is to control the amount of heat supplied to sample and reference so that their temperatures stay as nearly the same as possible. Using separate heaters for sample and reference allows measurement of the difference in power ΔP to be measured. With proper control and calibration, this will give the enthalpy change of the peak directly:

$$\Delta H = \int \Delta P \mathrm{d}t$$

This is known as **power-compensated DSC**. Essentially, it has been demonstrated that both strategies produce equivalent results with similar accuracy.

Instrumentation

The schematic diagram of the apparatus for DSC/DTA is shown in *Figure 1*. The temperature, both for the sample and the reference and also the furnace is measured by thermocouples, or resistance sensors. Higher sensitivity and greater stability are obtained if multiple sensors of inert material are used.

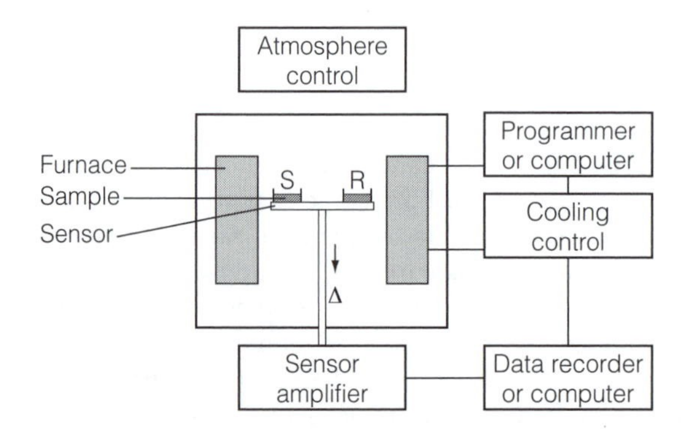

Fig. 1. Schematic diagram of DTA or DSC apparatus (Δ represents either the temperature or power difference).

The factors that influence the results of thermal methods (see SCRAM in Topic G1) are considered below.

- **The sample** is generally about 10 mg of powder, fibers or reactants such as monomers for plastic production. These are placed into the **crucible,** which should be unreactive and stable over the temperature range used. Platinum, aluminum, silica or alumina crucibles are commonly used. The sample and reference pans (either with alumina powder or sometimes an empty pan) are placed in their holders within the furnace, generally a wire-wound electrical heater controlled by the computer program.
- **The rate of heating** is user-determined, often about 10 K min^{-1}, but for the best approach to equilibrium, low heating rates are needed, and isothermal experiments may also be carried out. High heating rates save time, and can simulate situations like burning, but they tend to raise the temperature of recorded events.
- **The atmosphere** surrounding the samples can be controlled. A slow flow of nitrogen gas will give an almost inert atmosphere and sweep away harmful products. Oxygen may be used to study the oxidative stability of polymers. Carbon dioxide will react with some oxides to form carbonates.
- **The mass** of the sample, together with its volume and packing is important since these determine the heat transfer and the diffusion of gases across the sample.
- The computer records the values of ΔT or ΔP and of the temperature T and time t. Computer software has been designed to correct the temperature by calibration, to measure peak areas and onset points and to calculate reaction parameters.

Physical properties and changes

Typical physical changes in a sample (of a polymer) are shown in *Figure 2*. The first part of the curve shows a small deflection due to the heat capacity of the solid, glassy polymer. Around 80°C, the material changes to a rubbery nature, and its heat capacity increases. This is the **glass transition, T_g**. At about 120°C, the molecules in the polymer may move freely enough to form the crystalline polymer and so an **exothermic peak** for cold crystallization is observed. This form is stable until it melts at about 250°C giving an **endothermic peak**.

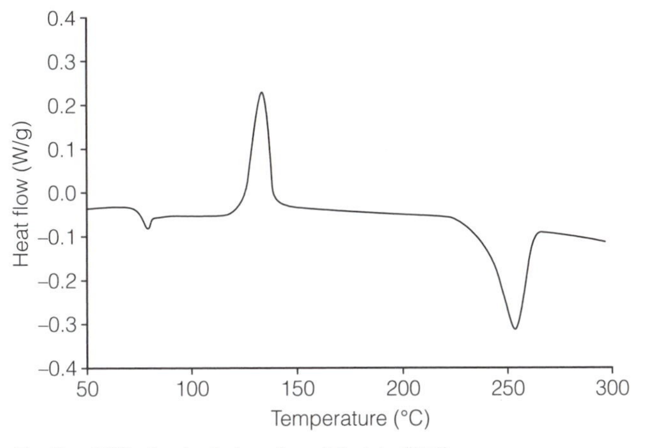

Fig. 2. DSC of polyethylene terephthalate (PET).

Calibration of DTA and DSC instruments is usually carried out using standards with well-characterized and tested transition temperatures and enthalpies of reaction; for example, the melting of indium occurs at 156.6°C and absorbs 28.7 J g^{-1}, while zinc melts at 419.4°C and absorbs 111.2 J g^{-1}.

Chemical reactions

The action of heat on samples has long been a useful 'dry test' to determine some of the qualitative characteristics of the material. Dehydration occurs with hydrates and with materials such as cellulose. Chemical decomposition and gas evolution are observed with carbonates, sulfates and nitrates. Occasionally, explosive reactions happen and all of these can be characterized by DTA and DSC.

Kaolinite is the pure form of white china clay. It is a hydrated aluminosilicate and found naturally. The small peak at low temperature (*Fig. 3*) is due to loss of hydrated moisture. Around 500°C the strongly bound hydroxyl groups release water producing a large, broad endotherm. When the temperature reaches 1000°C, the silica and alumina react exothermically to form crystalline mullite, $3Al_2O_3.2SiO_2$.

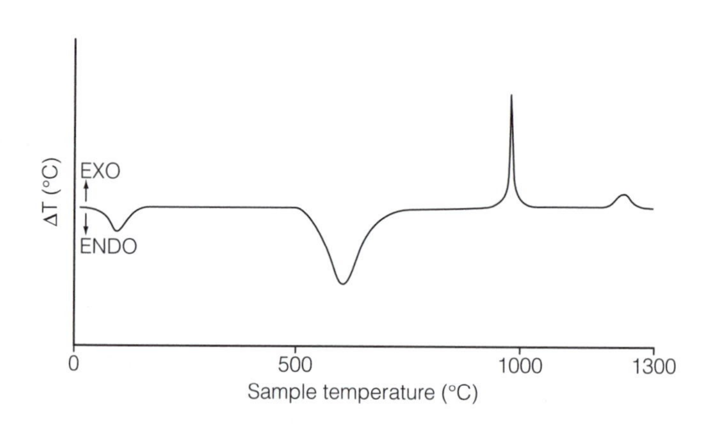

Fig. 3. DTA curve of kaolinite.

Applications

Although it must be noted that, like many thermal methods, DTA and DSC are *not* compound-specific, they are still most important test methods for a wide variety of disciplines and materials.

Inorganic materials, salts and complexes have been studied to measure their physical properties, chemical changes and qualitative thermal behavior. Minerals and fuel sources such as coal and oil have been examined and when new materials (e.g. liquid crystals) are discovered, DSC is frequently used to test them. However, by far the greatest use is made of these techniques in the pharmaceutical and polymer industries. One special use of DSC for physical changes is the determination of **purity**. While a pure substance melts sharply, perhaps over a few tenths of a degree near its true melting point, an impure sample may start to melt several degrees below this temperature, and will give a broad peak. Computer analysis of the shape of this peak allows an estimation of purity, but does not provide any information on the nature of the impurities.

Many studies of inorganic complexes, of polymer degradations and reactions between samples and reactive gases, have been followed by DTA and DSC. The oxidation of polyethene is tested by heating samples in oxygen or holding them isothermally at around 200°C and then changing the surrounding atmosphere to oxygen and noting the time at which oxidative reaction starts. This is a most useful test for blue polyethene water pipes.

G3 THERMOMECHANICAL ANALYSIS

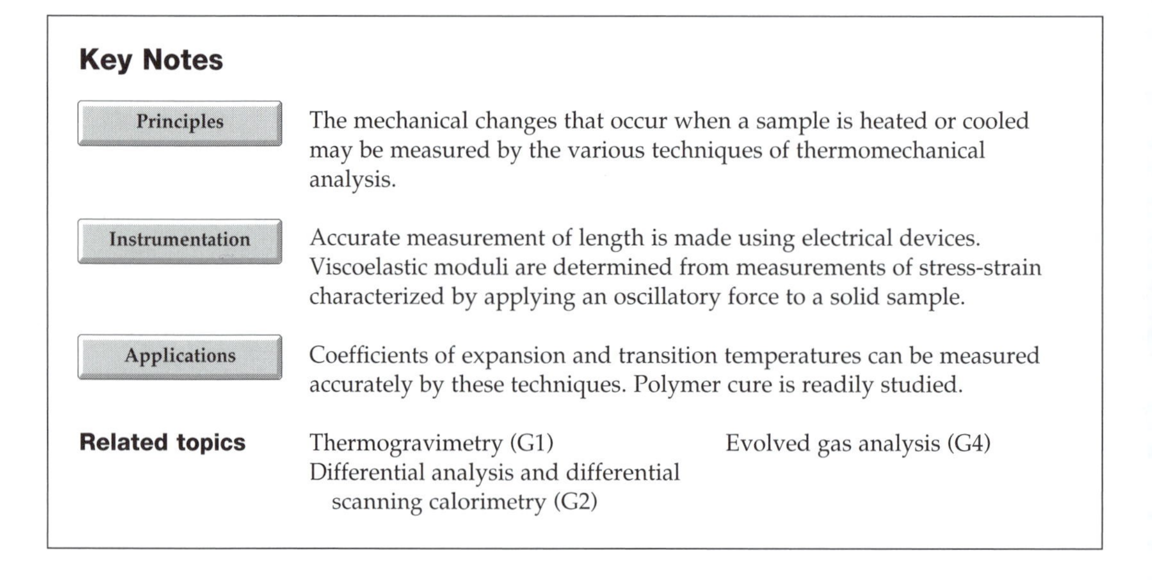

Key Notes

Principles	The mechanical changes that occur when a sample is heated or cooled may be measured by the various techniques of thermomechanical analysis.
Instrumentation	Accurate measurement of length is made using electrical devices. Viscoelastic moduli are determined from measurements of stress-strain characterized by applying an oscillatory force to a solid sample.
Applications	Coefficients of expansion and transition temperatures can be measured accurately by these techniques. Polymer cure is readily studied.
Related topics	Thermogravimetry (G1) Differential analysis and differential scanning calorimetry (G2) Evolved gas analysis (G4)

Principles

When any solid sample is heated, it will expand. The **coefficient of linear expansion**, α, is given by:

$$\alpha = (1/l)\mathrm{d}l/\mathrm{d}T \sim (l - l_o)/((T - T_o) \cdot l_o)$$

where l is the length, l_o the initial length and T and T_o the temperature and initial temperature. The coefficient of expansion is not constant with temperature, and whenever a phase change occurs, such as from one crystal form to another, it will change. This is very important with polymer and glassy samples, which are brittle below the **glass transition temperature**, T_g, but pliable and resilient above that temperature. This is shown in *Figure 2*. The measurement of length, or generally of dimensions such as volume, is also referred to as **dilatometry**.

When a sample is subjected to a force, F, it may behave in a variety of different ways. A large force, suddenly applied may break it, while smaller forces deform it, and liquid samples will flow. Elastic deformations are reversible and the sample returns to its original shape when the force is removed. Above the elastic limit it may undergo irreversible plastic deformation into a new shape.

For a true elastic material, the behavior is described by the elastic (or Young's) modulus, E,

$$E = (\text{stress}/\text{strain}) = (F/A)/(\Delta l\,/l)$$

where F is the applied force, A the cross sectional area, l the length and Δl the change in length measured.

Most materials, such as polymers, metals and glasses, possess some elastic and some viscous properties, and are described as **viscoelastic**. This causes the situation to become complex, and the measured modulus, E^*, also becomes complex:

$$E^* = E' + iE''$$

where E' is called the **storage modulus**, E'' the **loss modulus,** and their ratio is the **loss tangent** with $i = \sqrt{-1}$:

$$\tan(\delta) = E''/E'$$

Without concentrating on the involved nature of the mechanics and mathematics, the measurement of the moduli and loss tangent allows the study of many important commercial samples.

Instrumentation

The term **thermomechanical analysis (TMA)** is usually used when the force applied is constant, whereas if the force is made to oscillate, the technique is **dynamic mechanical analysis (DMA).**

The furnace, computer, atmosphere control and temperature sensors are all similar to those described in Topics G1 and G2. For mechanical measurements the sample holders and sensors are very different.

For TMA, the length of the sample and the changes in length that occur during heating are measured by a **linear variable differential transformer (LVDT)**. The movement of the transformer core produces an electrical signal, sensitive to direction, and this signal is transmitted to the data system.

The sample is placed on a support within the furnace and contact made with the LVDT through a rod. The force is applied either directly, by adding a weight to the rod, or electrically using the LVDT. If a zero load is applied to the sample, the expansion is measured. A schematic diagram of this system is shown in *Figure 1(a)*.

In a DMA, the system may operate in a number of different ways, to study compression, shear, bending or torsion. The oscillatory force is applied by the mechanism, and the sample behavior affects the response of the system. The frequency of the oscillation may be altered, but is often about 1–10 Hz. From the calibration of the apparatus, the data may be processed to give the storage and loss moduli. One typical sample holder for DMA in the 3-point bending mode is shown in *Figure 1(b)*.

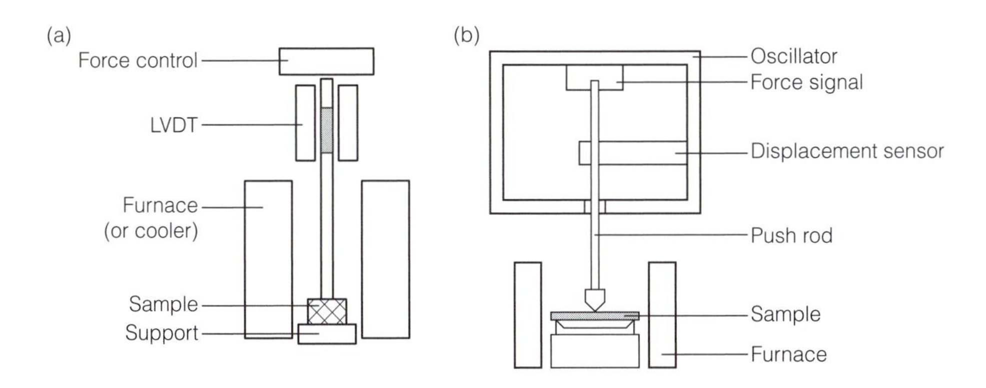

Fig. 1. (a) Schematic of a TMA system. (b) Schematic of a DMA system using 3-point bending.

Applications

The expansion of a plastic component may be studied by TMA. *Figure 2* shows that below 112°C, the coefficient of expansion is about one third of the value above this temperature. This measures the glass transition temperature, T_g, as 112°C. Note that the dimension of the sample is very small and the dimensional change even smaller.

Similar techniques can be used to study softening and penetration, swelling by solvent and sintering.

The DMA of a plastic component is illustrated in *Figure 3*. For many plastics, their structure causes characteristic low temperature transitions. Solvents or

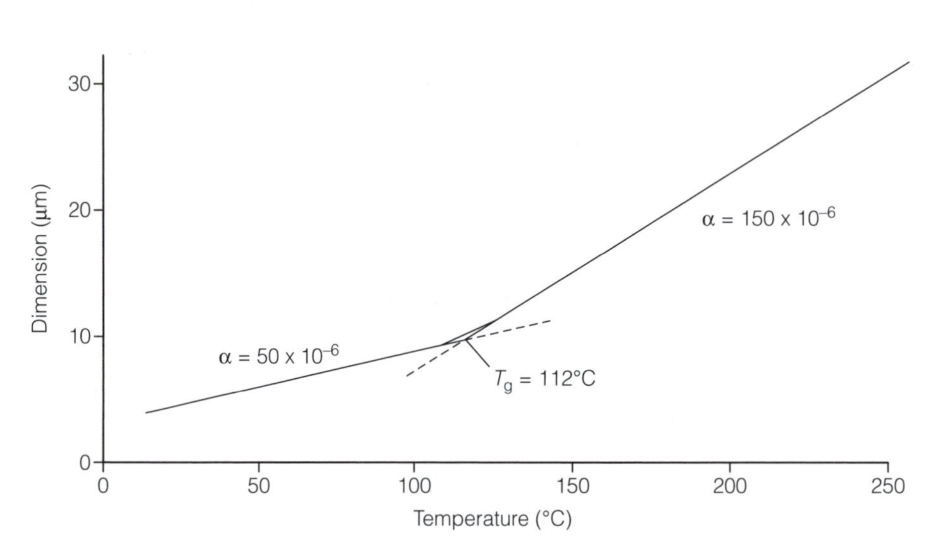

Fig. 2. TMA curve for printed circuit board.

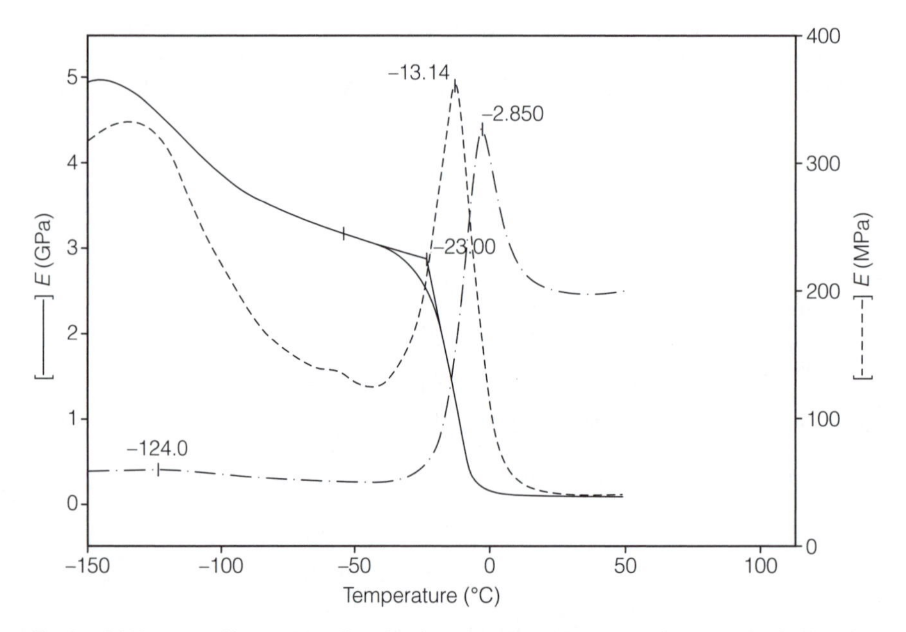

Fig. 3. DMA traces of latex rubber film showing a low-temperature transition at –124°C. The glass transition is at –13°C.

moisture may lower the temperatures of these transitions and this is easily observed by DMA experiments.

As the sample softens through the low temperature transition, the storage modulus decreases, while the loss modulus and loss tangent show broad peaks. At the glass transition, E' decreases greatly, while the loss modulus, E'' and $\tan(\delta)$ show a sharp peak showing the onset of irreversible viscoelastic behavior.

It should be noted that the sensitivity of the DMA technique is considerably greater than that of TMA or DSC in detecting transitions. The comparable techniques of **dielectric thermal analysis (DETA,** or **DEA)** produce similar results and cover a wider range of frequencies.

G4 Evolved gas analysis

Key Notes

Principles

When samples are heated, volatile products and gases may be evolved. The identification of these may be carried out by many analytical techniques. Solid products may be analyzed as well.

Instrumentation

Separation by chromatographic techniques and analysis by spectrometry, titration and potentiometry may all be used to identify and measure the evolved gases.

Applications

The decomposition of polymers, the distillation of oils and the evolution of gases from minerals, inorganics and complexes as well as the detection of products from catalytic reactions have all been studied by evolved gas methods.

Related topics

Infrared and Raman spectrometry: applications (E11)
Mass spectrometry (E14)
Thermogravimetry (G1)
Differential thermal and differential scanning calorimetry (G2)
Thermomechanical analysis (G3)

Principles

During a heating process, any reaction or decomposition may produce gases. Although the gases may sometimes be absorbed, the technique is generally referred to as **evolved gas analysis (EGA)**.

For example, the three-stage decomposition of calcium oxalate monohydrate gives first water vapor, then carbon monoxide, and finally carbon dioxide. Each of these can be identified, for example, by mass spectrometry (see Topic E14) as shown in *Figure 1*.

The first stage evolves water (m/z 18), the second carbon monoxide (m/z 28),

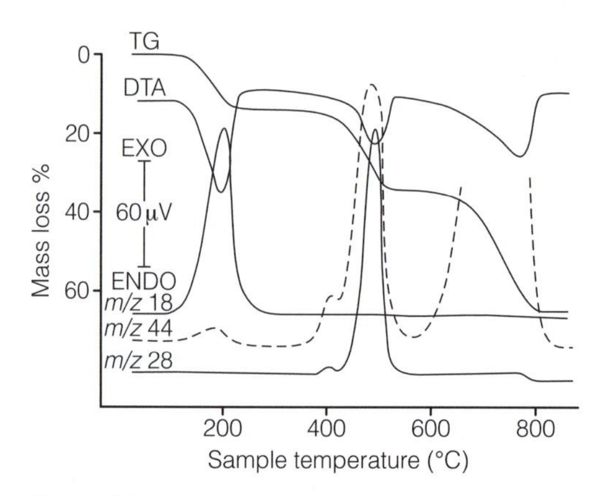

Fig. 1. Mass spectrometric detection and identification of gases evolved during the decomposition of calcium oxalate monohydrate.

which dissociates to give carbon dioxide (*m/z* 44). The third stage produces only carbon dioxide.

If several gases are evolved simultaneously, they should preferably be separated by chemical or chromatographic methods. Qualitative and quantitative methods for analysis of the gases may be classified as:

- physical – conductivity, density, absorption;
- chemical – titration, electrochemical, reaction;
- spectrometric – infrared and mass spectrometry, colorimetry.

These methods are described in previous sections. It is important to note that both the instrumentation and the analysis must be designed to prevent any further gas-phase reaction occurring.

Instrumentation

In order to make simultaneous measurements, the evolved gases are often detected in the carrier gas stream from thermogravimetric or other thermal analysis procedures. This provides programmed temperature control of heating and correct sample handling.

Physical detectors, such as those used for gas chromatographic detection are sometimes fitted to detect evolved gases, but they do not identify them. Moisture can be measured quantitatively using a capacitance moisture meter or by absorption and electrolytic determination.

Chemical methods may involve many of the techniques described in Section C. Any acidic or alkaline gas can be detected by absorption and titration, or potentiometry. Gas-sensing membrane electrodes and other ion-selective electrodes allow analysis of halide and sulfide ions.

Spectrometric methods, especially mass spectrometry (MS) and Fourier transform infrared spectrometry (FTIR) have been used, often coupled with thermogravimetry. For molecules that are polar and of low molar mass, FTIR is particularly useful. For nonpolar molecules and those of higher molar mass, MS is more adaptable. There are problems, however, in interfacing the thermal analysis instrument operating at atmospheric pressure to the MS operating under vacuum. This is discussed in Topic F3.

Special instrumentation allows study of gases for environmental investigations and it should be noted that **hot-stage microscopy** (or **thermomicroscopy**) and X-ray diffraction can be used to observe changes in the solid residues.

Applications

Acidic gases, such as the hydrogen chloride evolved from heating poly(vinyl chloride) can be measured by pH change, or by absorption in alkaline solution and back-titration, or using a chloride ion selective electrode. This is shown in *Figure 2*.

Carbon dioxide from heated concrete may be absorbed in barium hydroxide and measured conductimetrically. Sulfur dioxide from coal combustion can also be determined electrochemically. Ammonia evolved from ammonium aluminum compounds can be measured by pH change.

Ammonia evolved from minerals, and a complex collection of products from polymer decompositions can all be measured and identified by FTIR spectrometry. Mass spectrometric identification of gases, both polar and nonpolar (for example acidic gases, hydrocarbons and even homonuclear diatomics such as oxygen) provides a very versatile experimental method for analyzing the decomposition of materials, especially polymers. Phenol-formaldehyde resins

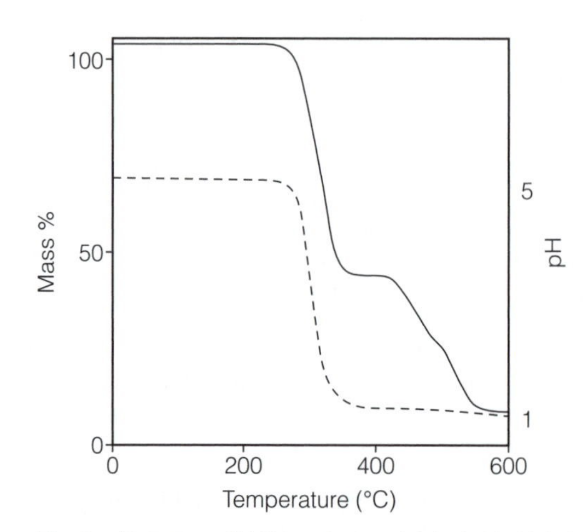

Fig. 2. Evolution of HCl gas from poly(vinyl chloride) detected by pH measurement (- - -) during TG (——).

break down in several stages over the temperature range 100 to 600°C and the evolution of water, formaldehyde and ammonia and, at higher temperatures phenol, can be detected by mass spectrometry, showing the course of the decomposition.

H1 Chemical sensors and biosensors

Key Notes

Sensors

Sensors are devices that respond to the presence of one or more analytes enabling qualitative and/or quantitative information to be obtained quickly and continuously.

Electrochemical sensors

Electrochemical sensors are based on potentiometric, amperometric or conductimetric cells that generate signals from electrochemical reactions of the analyte or in response to the presence of particular chemical species related to it.

Optical sensors

Optical sensors respond to the absorption or fluorescent emission of electromagnetic radiation by analytes, indicators or analyte–receptor complexes at characteristic wavelengths, principally in the visible region, but also in the ultraviolet and near infrared spectral regions.

Thermal- and mass-sensitive sensors

Thermal sensors measure heats of reaction generated by the oxidation of an analyte or its reaction with a reagent. Mass sensitive sensors are mechano-acoustic devices based on piezoelectric effects caused by the adsorption of analytes onto a crystal surface.

Sensor arrays

These are groups of sensors that allow simultaneous monitoring with different instrumental parameters so as to improve the selectivity of the system for a particular analyte or to monitor several analytes.

Related topics

Potentiometry (C3)
Voltammetry and amperometry (C9)
Conductimetry (C10)
Ultraviolet and visible molecular spectrometry: applications (E9)
Evolved gas analysis (G4)

Sensors

Sensors are small or miniaturized devices designed for the continuous monitoring of the physicochemical or biochemical properties of specific analytes so as to provide qualitative and/or quantitative analytical data. They comprise a thin layer of a chemically or biochemically sensitive substance (matrix and recognition element) in contact with a **transducer** (i.e. a means of converting the chemical or biochemical information into an electrical or optical signal) (*Fig. 1*). The transducer signal is then processed by suitable electronic circuitry incorporated into the device.

Chemical sensors react to the presence and concentration of an analyte by responding selectively to an electrical, optical, thermal or other property. **Biosensors** generally employ a thin layer of a substance incorporating an **immobilized reagent** that contains **biorecognition sites**. The biomolecules may be proteins, often enzymes, or other macromolecules that display specific interactions with an analyte species. Reagents are immobilized by entrapment or

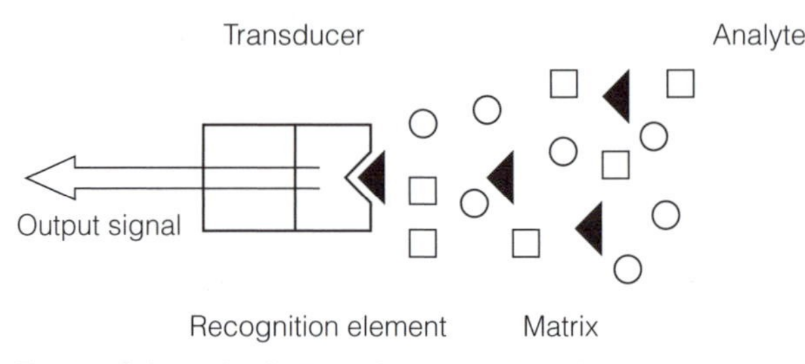

Fig. 1. Schematic diagram of a chemical or biosensor.

binding, methods used including physical adsorption onto or chemical bonding to the surface of the transducer, or entrapment in a cross-linked polymeric gel.

Sensors should be robust, have a rapid and reproducible response to the analyte(s), and have an appropriate selectivity/specificity and working range. They should be stable in operation, with minimal drift, and ideally be unaffected by changes in temperature, pressure and other adverse conditions. Sensors are used for industrial process stream analysis, monitoring air quality in urban and workplace environments, and hospital patients' body fluids for clinical purposes.

Electrochemical sensors

Potentiometric sensors are based on **ion selective electrodes, solid state redox electrodes** and **field effect transistors (FETs)**. Glass electrodes used for pH measurements can be incorporated into gas sensors for measuring levels of carbon monoxide, carbon dioxide, ammonia, oxides of nitrogen and sulfur and other gaseous analytes (Topic C3). Biosensors can be made by coating a glass electrode with a layer of an enzyme immobilized on the surface and which catalyzes a biochemical reaction (*Fig. 2*). For example, urease coated onto a gas-sensitive ammonia electrode (Topic C3) can be used to monitor urea which is hydrolyzed by the enzyme, the electrode responding to the ammonia produced at pH 7–8, i.e.

$$CO(NH_2)_2 + 2H_2O + H^+ \xrightarrow{\text{urease}} 2NH_4^+ + HCO_3^-$$

$$NH_4^+ + OH^- \rightarrow NH_3 + H_2O$$

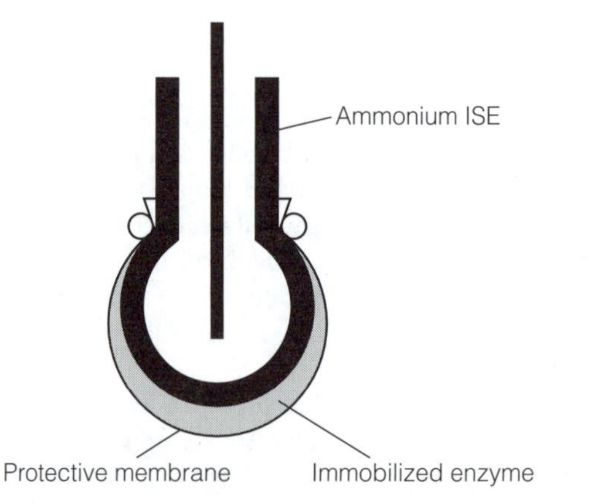

Fig. 2. Biosensor based on an enzyme-coated ammonium ion-selective electrode. Reproduced from R. Kellner et al., Analytical Chemistry, *1998, with permission from Wiley-VCH.*

Redox electrodes consisting of crystalline zirconium dioxide, doped with calcium or yttrium oxides to render it ionically conducting, are used to monitor oxygen in combustion gases and molten metallurgical samples. The redox equilibrium

$$O_2 + 4e^- \leftrightarrow 2O^{2-}$$

creates an electrode potential with a Nernstian response to concentration through the movement of O^{2-} ions through the crystal lattice.

Field effect transistors can be made sensitive to a range of ionic and gaseous analytes. Biosensors can be fabricated by incorporating gel or polymer layers containing immobilized enzymes, antigens or antibodies.

Amperometric and conductimetric gas sensors have been devised for the detection of oxygen, hydrogen, ammonia, sulfur dioxide and methane. Enzyme-based biosensors such as cholesterol or glucose electrodes for blood analysis depend on the oxidation of cholesterol or glucose oxidase, respectively, to produce hydrogen peroxide that is detected amperometrically (*Fig. 3*).

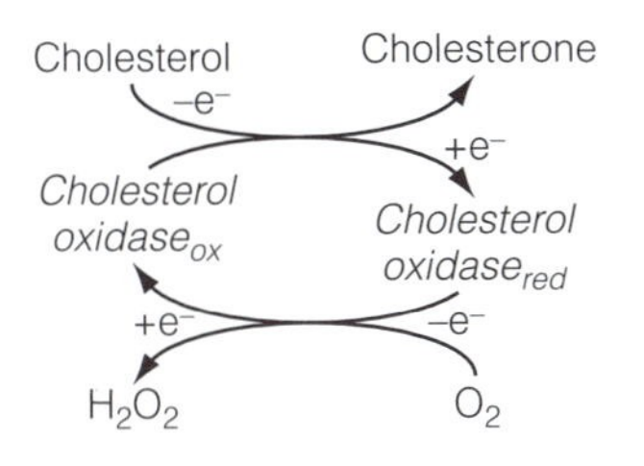

Fig. 3. Example of an oxidase-based sensor reaction: enzyme catalyzed oxidation of cholesterol. Reproduced from R. Kellner et al., Analytical Chemistry, *1998, with permission from Wiley-VCH.*

Optical sensors

These utilize glass, quartz or plastic **optical fibers** to transmit incident and attenuated or fluorescent radiation between a spectrophotometer, or **light-emitting diode (LED)** and detector, and the sample or sample stream. The ends of two separate fiber cables are positioned together in a y-shaped configuration a short distance above a reflector plate to form a sample cell (*Fig. 4*). Radiation emerging from one cable passes through the sample into which the cell is immersed and is reflected back through the sample into the end of the other cable. The path length of this optical cell is twice the distance between the ends of the cables and the plate. Fluorescent emission or reflected radiation from a solid surface can also be detected. Chromogenic reagents (e.g. pH or complexometric indicators) can be immobilized on the ends of the fibers in thin layers of supporting media such as cellulose and polyacrylamide. Sensors modified in this way are known as **optrodes**.

Home test kits, such as those for blood sugar or pregnancy, depend on visual observation of a developed color, but test strips can also be assessed by instrumental reflectance measurements. Enzyme-based redox reactions monitored by electrochemical biosensors may alter pH or generate products such as hydrogen peroxide that can react with a chromogenic reagent to form the basis of an optrode.

Optical sensors can be used to monitor pH, metal ions, dissolved gases and organic compounds down to ppm and ppb concentrations using radiation in the

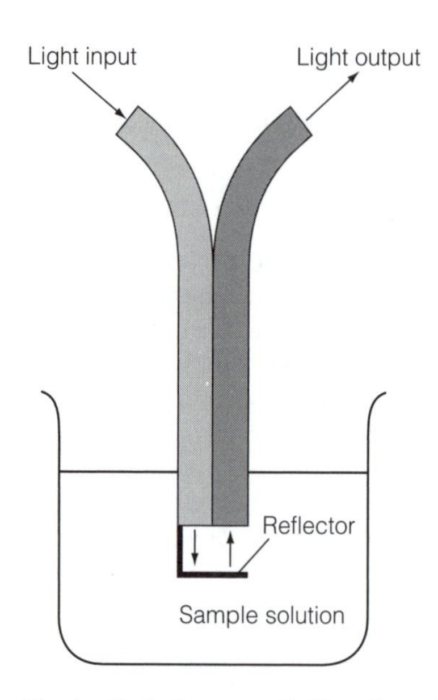

Fig. 4. Optical sensor with Y-configuration cell. Reproduced from R. Kellner et al., Analytical Chemistry, *1998, with permission from Wiley-VCH.*

visible, ultraviolet and near infrared regions by absorbance, fluorescence or reflectance. They have inherent advantages over electrochemical sensors in not requiring an electrode system and providing valuable spectral information over a range of wavelengths, but may suffer from ambient light interference, depletion of immobilized reagents, and slow kinetics of the reactions between analytes and reagents.

Thermal- and mass-sensitive sensors

Thermal sensors for oxidizable gases such as carbon monoxide and methane depend on measuring the change in resistance of a heated coil due to the **heat of reaction** resulting from oxidation of the gas by adsorbed oxygen. Thermal biosensors incorporating thermistors have been developed, which measure heats of reaction of enzymes in the detection of urea, glucose, penicillin and cholesterol.

Mass-sensitive devices are based on **piezoelectric quartz crystal resonators** covered with a gas-absorbing organic layer. Absorption of an analyte gas causes a change in resonance frequency that can be detected by an oscillator circuit and which is sensitive down to ppb levels.

Sensor arrays

Individual sensors are often nonspecific, but selectivity (Topic A3) can be improved by using groups or **arrays** of several sensors to monitor one or more analytes using different instrumental parameters and/or different sensor elements. Sensors in an array may be operated at different electrical potentials, frequencies or optical wavelengths. A sensor array for the simultaneous monitoring of pH, sodium and potassium levels in body fluids can be constructed from three ion-selective field effect transistors (*Fig. 5*), each with an appropriate sensitivity to one of the three analytes.

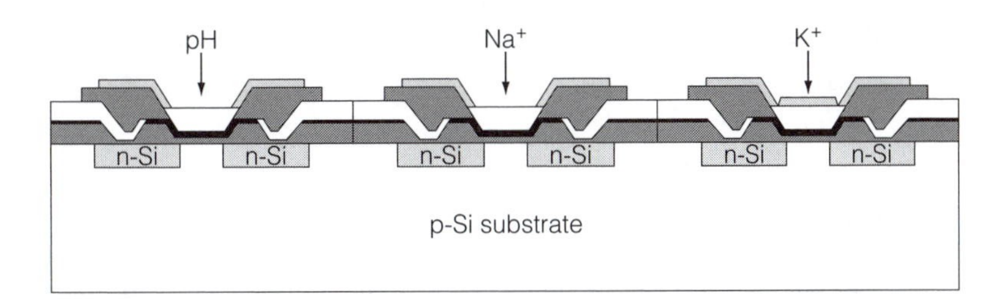

Fig. 5. FET sensor arrray for monitoring pH, sodium and potassium. Reproduced from R. Kellner et al., Analytical Chemistry, *1998, with permission from Wiley-VCH.*

H2 AUTOMATED PROCEDURES

Key Notes

Automation

The automation of some or all of the stages in an analytical procedure provides a number of advantages for busy laboratories or where hazardous samples are to be analysed. Automated sample processing and instrument control with the aid of robots is increasingly commonplace.

Laboratory robots

Robots are mechanical devices capable of performing both simple repetitive tasks and complex operations unattended. These include weighing, dispensing reagents, dilutions, extractions, movement of samples and instrument control.

Autosamplers

Multiple samples and standards prepared for routine titrimetric, chromatographic, spectrometric and other types of instrumental analysis can be loaded into autosamplers that transfer them one at a time to the instrument for analysis in a pre-determined sequence.

Related topics

Chemical sensors and biosensors (H1)

Computer control and data collection (H3)

Automation

The partial or complete automation of analytical procedures offers significant advantages in cost savings and increased sample throughput for busy laboratories. Personnel can be released for more demanding work, and the elimination of human error in repetitive operations leads to improved precision and accuracy. The handling of toxic or radioactive samples by remote control, the use of laboratory robots and computer-controlled operations are all important features of modern analytical laboratories. Automated analyses frequently make use of chemical and biosensors (Topic H1).

Many and sometimes all of the practical steps in an analytical procedure can be automated. These include:

- sample preparation by dissolution;
- addition of reagents, mixing, digestions, filtrations, dilutions;
- liquid or solid-phase extractions;
- titrations;
- setting and monitoring instrument parameters;
- presenting samples to instruments;
- chromatographic separations;
- spectrometric measurements;
- electrochemical, thermal or radioactivity measurements.

A schematic diagram of a potentiometric autotitrator is shown in *Figure 1*.

Laboratory robots

Robots are programmable mechanical devices that are the central components of laboratory work stations. They can be made to perform a variety of manipula-

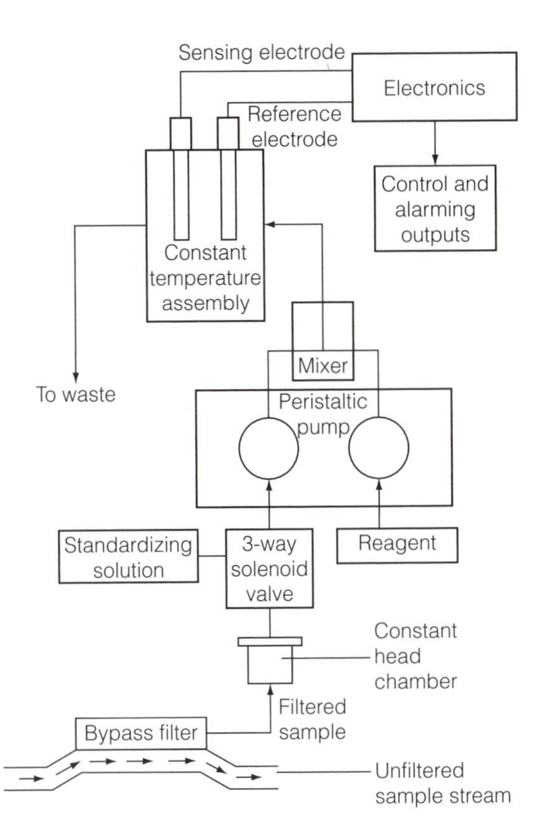

Fig. 1. Schematic diagram of a potentiometric autotitration system. From Principles of Instrumental Analysis, *2nd edn, by D.A. Skoog & D.M. West © 1980. Reprinted with permission of Brooks/Cole, an imprint of the Wadsworth Group, a division of Thomson Learning.*

tive and repetitive tasks ranging from simple operations, such as weighing samples, adding reagents or filtrations to multistep procedures for sample cleanup by solvent or solid phase extractions. They are computer-controlled and can be programmed and reprogrammed to perform sequences of operations according to specific analytical requirements.

The spatial geometry of robots may be cylindrical, cartesian or anthropomorphic (mimicking human movements), the first form being the most common. A typical laboratory work station with a computer-controlled cylindrical robot is shown in *Figure 2*. Typical operations include:

- manipulation of glassware and other apparatus;
- weighing and dissolution of samples;
- addition of reagent solutions and solvents;
- control of heating, cooling and mixing;
- filtrations and extractions;
- instrument operation and control.

As for automation in general, robots release laboratory staff for more demanding and nonrepetitive tasks, increase sample throughput, and contribute to improved analytical precision and accuracy. They can also be designed to work with hazardous materials so as to protect laboratory personnel from direct contact with toxic or radioactive substances.

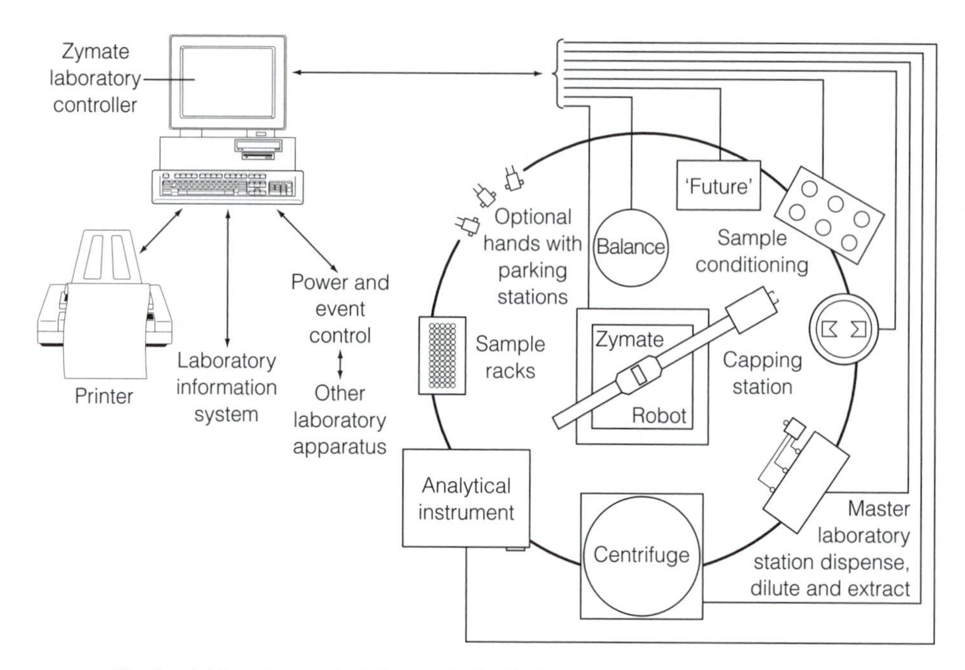

Fig. 2. A laboratory work station and cylindrical geometry robot arm. Reproduced with permission from Zymark Co.

Autosamplers

Autosamplers are used to enable a series of samples and standards to be taken from pre-loaded vials and analyzed in sequence, normally under computer control. They consist of a rectangular or circular array of sample vials in a rack or turntable, and a means of transferring measured volumes to an analytical instrument or another apparatus. They may be used in conjunction with a laboratory work station or dedicated to a particular instrument, such as a gas or liquid chromatograph, mass spectrometer or flow injection analyzer. Turntable autosamplers have a fixed sampling device consisting of a hollow stainless steel needle pipet that dips into each sample vial as it is rotated into position, the liquid being drawn into the pipet under vacuum. More versatility is available with xyz autosamplers because they automatically adjust the height of the pipet (the z-direction) to take account of sample volume as well as being able to move to any sample vial (x and y directions) following any predetermined sequence.

H3 Computer control and data collection

Key Notes

Microprocessors and microcomputers

Digital computers consist of hardware and software components that use binary code for data and word processing. The microprocessor, an integrated circuit chip that performs all the operations and computations, is at the heart of a microcomputer, which also includes various forms of memory for program and data storage, and input/output devices.

Computer–instrument interfacing

Computer control of operating parameters and the digitizing of analog detector signals for storage and processing are important aspects of interfacing computers with analytical instruments.

Networking and laboratory management

The electronic transfer of data and other information between instruments and laboratories is facilitated by connecting them together in a network. The analytical work and overall management of one or more laboratories can be controlled and monitored through the use of specifically designed software packages.

Related topic

Data enhancement and databases (H4)

Microprocessors and microcomputers

A digital computer consists of four principal hardware components.

(i) The **central processing unit** (**CPU**), which includes the **microprocessor integrated circuit chip**, registers for the temporary storage of data, and a **high frequency clock** to synchronize all operations. Clock frequencies, which determine computing speed, are steadily increasing and are currently approaching or exceeding 1.5 GHz.

(ii) **Random access memory** (**RAM**) for the temporary storage of current programs and data, **read only memory** (**ROM**) for the storage of data for reference and frequently used routines, and long term memory on magnetic and optical disks (CD and DVD), capable of storing from one or more megabytes (Mbytes) of information up to several gigabytes (Gbytes). (1 byte, or 8 bits (**bi**nary digi**ts**) defines an alphanumeric character.)

(iii) **Input** and **output** (**I/O**) devices, i.e. keyboards, VDU screens, printers, plotters, scanners, instrument interfaces, modems and digital cameras.

(iv) Parallel or serial transmission lines (**buses**) for internal and external transfer of program instructions and data.

Computer–instrument interfacing

Computers can be programmed to set and monitor instrument parameters to ensure stable and reproducible operation. Groups of parameters can be stored and retrieved as standard methods for routine use. **Self-diagnostic** routines to test the condition of instrument components and locate faults are a common feature of many computer–instrument software packages.

Most analytical instruments generate analog detector signals in the form of a varying voltage or current. To store and process the signal, it must first be digitized at an interface between the instrument and the computer using an **analog-to-digital converter** (**ADC**). The detector signal may vary relatively slowly with time, as with an autotitrator or a UV/visible spectrometer, or very rapidly, as with a capillary gas chromatograph linked to a mass spectrometer. An ADC must be capable of sampling and converting the sampled signal to a digital value at an appropriate rate (in as little as a few microseconds) so as to be ready for the next sample, and the digitized record should be as accurate a version of the original analog signal as possible. This normally requires an ADC with between 10 and 16 bit resolution. For example, an analog signal that varies between 0 and 1 V and is digitized with a 12-bit ADC would produce corresponding digital values in the range 0 to 4095, giving a resolution of 1 in 4096 ($2^{12} = 4096$), or voltage increments of 2.44×10^{-4} (~0.2%).

If the computer is to control instrumental operating parameters, or if an analog data output is required, the reverse process of **digital-to-analog conversion** (**DAC**) is used. As for ADC, a DAC should have a resolution of 10 to 16 bits to ensure the generation of an acceptable analog signal.

Networking and laboratory management

Electronic connections beween analytical instruments, databases, storage media and other devices to facilitate the transfer, archiving and retrieval of results and other information are known as **networks**. Customized software enables reports to be generated in a desired format, and library databases to be accessed to assist in the interpretation of results. Samples passing through the laboratory can be logged, and other information concerned with its organization, such as statistical assessments of the workload, and monitoring the performances of individual instruments can also be controlled by a Laboratory Information and Management System (**LIMS**) software package. Networks may be localized in one or a small group of laboratories (local area networks, **LANs**) or can extend throughout a national or international organization by use of the internet or Compuserve. *Figure* 1 is a diagram of a typical LAN.

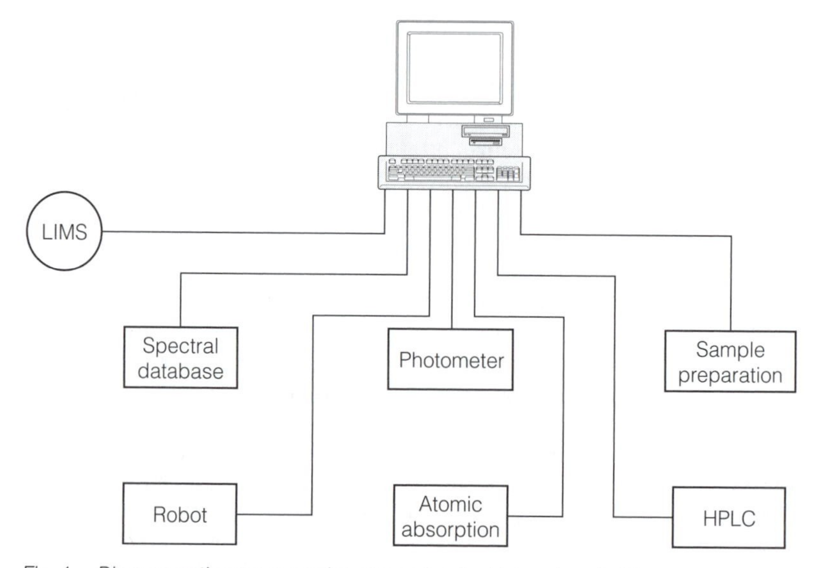

Fig. 1. Diagrammatic representation of a networked laboratory (LAN).

H4 DATA ENHANCEMENT AND DATABASES

Key Notes

Data processing Digitized raw analytical data can be computer-processed to extract the maximum amount of useful information. This includes noise reduction, signal enhancement, calibration and quantitation, the identification of unknown analytes, and the characterization of materials.

Databases Large amounts of analytical and other chemical data can be stored in digital form for access when required. This can range from small locally generated databases for use with a single instrument to large ones compiled by national or international organizations for general availability.

Library searches Search software enables data libraries to be searched for specific information, including physicochemical data on elements, compounds, materials and products. The ability to identify or classify unknown analytes or substances is an important facility of library search software.

Related topic Computer control and data collection (H3)

Data processing

Raw analytical data often contains electronic noise and other spurious signals such as detector responses from sample components other than the analyte(s) of interest. **Digitizing** and **computer processing** enable both the noise and interfering signals to be reduced or eliminated using chemometric routines (Topic B5). Additional software procedures can be used to process calibration and sample data for quantitative analysis by establishing detector responses, computing results and applying statistical tests. Quantitative data is conveniently handled by **spreadsheets**, such as Microsoft Excel and statistics packages such as Minitab. Tabulated results can be incorporated into reports generated in any required format by a word processor. Stored information can be manipulated and presented in tabular or graphical forms to aid interpretation. The processing of spectrometric and chromatographic data provide good examples of data enhancement and quantitative computations, for example:

- smoothing to reduce noise and co-adding spectra from multiple scans to increase the **signal-to-noise (S/N)** ratio of weak spectra;
- spectral subtraction to remove background interference or contributions from components of a sample that are not of interest to reveal the presence of peaks that were previously obscured;
- scale expansion to show details in a particular region of a spectrum or to increase sensitivity;
- flattening a sloping baseline that may be distorting spectral peaks;
- processing calibration and sample data for quantitative analysis;

- the application of statistical tests and chemometric procedures to assess quantitative results and extract additional information from complex data;
- calculation of chromatographic parameters such as efficiency, resolution, peak assymetry and detector response;
- comparison of chromatographic retention data for standards and samples to enable unknown analyte peaks to be identified;
- measurement of chromatographic peak areas using a range of options for defining baselines and separating overlapping peaks;
- display of developing chromatograms in real time, including scale and sensitivity changes;
- processing and presentation of complex data from hyphenated techniques such as gas or liquid chromatography-mass spectrometry and inductively coupled plasma mass spectrometry (ICP–MS).

Important areas of data processing include the use of chemometrics (Topic B5) to simplify complex data for characterizing materials, quantitative spectrometric analysis using multiple wavelengths, and routines to optimize experimental conditions for high-performance liquid chromatography.

Databases

A **database** may consist of a simple look-up table relating two variables, such as solvent composition and polarity, tables of statistical factors for tests of significance (Q, F and *t*-tests), or archived sample data and analytical procedures. Larger compilations of data may list chemical formulae and structures or characteristic properties of elements, compounds, materials and commercial formulations (e.g., boiling point, viscosity, dielectric constant, hardness and toxicity). Spectrometric and chromatographic analytical databases are of particular value in the characterization and identification of unknown substances, and some examples are given in *Table 1*. The data is usually **compressed** and **encoded** to maximize the amount of information that can be stored.

Table 1. Some analytical spectrometric and chromatographic databases

Technique	Data base/source	Data
Atomic emission spectrometry	Plasma 2000/Perkin Elmer	50 000 atomic lines
Gas chromatography	Sadtler	Retention indices
Infrared spectrometry	Aldrich/Nicolet	>100 000 spectra
Mass spectrometry	NIST/EPA/MSDC	50 000 spectra
Nuclear magnetic resonance spectrometry	Bruker	19 000 spectra

The formatting of a database involves the creation of several types of files that are manipulated with specialized software. A **source file** containing raw analytical data is converted to a **library file** by reducing noise, eliminating unimportant data and compression. Associated **exchange files** enable data to be transferred in a standard format such as **JCAMP/DX** for spectrometric data and **JCAMP/CS** for chemical structures.

Library searches

Search algorithms are used to retrieve information from databases as quickly as possible, often within a few seconds. User-defined criteria can be selected to direct the search and/or to limit the amount of data retrieved and specify the mode of presentation. Search algorithms are based on **multivariate chemo-**

metric procedures such as **cluster analysis** and **similarity measures** (Topic B5), and may use **sequential searches** or **hierarchical search trees**.

A sequential search of a spectral library, for example, involves the comparison of every part of a sample spectrum with library spectra, and is suitable only for small libraries. A hierarchical search involves comparing groups (families) of spectra having the same set of key features as the sample spectrum, enabling large libraries classified in a tree-like structure to be searched very efficiently.

Searching spectral libraries may involve the use of **inverted lists**. These consist of each characteristic absorption band or emission line along with a list of corresponding numbered library spectra that include that particular band or line. A list for the spectrum of an unknown analyte can then be rapidly checked against the library lists. An example of part of an inverted list for an infrared spectral library is shown in *Figure 1*. It includes spectrum No. 66 among those listed with an absorbance band at 1220 cm^{-1} and spectrum No. 105 among those listed with an absorbance band at 2730 cm^{-1}.

Correlation coefficients that define the quality of a match (a perfect match corresponds to a value of 1.000) between an unknown and a library spectrum or structure) are used to compile a **hit list** that places possible identities in descending rank order. A hit list shows the five to ten most probable identities of an unknown based on the selected search criteria, but it is limited by the content of the library and may include unlikely or impossible identifications.

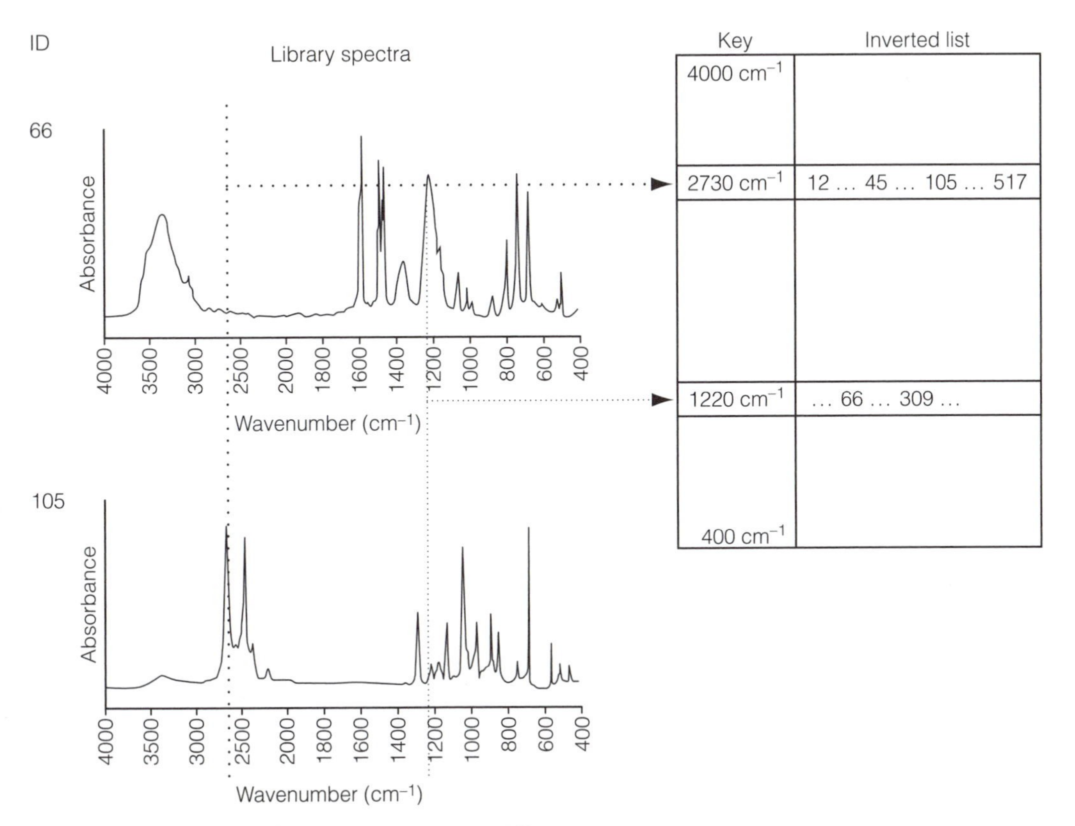

Fig. 1. Part of an inverted list for an infrared spectral library.

Table 2 shows an example of a hit list generated for identifying a benzodiazepine drug separated by high-performance liquid chromatography using a library search of UV spectra recorded by a diode-array detector (DAD).

Table 2. Hit list compiled from a library of UV absorption spectra for identifying a benzodiazepine drug separated by HPLC using a DAD

Compound	Correlation coefficient
Oxazepam	0.999
Chlordiazepoxide	0.966
Nordiazepam	0.960
Diazepam	0.842
Fluorazepam	0.753

FURTHER READING

General Reading

Atkins, P.W. (1998) *Physical Chemistry*, 6th edn. Oxford University Press, Oxford, UK.

Fifield, F.W. and Kealey, D. (2000) *Principles and Practice of Analytical Chemistry*, 5th edn. Blackwell Science, Oxford, UK.

Harris, D.C. (1998) *Quantitative Chemical Analysis*, 5th edn. Freeman, USA.

Kellner, R., Mermet, J-M., Otto, M. and Widmer, H.M. (eds) (1998) *Analytical Chemistry*. John Wiley & Sons, Chichester, UK.

Skoog, D.A., Holler, J.F., Nieman, T.A. (1997) *Principles of Instrumental Analysis*. Thomson, New York.

Whittaker, A.G., Mount, A.R., Heal, M.R. (2000), *Instant Notes Physical Chemistry*. Bios, Oxford, UK.

Willard, H.H., Merritt, L.L. Jr., Dean, J.A. and Settle, F.A. Jr. (1998) *Instrumental Methods of Analysis*, 7th edn. Wadsworth, USA.

More advanced reading

Section A

Kenkel, J. (1999) *A Primer on Quality in the Analytical Laboratory*. CRC Press, UK.

Section B

Miller, J.C. and Miller, J.N. (2000) *Statistics and Chemometrics for Analytical Chemistry*, 4th edn. Ellis Horwood PTR Prentice Hall, UK.

Section C

Bard, A.J. and Faulkner, L.R. (2000) *Electrochemical Methods: Fundamentals and Applications*. J. Wiley & Sons, Chichester, UK.

Kissinger, P.T. and Heineman, W.R. (eds), (1995) *Laboratory Techniques in Electroanalytical Chemistry*. M. Dekker, New York.

Monk, P.M.S. (2001) *Fundamentals of Electroanalytical Chemistry*. J. Wiley & Sons, UK.

Wang, J. (2000) *Analytical Electrochemistry*. J. Wiley & Sons, Chichester, UK.

Section D

ACOL book on *Gas Chromatography*, 2nd edn (1995). John Wiley & Sons, Chichester, UK.

Anderson, R. (1987) *Sample Pre-treatment and Separation*. John Wiley & Sons, Chichester, UK.

Baker, D.R. (1995) *Capillary Electrophoresis*. John Wiley & Sons, Chichester, UK.

Braithwaite, A. and Smith, F.J. (1996) *Chromatographic Methods*, 5th edn. Chapman and Hall, UK.

Lindsay, S. (1992) *High Performance Liquid Chromatography*, 2nd edn. John Wiley & Sons, Chichester, UK.

Section E, F

Barker, J. (1998) *Mass Spectrometry*, 2nd edn. John Wiley & Sons, Chichester, UK.

Pavia, D.L., Lampmann, G.M. and Kriz, G.S. Jr. (2001) *Introduction to Spectroscopy*, 3rd edn. Harcourt, USA.

Silverstein, R.M. and Webster, F.X. (1997) *Spectrometric Identification of Organic Compounds*, 6th edn. John Wiley & Sons, New York, USA.

Williams, D.H. and Fleming, I. (1995) *Spectroscopic Methods in Organic Chemistry*, 5th edn. McGraw Hill, UK.

Section G

Haines, P.J. (1995) *Thermal Methods of Analysis: Principles, Applications and Problems.* Blackie, UK.

Hatakeyama, T. and Quinn F.X. (1999) *Thermal Analysis: Fundamentals and Applications to Polymer Systems.* John Wiley & Sons, Chichester, UK.

Section H

Cattrall, R.W. (1997) *Chemical Sensors.* Oxford University Press, UK.

INDEX